Copyright © 2012, The Geological Society of America (GSA), Inc. All rights reserved. Copyright is not claimed on content prepared wholly by U.S. government employees within the scope of their employment. Individual scientists are hereby granted permission, without fees or further requests to GSA, to use a single figure, a single table, and/or a brief paragraph of text in other subsequent works and to make unlimited photocopies of items in this volume for noncommercial use in classrooms to further education and science. Permission is also granted to authors to post the abstracts only of their articles on their own or their organization's Web site providing the posting cites the GSA publication in which the material appears and the citation includes the address line: "Geological Society of America, P.O. Box 9140, Boulder, CO 80301-9140 USA (http://www.geosociety.org)," and also providing the abstract as posted is identical to that which appears in the GSA publication. In addition, an author has the right to use his or her article or a portion of the article in a thesis or dissertation without requesting permission from GSA, provided the bibliographic citation and the GSA copyright credit line are given on the appropriate pages. For any other form of capture, reproduction, and/or distribution of any item in this volume by any means, contact Permissions, GSA, 3300 Penrose Place, P.O. Box 9140, Boulder, Colorado 80301-9140, USA; fax +1-303-357-1073; editing@geosociety.org. GSA provides this and other forums for the presentation of diverse opinions and positions by scientists worldwide, regardless of their race, citizenship, gender, religion, sexual orientation, or political viewpoint. Opinions presented in this publication do not reflect official positions of the Society.

Published by The Geological Society of America, Inc.
3300 Penrose Place, P.O. Box 9140, Boulder, Colorado 80301-9140, USA
www.geosociety.org

Printed in U.S.A.

GSA Books Science Editors: Kent Condie and F. Edwin Harvey

Library of Congress Cataloging-in-Publication Data
Burchfiel, B.C.
  Tectonics of the southeastern Tibetan Plateau and its adjacent foreland / by B.C. Burchfiel, Chen Zhiliang.
      p. cm. — (Memoir / Geological Society of America ; 210)
  Includes bibliographical references.
  Summary: "This volume provides a summary of the geology of Eastern and Southeastern Tibet and its foreland. It covers an area of approximately 1.5 million square kilometers in 15 chapters on tectonic units that the authors recognize during 25 years of both field and laboratory study. Each chapter discusses the authors' understanding of the geology and offers interpretations of special geological relations, both local and regional, as well as currently unresolved problems of which there are many in this vast and poorly known region. Chapter 16 summarizes and interprets the preceding chapters. The volume is accompanied by a DVD containing four plates: two tectonostratigraphic maps, a map of unconformities and a plate of cross sections, in both Illustrator and ArcGIS formats. This is a unique map presentation and one that the authors suggest as a model for all geological maps"—Provided by publisher.
  ISBN 978-0-8137-1210-9 (cloth)
 1. Geology, Structural—Tibet, Plateau of. 2. Tibet, Plateau of.  I. Burchfiel, B. C., 1934– . II. Chen, Zhiliang. III. Geological Society of America. IV. Series: Memoir (Geological Society of America) ; 210.
  QE634.C6T43 2012
  555.1'8—dc23

2012031754

Cover: Looking east along the Bangong suture belt about 10 km east of Dengqen. Large white outcrops are marble and locally limestone blocks within the Bangong Middle and Upper Jurassic mélange. The scaly mélange matrix is visible along the road where the foliation is vertical. Photo courtesy B. Clark Burchfiel.

10 9 8 7 6 5 4 3 2 1

# *Dedication*

This work is dedicated to the thousands of Chinese geologists who worked for several generations to produce the geological maps on which this volume is largely based. Much of the geological mapping was done prior to the mid-1980s, however, we have found through our fieldwork that the maps illustrate the geology with surprising but variable accuracy. These geologists worked under very difficult conditions through very unstable times during the Second World War and for several decades afterward. More recent work may have changed the ages assigned to some of the stratigraphic units, but the structures shown on these maps remain a fair representation of what can be observed in the field. We have reinterpreted many of the structures and some of the contacts, and such changes will be discussed in the text. However, our efforts in no way subtract from the incredible contribution made by Chinese geologists who have produced the maps in this terrane of immense relief, difficult access, and highly complex geology.

# Contents

*Acknowledgements* ........................................................................... vii
*Abstract* ............................................................................................ 1

    1. Introduction ................................................................................ 3
    2. Yangtze Unit ............................................................................... 9
    3. Transitional Unit ....................................................................... 37
    4. Songpan Ganze Unit ................................................................ 53
    5. Yidun Unit ............................................................................... 63
    6. Qiangtang Unit ......................................................................... 77
    7. Lanping-Simao Unit ................................................................ 97
    8. Lhasa Unit .............................................................................. 117
    9. Bangong Suture Zone ............................................................ 127
  10. Syntaxial Unit ......................................................................... 131
  11. Tengchong Unit ...................................................................... 135
  12. Baoshan Unit .......................................................................... 139
  13. Changning-Menglian Unit ..................................................... 149
  14. Lin Cang Unit ........................................................................ 155
  15. Ailao Shan Unit ..................................................................... 161
  16. Synthesis ................................................................................ 165

*Epilogue* ........................................................................................ 207
*Appendix 1. Notes on map file formats* ........................................ 209
*Appendix 2. List of layers on Plates 1 and 2* ................................ 211
*Glossary of igneous rock symbols and names, in English and Chinese* ........................ 223
*References cited* ............................................................................ 225

# *Acknowledgments*

This work is the outgrowth of more than 25 years of cooperative work between MIT (Massachusetts Institute of Technology) geologists and geophysicists and geologists from many different Chinese agencies. Most consistently the work has been a cooperative effort between MIT and the Chengdu Institute of Geology and Mineral Resources, but it has also included work with the Yunnan Institute of Geological Sciences, the Geological Institute of the Chinese Academy of Sciences, and the former Chinese Ministry of Geology (now the Ministry of Land and Natural Resources). Strong support has been given to this work by these agencies, in particular the Institute of Geology and Mineral Resources, Chengdu, P.R. China. Support from the American side has been mainly through National Science Foundation grants EAR-9614970, EAR-9814303, EAR 0003571, EAR 8904096, EAR 9706300, National Aeronautics and Space Administration grant NAGW-2155, and the Schlumberger Chair at MIT. The coauthors specifically acknowledge the personal support over the years of Professor Liu Baojun and Drs. Ding Jun, Li Tingdong, and Xiao Xuchang.

Beehive Mapping of Watertown, Massachusetts, placed Plates 1, 2, and 3 into Illustrator and GIS formats from our drafted versions.

*Publication of this volume was made possible by support from*

*Director Ding Jun*
*Chengdu Institute of Geology and Mineral Resources*

*Department of Earth, Atmospheric and Planetary Sciences*
*Massachusetts Institute of Technology*

## Tectonics of the Southeastern Tibetan Plateau and Its Adjacent Foreland

**B.C. Burchfiel**
*Earth, Atmospheric and Planetary Sciences, Massachusetts Institute of Technology, 77 Massachusetts Avenue, Cambridge, Massachusetts 02139, USA*

**Chen Zhiliang**
*Ministry of Land and Resources, Chengdu Institute of Geology and Mineral Resources, 2 Beisanduan Yihuanlu, Chengdu, Sichuan 610081, People's Republic of China*

### ABSTRACT

The region of southeastern Tibet and its adjacent foreland presently lie north and east of the Eastern Himalayan syntaxis and consist of 14 tectonic units of which only five (Yangtze, Transitional, Baoshan, Lhasa, and Tengchong units) can be considered to be large continental fragments with a coherent Precambrian basement. The others are underlain by either thinned continental crust, or fragments of continental crust, arcs, and oceanic rocks swept together as a collage and forming a basement for late Paleozoic, Mesozoic, and Cenozoic arcs and sedimentary sequences that make up current tectonic elements. The history of fragmentation, accretion, and deformation has been complex and related to tectonic activity at several plate boundaries that have been active at different times from Paleozoic to Recent and have been reactivated at numerous times during the evolution of the region. Here we focus on the evolution from mid-Paleozoic to Recent with the major emphasis on Mesozoic and Cenozoic time, discussing the Yangtze, Transitional, Songpan Ganze, Yidun Arc, Qiangtang, Lanping-Simao, Lhasa, Bangong Suture Zone, Syntaxial, Tengchong, Baoshan, Changning-Menglian, Lin Cang, and Ailao Shan units, and synthesizing their relationships.

The area contain examples of the complex heterogeneous evolution of regions where there was early accretion of fragments forming a collage, later modified by rifting with the deposition of thick sequences of sedimentary rocks to form some tectonic units later deformed by intracontinental shortening and disruption by large strike-slip faults. Most of the folded belts of several ages must be basement involved and probably decoupled from the deeper crust, but the nature of their relations to the middle and lower crust remains unproven. The major early Cenozoic shear zones have large-scale displacements related to southeastward extrusion and often contain mylonitic rocks that suggest they were not only zones of dominantly lateral shear, but also had components of vertical extrusion of heated ductile lower crust during their formation in a transpressional tectonic environment. Numerous late Cenozoic shear zones bound crustal fragments that rotate clockwise around the Eastern Himalayan syntaxis. Many of these shear zones reactivate older sutures or shear zones, however, some formed by breaking though previously unfaulted areas. Superposed deformations in different tectonic settings are common, indicating a temporal and spatial pattern of changing rheology and strain. The Cenozoic deformation is the result of the interaction between two dynamic systems: the India-Eurasia intracontinental convergence system and the Western Pacific–Indonesian subduction system. Unfortunately, southeastern Tibet and its foreland is a region of mostly poorly known geology for which unraveling its detailed evolution will require extensive future research, although many of the important areas for investigation are discussed herein.

MANUSCRIPT ACCEPTED 18 MAY 2012

CHAPTER 1

# Introduction

The southeastern part of the Tibetan Plateau and its adjacent foreland (Fig. 1-1 and Plates 1 and 2 [available on DVD accompanying this volume and in the GSA Data Repository[1]]) are a complex region underlain by rocks that record a geological evolution from the Precambrian to Recent. Here we describe the general geology and structure of the region with the main emphasis on its evolution from late Paleozoic to the present. During this period all the tectonic elements were assembled to form the underpinnings of the Tibetan Plateau and its eastern and southeastern foreland. Not only were the tectonic units accreted, but they were also strongly modified by post-accretionary processes that are still continuing today. During the accretionary process, several major and some minor oceanic tracts were subducted, leaving only scattered ophiolitic remnants marking sutures that are sometimes cryptic. The last lithospheric fragment to be accreted was the Indian subcontinent, whose initial contact with Eurasia took place ca. 50 Ma (Garzanti and Van Haver, 1988; Green et al., 2008; Rowley, 1996, 1998; Zhu et al., 2005), but the timing at the eastern Himalayan syntaxis that lies in the northwestern part of Plate 1 is poorly constrained. Recent data presented by Aikman et al. (2008) suggest that the collision probably occurred here at about the same time. The region discussed lies entirely within China and north and east of the eastern Himalayan syntaxis, where ongoing indentation of India into Eurasia produced one of the most tectonically active areas in the world. The region discussed here adjoins that of a previous work published in 1995 (Burchfiel et al., 1995) on the geology and structure of the Longmen Shan and adjacent regions in the eastern part of the Tibetan Plateau and its foreland. The map in that publication has been redrawn as Plate 2 and adjoins the northeast corner of the map presented here as Plate 1. Since initial contact, India has continued to move northward relative to Eurasia ~3600 km at the eastern Himalayan syntaxis based on seafloor spreading data (Molnar and Stock, 2009). This large-magnitude intracontinental deformation has strongly deformed and continues to deform all tectonostratigraphic units in the region. One of the major questions on which we will focus is how this immense amount of convergence was accommodated.

The Tibetan Plateau is the largest area of high topography and thick crust on the Earth today. It is also an area where the rates of deformation during Cenozoic intracontinental convergence have been rapid since initial lithospheric contact between the continental lithosphere of India and Eurasia, and they continue to be rapid today; Late Cretaceous and early Cenozoic convergence between India and Eurasia was ~118 mm/a, and at ca. 50 Ma it decreased to 83 mm/a up to ca. 20 Ma, when it decreased again to its present value of 44 mm/a at the eastern Himalayan syntaxis (Molnar and Stock, 2009). Some of the convergence has been absorbed by shortening of the northern part of the Indian plate to form the Himalayan mountain chain, but most of it, perhaps more than two-thirds, has been absorbed by deformation within the Eurasia plate (Royden et al., 2008). These facts have attracted many geologists to this region, because the processes of intra-continental deformation should be abundantly obvious and decipherable. However, because numerous complexities and uncertainty in the geology hinder development of interpretations and include (1) the inherited complex anisotropy of the Tibetan crust developed during the late Paleozoic to early Cenozoic assembly of crustal elements, (2) the evolution of crustal rheology during crustal thickening, (3) differences in interpretation of the geology and major structures within the region, (4) difficulty in mapping within a region of great topographic relief and poor access, (5) surprisingly poor outcrops in many areas because of thickly cemented scree and soil cover even in areas of extreme mountain slopes, (6) strong deformation and rotation of older structural features, (7) difficulty in determining the age of individual structures because of the lack of rocks of the appropriate age or of poorly constrained ages on abundant non-marine sedimentary rocks, (8) inherited crustal and lithospheric anisotropy, and (9) complex temporal and spatial evolution of crustal rheology. Thus, as the task of unraveling how intracontinental convergence was absorbed, time and space relations between geological elements have remained very difficult to resolve. We first describe the general geology of the tectonic units we identify within the region as we understand it, realizing that much of the geology remains poorly known. This discussion is followed by our interpretation of its tectonic development, and finally we focus on many of the major unresolved tectonic problems that remain to constrain interpretations of how the post-collision convergence was absorbed in SE and east Tibet.

We presently cannot develop a well-constrained interpretation of the post-collisional evolution of the southeastern part of the Tibetan Plateau and its foreland, but we hope the discussions and maps presented below will show where we stand today and point to problems of regional and local importance. Not only are space-time relations of deformation difficult to determine, but the evolving rheology of the crust and lithosphere has caused changes in the nature of the deformation within the SE part of

---

[1]GSA Data Repository Item 2012297, Plates 1–4, is available at www.geosociety.org/pubs/ft2012.htm, or on request from editing@geosociety.org or Documents Secretary, GSA, P.O. Box 9140, Boulder, CO 80301-9140, USA.

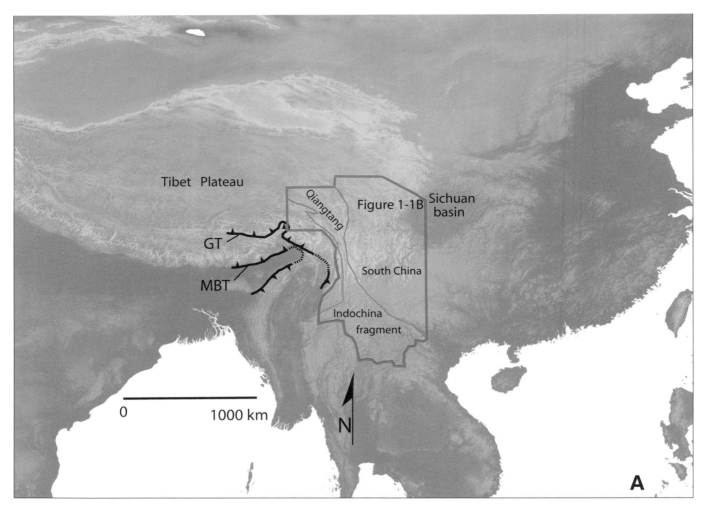

Figure 1-1 *(Continued on following pages).* (A) Location of southeastern Tibet and adjacent regions. The red outline shows area of generalized tectonic map of Figure 1-1B. Blue lines show commonly accepted boundaries of Indochina fragments extruded from SE Tibet during early Cenozoic time. Also shown is the boundary of the Qiangtang Unit of SE Tibet. GT—Gangdese thrust; MBT—Main Boundary thrust and border thrust of the Indoburman Ranges.

the Tibetan Plateau. As in all mountain ranges, even though the processes of crustal deformation and their crustal manifestations may be broadly similar, there remain geological features within the SE Tibetan Plateau that may be somewhat unique because of the early history of crustal assembly and the large magnitude of deformation.

The basis for the discussion is the tectonostratigraphic maps (Plates 1 and 2), cross sections (Plate 4), an unconformity map (Plate 3), and stratigraphic sections presented as individual figures accompanying each tectonic unit. All these plates have been derived mainly from one basic source, the 1:200,000-scale geological maps of the region (Bureau of Geology and Mineral Resources of Xizang Autonomous Region; Bureau of Geology and Mineral Resources of Sichuan Province; Bureau of Geology and Mineral Resources of Yunnan Province; see in References Cited). These maps and their accompanying explanatory texts are old, with published dates from the 1960s to the 1990s, and they form a body of data of varying quality; but they contain much detail that more recent compilations lack. Our compilations are based on the 1:200,000-scale map sheets, modified by our generalizations of their tectonostratigraphy by our own field studies and analysis over much of the area during the past 25 years. We have made both structural and stratigraphic interpretations of relations that are not shown on the original maps. Because these maps were made at different times by different geological groups, there are often problems in connecting geology between maps, a fact that cannot be easily rectified. Such difficulties are fairly easy to recognize on compilations in Plates 1 and 2 and will be mentioned in the text where such problems are important. Also, ages of plutonic rocks shown on Plate 1 are those assigned by the geologists who did the mapping, but very few of the plutons are dated. Their compositions and ages are shown by the symbols whose explanations are presented in Appendices 1 and 2. Because so few plutons are dated, their

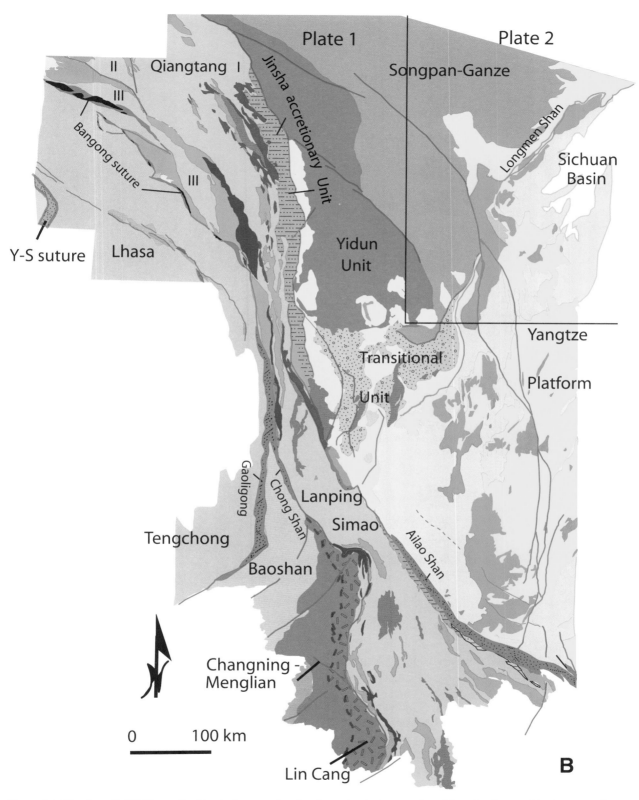

Figure 1-1 *(Continued)*. (B) Generalized tectonic map, showing the major tectonostratigraphic units discussed in the text and some structural elements that bound these units. The area of Plate 1, the tectonostratigraphic map that accompanies this report, does not include the northeast corner of Figure 1-1, but that area is covered by Plate 2. The Qiangtang Unit has been divided into three subunits (Q I, Q II, Q III). Y-S—Yarlang-Zangbo.

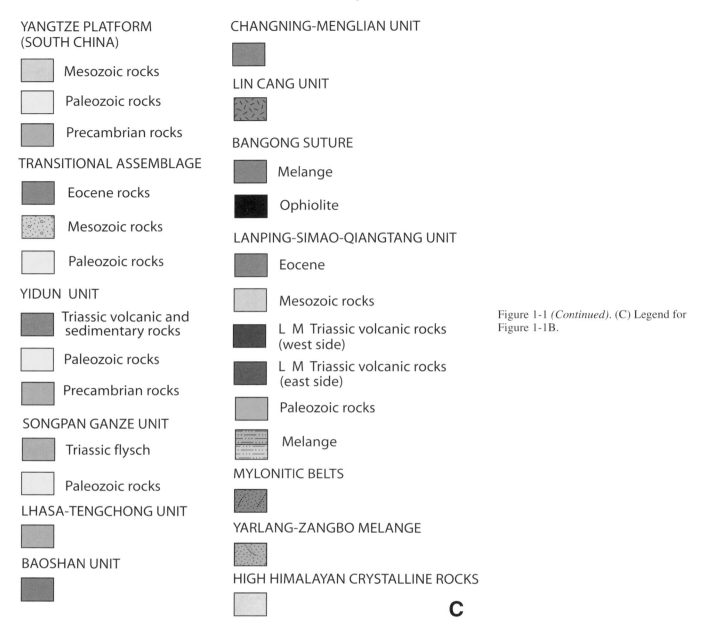

Figure 1-1 *(Continued)*. (C) Legend for Figure 1-1B.

assigned ages must be taken as of limited reliability. The plutons on Plate 2, however, have been extensively dated, and references are presented in the text.

In the following discussions we will try to present the geology and structure in as unbiased a way as possible. However, there remain many different interpretations, not only in the literature but also among the authors of this book. Thus we present the differences in interpretations where necessary. In essence we will present a picture of what is known and what is not known. We also will point to areas where new work is needed to resolve critical geological features and relationships. In the final chapter we try to offer regional interpretations for the development of the SE Tibetan Plateau, and the broader regional framework of SE Asia where necessary and how they relate to intracontinental deformation during post-collisional convergence of India relative to Eurasia. Our analyses have led us to interpretations that are to some degree unconventional but testable.

## MAPS THAT ACCOMPANY THIS VOLUME

### Plate 1: Tectonostratigraphic Map

Plate 1 is a tectonostratigraphic map that combines individual stratigraphic units into larger units interpreted to divide the rocks into simplified assemblages that have tectonic significance as discussed in the text. These assemblages of units are grouped into the 14 major tectonic units of Plate 1. Greater detail for the stratigraphic units is discussed in the text and is shown on both the stratigraphic sections in the text and on the geological cross sections presented on Plate 4.

The map was constructed by redrawing from 96 1:200,000-scale Chinese map sheets. These map sheets were used as the basis of the geology shown on Plate 1, but with additions of folds and interpretations of faults shown by geological relations on the maps and observation that we have made in the field. These interpreted map sheets were then compiled into a single map sheet (Plate 1). However, the compilation is not well geographically registered because of distortions induced by the non-stable base of the redrawn maps, a problem that was unavoidable. Attempts to geographically register the maps were made but were not completely successful. Geographic coordinates are shown on Plate 1, but they must be taken as approximate.

The color scheme presented on Plate 1 was designed to bring out the distinctions between the major tectonic units. Certain colors were used to keep large time intervals uniform across the map, but different color hues were used within the different tectonic units. For example, blue colors were used for Mesozoic rocks but were assigned different hues between different tectonic units. These color assignments can be seen quickly by reference to the Legends for the different tectonic units on Plate 1.

All the features and units on Plate 1 are in different layers so that the digital map can be fully manipulated. The reader can upgrade the map as new data become available by changing different areas on the map by removing and adding only the data they deem to be necessary on the original map: One needs only to open the layers necessary to make changes, locking the remaining layers. For interpretive purposes, any of the units or linework for structures and contacts can be highlighted for study by hiding all unwanted layers and leaving only those layers of interest displayed; e.g., if the structures that involve Paleogene rocks are of interest, the distribution of Paleogene strata can be highlighted, hiding all other tectonostratigraphic units; then the relation of these rocks to mapped structures can be examined. The map can be used as a base map for specific areas by outlining areas of interest, and then copying and transferring the area to another file that can then be modified for a specific purpose. New layers can be added, overlaid on the map, if other data not shown are useful.

**Plate 2: Tectonostratigraphic Map of the Longmen Shan and Regions to the East**

Plate 2 is a colored, larger scale version of the tectonostratigraphic map of the Longmen Shan region from Burchfiel et al. (1995) and is designed to be placed in the upper right corner of Plate 1. It is necessary to include this map, because many of the major structures have been better studied in this region and continue onto Plate 1 so that the entire western margin of the Yangtze Unit can be integrated. Some of the major structural boundaries that are relevant in the text-discussions of Plate 1 are highlighted: The eastern boundary faults of the early Mesozoic Longmen Shan thrust belt are highlighted with a broader white overlay, similar to the boundaries of the tectonic units on Plate 1. The eastern limit of the early Mesozoic thrust faults has been strongly modified by younger thrust, normal, and strike-slip faults; thus the boundary shown is irregular. Within Plate 2, where it adjoins Plate 1 in the south, the two thrust faults of the Longmen Shan thrust belt—the Jinhe-Qinhe and Xiaojinhe thrust faults—are shown separately.

**Plate 3: Unconformity Map**

Plate 3 shows color codes for all the major unconformities by age; it color-codes the age at the base of the first strata lying above the unconformity. The color code is shown on the map legend. The map is useful because it shows the areal distribution of major unconformities that display the distribution, both areally and locally, where major deformations ceased. For example, in the south-central part of Plate 1, within the Lin Cang and Changning-Menglian units and adjacent parts of the Baoshan and Lanping-Simao units, it shows that major deformation ceased in this region in pre–Middle Jurassic time, distinguishing this region from other tectonic units. Like all the digital maps in this report, this map can be upgraded or used for interpretation by highlighting areas of distribution of specific unconformities and hiding the others.

**Plate 4: Geological Cross Sections on Plate 1**

Plate 4 contains cross sections on Plate 1 for two layers. Layer 1 contains the cross sections as only slightly changed from the original cross sections on the 1:200,000-scale Chinese map sheets. Layer 2 contains a generalized interpretation of the deeper structure made by the authors of this work.

Layer 1 contains more detail for the stratigraphic units than is shown on Plate 1 and uses the international color designations, the same color scheme as shown on the stratigraphic sections presented in the figures within the text. Stratigraphic details are generalized from those given on the Chinese maps for both the cross sections and stratigraphic sections.

Layer 2 can only be considered as simplistic interpretations of the deeper structure. The Chinese cross sections in layer 1 were made prior to many geological advances in the interpretation of thrust and fold belts. Even at present there is essentially no direct information for the subsurface structure below the region of Plate 1. Some of the problems in interpreting the deeper structure shown on layer 2 include the following: (1) surface geology confirms that at least some, if not most, of the structure on the cross sections involves basement rocks, but the depth and details remain unknown; (2) faults are of many different ages, but details of their ages are poorly constrained or unknown so that cross-cutting relations at depth also remain unknown; (3) some faults exhibit recurrent movement but not in enough detail to be shown; (4) geological maps are not detailed enough for quantitative assessment of fault geometry to be accurately made; and (5) from field observation there is a tendency for the geological maps to overestimate dips on many faults, and low-angle faults on the maps are extremely rare, even in areas where we have observed them.

Because Plate 4 presents data in two layers, the deeper structural interpretation can be hidden in order to see only the cross sections shown on the original geological maps.

## MAJOR TECTONIC UNITS OF THE SE TIBETAN PLATEAU AND ITS FORELAND

We divide the southeastern Tibetan Plateau and adjacent foreland regions into 14 tectonic units (Fig. 1-1 and Plates 1 and 2). They represent the major crustal-lithospheric units formed and assembled during late Paleozoic and Mesozoic time and modified during Cenozoic post-collisional convergence in the region north and east of the eastern Himalayan syntaxis. The tectonic units are, in order of presentation:

I. Yangtze Unit
II. Transitional Unit
III. Songpan Ganze Unit
IV. Yidun Unit
V. Qiangtang Unit
VI. Lanping-Simao Unit
VII. Lhasa Unit
VIII. Bangong Suture Zone
IX. Syntaxial Unit
X. Tengchong Unit
XI. Baoshan Unit
XII. Changning-Menglian Unit
XIII. Lin Cang Unit
XIV. Ailao Shan Unit

Other tectonic units could be delineated. However, these are the ones we recognize, and other possible units not shown in Figure 1-1 will be mentioned where necessary.

# CHAPTER 2

# *Yangtze Unit*

The Yangtze Unit, often referred to as a *paraplatform,* consists of Paleozoic through Cenozoic strata in the southeastern and eastern parts of the map area (Figs. 1-1B and 2-1; Plates 1 and 2 [available on DVD accompanying this volume and in the GSA Data Repository[1]]). Mesozoic strata define the Songpan Ganze, Transitional, and Yidun Units to the west, but upper Neoproterozoic to Permian strata that at least locally underlie them is of the Yangtze type and will be discussed with those units. This area lies north of the Ailao Shan metamorphic belt, whose original position relative to the Yangtze Unit remains uncertain, but it may be a displaced and metamorphosed part of the Yangtze Unit. The Zongza Unit, a narrow, NNW–trending belt of Paleozoic strata that lies east of the Qiangtang Unit and underlies the western part of the Yidun Unit, contains a Yangtze stratigraphic succession and forms the basement rocks for the western part of the Yidun Unit (Fig. 2-1). The Yangtze Unit is underlain by Precambrian rocks, mainly Proterozoic metasedimentary, metamorphic, and igneous rocks, and is overlain by a few kilometers to >10 km of upper Neoproterozoic to Mesozoic strata of mainly marine shallow-water or non-marine rocks with variable lateral development. The Yangtze Unit is often called a *paraplatform* because of its relatively tectonically stable character throughout most of the Paleozoic with identifiable continental margins in the area of the Longmen Shan to the northwest and the Qingling to the north, and it extends to the east across much of southeast China beyond the limits shown on Plates 1 and 2. Its original margins to the south and southwest have been tectonically modified, and the difficulty in defining these margins will be discussed below.

The platform designation for this tectonic unit is also derived from its sequence of shallow-marine and non-marine sedimentary rocks with numerous unconformities. It is often referred to in the literature as the *South China block,* but we reject the term *block,* as it implies a structural rigidity that was lacking, particularly during Mesozoic and Cenozoic time. Rocks of the Yangtze Unit are deformed throughout the area of Plate 1 and in the western and eastern parts of that shown on Plate 2, but are only mildly deformed within the Sichuan basin, seen on Plate 2. Whereas the sedimentary succession varies considerably across the Yangtze Unit, our interest lies mainly in sedimentary rocks that overlie a complex Proterozoic basement and their tectonic history; thus basement rocks are discussed only briefly below.

The Yangtze platform can be subdivided into four overlapping subunits: the South China fold belts, the Kungdian high, the eastern Longmen Shan, and the Chuxiong basin (Fig. 2-1). The eastern part of the Yangtze Unit (Plates 1 and 2) has been deformed into generally NE- to E-W–trending folds—the South China fold belts—which over much of the area probably involve Precambrian basement. Toward the west the folds become more irregular in trend and increasingly involve more basement rocks, which can be seen by more extensive exposures of basement rocks (Plates 1 and 2). On Plates 1 and 2 only a small part of a broad belt of Mesozoic deformation covers much of the eastern part of southeastern China, where folding occurred at several times and so cannot be regarded simply as a single fold belt; thus the designation *South China fold belts* (see below). The Kungdian high is a north-south–trending region underlain by basement rocks of the Yangtze Unit that were intermittently positive during the Paleozoic and where Paleozoic strata are commonly incomplete, thin, wedge out, or faulted out on both its eastern and western flanks. Stratigraphic relations within the Paleozoic strata indicate that the high was structurally active at various times during the Paleozoic. Along its crest, Upper Triassic strata lie directly on Proterozoic basement rocks. Within the area of Plate 1 this is shown on the unconformity map, Plate 3 [see footnote 1], by the unconformity that underlies Upper Triassic [T3] rocks in the central part of the Yangtze Unit and can be followed northward into the Longmen Shan region shown on Plate 2. To the west the high underlies the eastern part of the Chuxiong basin, which contains up to 10 km of Triassic to early Cenozoic, dominantly non-marine red beds. The Kungdian high is well displayed by the north-south–trending belt of Precambrian basement rocks exposed within in the central part of the Yangtze Unit (Plate 1).

## BASEMENT ROCKS OF THE YANGTZE UNIT

Basement rocks of the Yangtze Unit shown on Plates 1 and 2 are the western part of a more extensive region composed of two accreted continental fragments, the Yangtze and Cathaysian fragments, to the northwest and southeast, respectively; the boundary between these two fragments is eastward, beyond the limits of Plates 1 and 2. The time of suturing of these two fragments remains somewhat controversial, regarded by Li et al. (2002) as latest Precambrian, by Lingzhi et al. (1989) as early Paleozoic, and by Wang et al. (2007), Yan et al. (2006), and Hsü et al. (1987, 1988) as late Paleozoic and early Mesozoic. Recent studies by Liang et al. (2004) and Xiao and He (2005) have supported the collision of these two fragments as Late Permian and Early

---

[1]GSA Data Repository Item 2012297, Plates 1–4, is available at www.geosociety.org/pubs/ft2012.htm, or on request from editing@geosociety.org or Documents Secretary, GSA, P.O. Box 9140, Boulder, CO 80301-9140, USA.

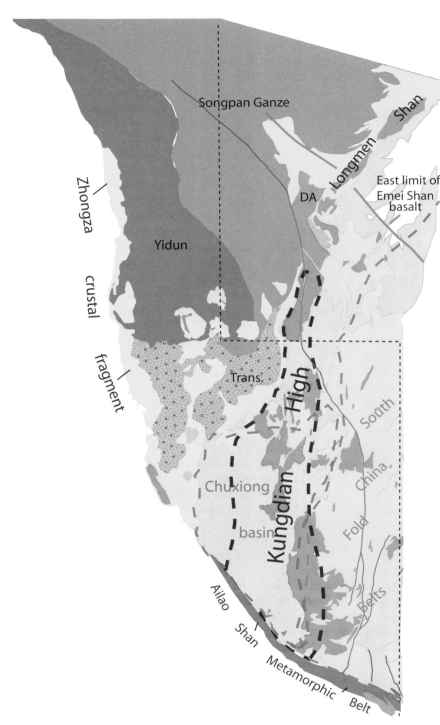

Figure 2-1. Location of Yangtze Platform Unit (red, Precambrian; yellow, Paleozoic; light blue, Mesozoic) rocks in the southeastern parts of Plates 1 and 2. The late Neoproterozoic–Paleozoic rocks of the Yangtze Unit in the western part of Plates 1 and 2 underlie the Mesozoic strata of the Songpan Ganze and Transitional (Trans.) Units, and locally, the Yidun Unit. These units are defined on the basis of their Mesozoic strata, which are different from those of the Yangtze Unit. Along the western side of the Yidun and Transitional Units is a narrow belt, the Zhongza fragment, which contains upper Neoproterozoic and Paleozoic strata of the Yangtze type and which underlies Mesozoic strata that define these two units. It is bounded on the south by the Ailao Shan metamorphic belt. Interior subdivisions of the Yangtze platform unit, the Chuxiong basin, the Kungdian high, and the South China fold belts are also shown. DA—Danba antiform. Dashed line is boundary between Plates 1 and 2.

Triassic, although a Neoproterozoic to early Paleozoic accretion is more likely (Wang and Li, 2003a, 2003b). Although the boundary between these two fragments lies to the southeast of the area of Plate 1, and the following discussion of the Yangtze Unit is focused primarily on the area within Plate 1, there is a thick succession of Proterozoic clastic strata in the southeasternmost part of Plate 1 that is unlike the characteristic shallow-water strata of the remainder of the Yangtze Unit (Figs. 2-2 and 2-3). These rocks may represent the southernmost exposures of the passive margin of the Yangtze platform within the map area or the southern extent of a NE-trending Neoproterozoic basin (Nanhua basin) within the Yangtze-Cathaysian craton (see Li et al., 2003,

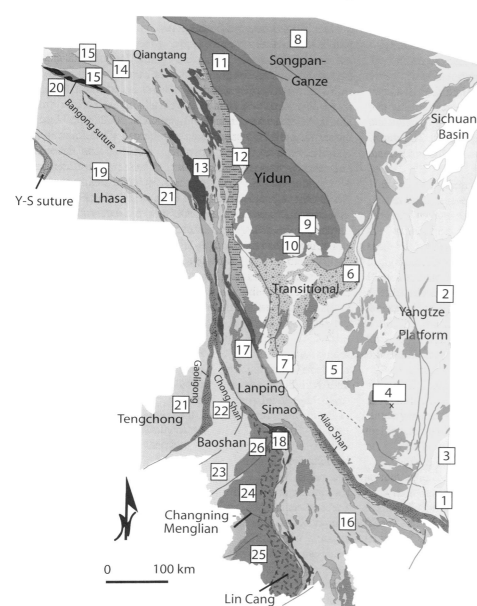

Figure 2-2. Locations of stratigraphic section figures throughout the text by figure number and designation (e.g., no. 1 for location). X below location no. 4 is the location of Figure 2-7.

and Wang and Li, 2003a, 2003b). Also the tectonic activity in the eastern part of Plate 1 is the western and marginal part of the broadly deformed region in southeastern China that consists of locally metamorphosed rocks farther east with abundant magmatic activity that must be considered to gain an understanding of the tectonic setting shown on Plate 1.

Basement rocks of the Yangtze Unit are shown as one unit in Plates 1 and 2 (unit pt), although in detail they consist of two major rock groups of Mesoproterozoic and of older Neoproterozoic age. They are mostly covered by upper Neoproterozoic and Paleozoic strata, and the most extensive outcrops are in the middle of the Yangtze Unit (Plate 1) and within the eastern part of the Longmen Shan (Plate 2). They are more rarely exposed within some folds and thrust faults in the eastern part of the area of Plate 1, referred to below as the South China fold belts. In some parts of the Yangtze Unit, although not in the area of Plate 1, are exposures of Archean granitic gneiss that has yielded U-Pb ages of 2.90–2.95 Ma (Zheng et al., 2006). Studies by Zheng et al. (2006) of zircons from lamporite diatremes within Proterozoic rocks in several areas have yielded ages of 2.9–2.8 Ga and 2.6–2.5 Ga, and Hf model ages of 2.6 to ca. 3.5 Ga or older, suggesting that Archean rocks may be widespread beneath the younger cover of the Yangtze Unit. Mesoproterozoic and older Neoproterozoic rocks form extensive outcrops (Plate 1). The Mesoproterozoic rocks (pt) crop out mainly along the Kungdian high, paleo-positive area that trends north-south within the central part of the

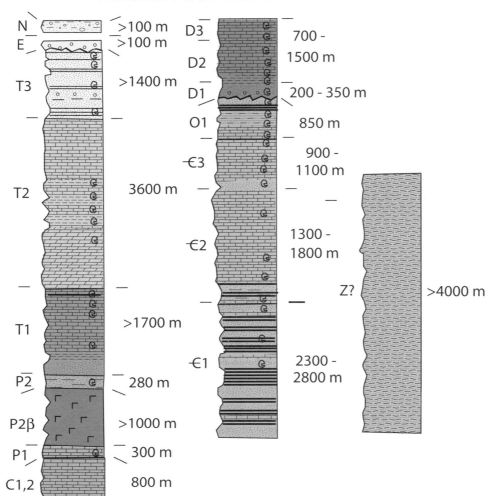

Figure 2-3. Stratigraphic section for rocks in the southeast corner of the Yangtze platform (for location, see Fig. 2-2, no. 1). All stratigraphic sections in the text show international designations for ages of rocks and color scheme, and the stratigraphic units shown on the cross sections use the same scheme, but the details of the stratigraphic units are generalized on the cross sections in Plate 4.

Yangtze Unit (Plate 1) and extend north, forming the Danba antiform (DA) in the southern Longmen Shan region (Fig. 2-1; Plate 2; Li et al., 2003; see also Burchfiel et al., 1995). To the east, rare exposures of these rocks occur within folds and along thrust faults that involve Paleozoic and Mesozoic strata of the South China fold belts. The older Neoproterozoic and Mesoproterozoic rocks are characterized by unmetamorphosed or low- to medium-grade meta-sedimentary and meta-igneous rocks that were intruded by large volumes of mafic to felsic plutonic rocks that commonly date from 950 to 850 Ma (Yao et al., 2012: shown as pt in Plate 1, where the meta-igneous and meta-sedimentary rocks have not been separated). The metamorphic rocks have been strongly faulted and folded prior to deposition of the overlying younger Neoproterozoic sequences (Za and Z2). This N-S belt of Precambrian rocks contains numerous N-S–striking faults, some of which can be shown to have Proterozic displacements, and many of which have been reactivated at a later time. This belt is often referred to as the *Panxi rift* (see Li et al., 2003).

## LOWER PART OF THE YOUNGER NEOPROTEROZOIC ROCKS OF THE YANGTZE UNIT

The Neoproterozoic rocks of this unit consist of two major sequences, an older sequence of volcanic and clastic rocks of variable thickness (Plates 1 and 2, Za), and a younger unconformable or disconformable overlying sequence of shallow-marine to nonmarine strata that form the base of the platform cover sequence of Paleozoic and Mesozoic age (Plates 1 and 2, Z2; see below). The older sequence of younger Neoproterozoic rocks unconformably overlies the older Neoproterozoic and/or Mesoproterozoic metamorphic and plutonic rocks (see distribution of exposed Za unconformity on Plate 3). They consist of a thick sequence of largely non-marine maroon, red, tan, and white coarse-grained sandstone and conglomerate that may reach >3 km in thickness (Za: often referred to as the lower *Sinian System* on Plates 1 and 2). Some of the conglomerate is interpreted to be of glacial origin. The sections locally contain thick sequences of felsic and mafic volcanic rocks. The sedimentary rocks are largely uncleaved, except locally where affected by strong Mesozoic and Cenozoic deformation. This older sedimentary sequence is less extensively exposed than the underlying metamorphic and igneous rocks, a distribution at least partly related to their deposition in grabens or half grabens formed during Neoproterozic rifting. Because of the rift depositional setting, facies and thickness changes were rapid, and the units are commonly only locally developed. Wang and Li (2003a, 2003b) have reviewed the stratigraphy of these rocks and suggested that they were deposited during four different rift phases that began ca. 820 Ma and ended ca. 690 Ma. The exposures of these rocks are most extensive along the Kungdian high, which has led to the interpretation that this was the site of a tectonic element, the Panxi rift, during Neoproterozic time (Li et al., 2003); limited exposures, however, make their regional distribution and limits of the Panxi rift hard to reconstruct, and correlative rocks are also present locally within the fold belt to the east (Plate 1).

# LATE NEOPROTEROZIC TO CENOZOIC SEDIMENTARY SUCCESSIONS IN THE YANGTZE UNIT

The Yangtze Unit in general is a region characterized by widespread deposition of strata in shallow-marine and non-marine environments from latest Neoproterozoic to early Cenozoic time, but there is variability in the deposition of these rocks within the major subdivisions of the Yangtze Unit (Plates 1 and 2). Because of these variations the important stratigraphic successions will be discussed for the major structural subdivisions of the Yangtze Unit (Fig. 2-1).

## General Distribution

The South China fold belts form the eastern part of the Yangtze Unit in Plates 1 and 2 (Fig. 2-1). Even though fold and fault trends to the northeast or east are commonly parallel, different parts of the region were interpreted to have been deformed at different times, perhaps from Late Triassic to Early Cretaceous time in its eastern part, but into early Cenozoic time in its western part. Many parts were deformed more than once; thus the plural designation of the fold belts. The geology of the fold belts contrasts with the western part of the Yangtze Unit, such as in the Kungdian high, where deformation during Paleozoic time generally consisted of faults that influenced lateral facies distributions and thicknesses of strata. Numerous unconformities are present in the succession from latest Neoproterozoic to early Cenozoic time (see Plate 3). During early Late Permian time there was widespread mafic igneous activity, both intrusive and extrusive, centered in the Transitional Unit, which produced the voluminous Emei Shan basalts (Pβ in Plate 1) that are present across the entire southern part of the Yangtze Unit, and reaching thicknesses >4 km in their western part and >5 km thick in the adjacent Transitional Unit. These basalts have been interpreted to have been extruded above a mantle plume (see below; e.g., Xu and Chung, 2001; Shellnutt et al., 2008). Folds in the South China fold belts extend into the area of the Kungdian high, and rocks of the Chuxiong basin to the west of the Kungdian high are also folded, but their folding is largely unrelated to the folds in the South China fold belts farther east (see below).

## Sedimentary Succession of the South China Fold Belts

The sedimentary succession of the South China fold belts is characterized by mostly shallow-marine, and rarely non-marine, uppermost Neoproterozoic, Paleozoic, and early Mesozoic deposits (Figs. 2-3 through 2-6; see Fig. 2-2 for locations: Note that all stratigraphic sections are more detailed than the combined units shown on Plates 1 and 2, and that international symbols and colors are used). Plate 1 shows well the general distribution of rocks in the South China fold belts that crop out in four broad, NE-trending belts. In the north is a belt of Paleozoic and Mesozoic strata with extensive outcrops of Emei Shan basalt, followed to the south by a belt of primarily Paleozoic strata that lie below the Emei Shan basalt and expose some large areas of basement rocks. To their south is a broad belt of Triassic age, followed to the southeast by Paleozoic strata that contain very thick sections of Cambrian and latest Neoproterozoic rocks. Basement rocks (both Za and pt) are exposed in a few folds or faulted folds in the northern two belts, and are not exposed in the southern two belts, but they become more widely exposed to the west where the folds of the northern two belts merge into the Kungdian high. In the northwestern part of the Yangtze Unit, Jurassic and Cretaceous rocks thicken to the west, forming parts of a foredeep for major east-vergent thrusting and folding along the southern continuation of the Longmen Shan thrust belt that is unrelated to formation of the South China fold belts (e.g., Plate 4, sections 5 and 6 [see footnote 1]; Plate 2; and see below, Burchfiel et al., 1995; Chen and Chen, 1987).

Two representative sections for the rocks of the South China fold belts are shown in Figures 2-3 and 2-4. The basal uppermost Neoproterozoic unit (Z2 on Plate 1, referred to in Chinese literature as Sinian in age; for additional details, see also Chen and Chen, 1987; Bureau of Geology and Mineral Resources of Sichuan Province, 1991; Bureau of Geology and Mineral Resources of Yunnan Province, 1990) begins with a basal conglomerate and sandstone, which unconformably overlie both the upper Neoproterozoic coarse-grained sedimentary and volcanic rocks (Za) and more commonly the lower Neoproterozoic and/or Mesoproterozoic metamorphic basement rocks (pt on Plate 1; Plate 3). Above is a unit of thin-bedded, commonly nodular dolomite and limestone that varies in thickness from a few meters to >1600 m and that forms the characteristic basal unit of the Yangtze Unit; however, it is not present everywhere, as younger Paleozoic strata locally rest on older rocks. These rocks are known for their interbeds of phosphatic shales with abundant Ediacaran fauna in some areas (e.g., Luo and Jiang, 1996; Xiao et al., 1998; Xing and Luo, 1984; Zhang and Babcock, 2001). Above lies a Cambrian to Permian sequence of sedimentary rocks that contain disconformities and low-angle unconformities that become more common to the west along the eastern part of the Kungdian high (Plate 3, and also, e.g., see Plate 4, sections 2, 3, 4, 5). Cambrian to Middle Devonian (Pz) rocks are generally interbedded units of sandstone, shale, and limestone, whereas from Middle Devonian to Permian the sections are dominated by limestone. The entire Paleozoic section is several kilometers thick. The most important unconformity is below Devonian rocks, and in places Devonian strata rest directly on basement rocks (see Plate 4, section 3). Such relationships appear to have developed where major faults brought up older rocks prior to Devonian sedimenation. Locally, Ordovician and/or Silurian rocks are missing at unconformities. Toward the Kungdian high to the west the unconformities become more numerous and obvious, and more of the section is

## NORTHEAST YANGTZE UNIT

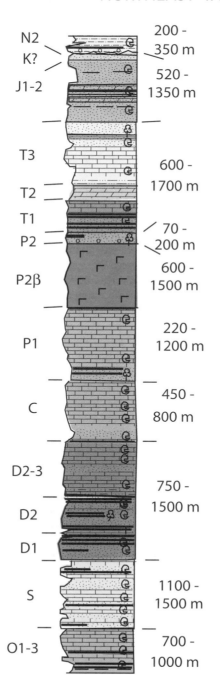
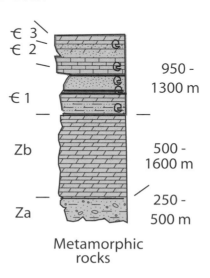

Figure 2-4. Stratigraphic section of rocks in the northeast corner of the Yangtze platform of Plate 1 (see Fig. 2-2, no. 2, for location).

missing (e.g., Fig. 2-5). Many of the Paleozoic units thin as they approach the Kungdian high, and different parts of the Paleozoic section rest directly on the Precambrian basement rocks, demonstrating that the high was an active positive area at several times during Paleozoic time (Plate 3 and Plate 4, sections 5, 6, 7). One of the most widespread, and perhaps the most important, Paleozoic units is the Emei Shan basalt of late, but not latest, Permian age (ca. 260 Ma), which is present throughout most of the South China fold belts, extending across the entire Yangtze Unit to the west, and shown as a separate unit on Plates 1 and 2 (Pβ). It varies in thickness from ~100 m to >1000 m over most of the area, where it consists of numerous basalt flows, but the unit is locally cut out to the west and is missing over the Kungdian high where Triassic strata lie on the basement rocks (Plate 3,

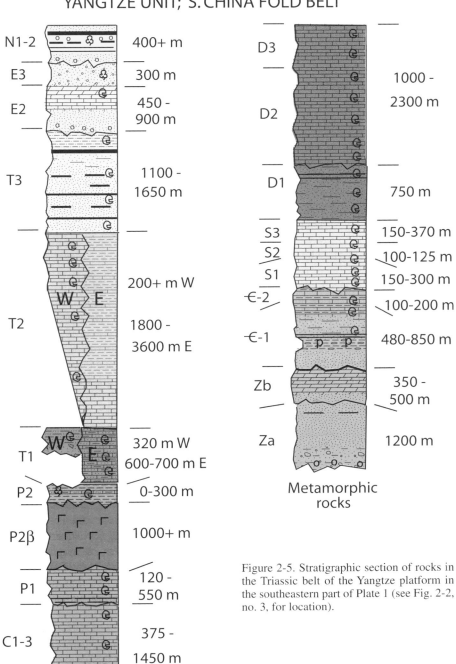

Figure 2-5. Stratigraphic section of rocks in the Triassic belt of the Yangtze platform in the southeastern part of Plate 1 (see Fig. 2-2, no. 3, for location).

and Plate 4 sections including the eastern end of section 32; also Fig. 2-6). The Emei Shan basalt is missing from an area shown in the northeastern part of Plate 2, along a generally NW-trending line (Fig. 2-1).

In the southeastern part of the South China fold belts is an area underlain by an anomalously thick sequence (8+ km thick, probably a structural thickness) of Paleozoic and latest Neoproterozoic strata (Plate 1 and Fig. 2-3). The oldest rocks conformably underlie dated Lower Cambrian strata and consist of a thick succession (4+ km; probably a structural thickness) of low-grade phyllitic meta-sandstone and shale that is assigned a Neoproterozoic age (Z2, Plate 1). These rocks are lithologically unlike the Neoproterozoic strata of the typical Yangtze succession to the north, from which they are separated by a broad belt of younger strata so that their lateral relations cannot be established. Conformably above them are well-dated (trilobite and brachiopod) strata that range from Early to Late Cambrian in age. Cambrian strata consist of a lower part dominated by sandstone and shale and an upper part that is dominantly limestone. These Cambrian strata are similar to the strata in the Yangtze platform, but they are much thicker (compare Fig. 2-3 with Fig. 2-4; Plate 4, sections 1, 2). These strata are overlain by Ordovician strata that are unconformably overlain by Devonian to Permian strata, which include the upper Permian Emei Shan basalt. The Devonian to Permian rocks are similar in lithology and thickness to the Yangtze succession. This area of anomalously thick Cambrian and Neoproterozoic strata is faulted on the SW side against Triassic rocks along the NW-striking Nanxihe fault and are overlain on the north side by a broad area of late Paleozoic and Triassic rocks in continuity with the Yangtze platform succession. Although west of the Nanxihe fault no rocks older than Permian Emei Shan basalt are exposed, and because the post-Devonian rocks are similar to the Yangtze rocks to the north, the Nanxihe fault may have a large pre-Devonian displacement. Eastward the Cambrian and Neoproterozoic rocks extend into South China and northwestern Vietnam beyond the area of Plate 1. The thick section of Cambrian and Neoproterozoic rocks indicates greater subsidence and deeper water deposition than the more characteristic Yangtze platform succession, which may suggest an approach to a continental margin or basinal environment (perhaps the southern extent of the NE-trending Nanhua basin in Southeast China; Wang and Li, 2003a, 2003b); but their depositional setting remains unclear (see, e.g., Wang and Li, 2003a, 2003b). Importantly, along their south side they are truncated by mylonitic metamorphic rocks of the Day Nui Con Voi belt that are often correlated with the Ailao Shan mylonite belt but separated from it by a narrow belt of Triassic rocks (see below, and Burchfiel et al., 2008a). The Day Nui Con Voi mylonitic rocks

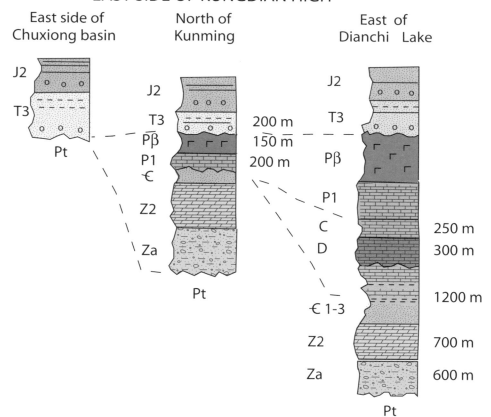

Figure 2-6. Stratigraphic sections showing the thinning and wedging out of units onto the eastern side of the Kungdian high. The way the units are cut out to the west indicates that the high was positive at different times during the Paleozoic (see Fig. 2-2, no. 4, for location).

are interpreted to be an important Cenozoic left-lateral shear zone (Tapponnier et al., 1982), to be discussed below. Large-magnitude displacement on either or both the Nanxihe fault or the Day Nui Con Voi shear zone may have caused removal of the southwestward continuation not only of the typical rocks of the Yangtze Unit but also of the thick succession of pre-Devonian strata, a problem discussed below.

The Mesozoic rocks (Mz on Plate 1) of South China fold belts consist of shallow-water marine and non-marine rocks, with the non-marine rocks dominant in the upper part. The characteristic platform succession of the South China fold belts consists of 3–4 km of Lower Triassic to Jurassic strata. Locally, however, particularly to the northwest, the section continues into the Cretaceous (see NE corner of Plate 1 labeled Mz2). Lower Triassic strata overlie both conformably and unconformably (mostly disconformably) Paleozoic rocks, and in places Middle Triassic rocks are at the base of the succession. Over much of the unit, particularly to the east and southeast, Triassic strata are 1–4 km thick and consist of a basal section of Lower Triassic sandstone and shale, overlain by a thick succession of limestone that began either in the late Early Triassic or Middle Triassic, but which is largely Middle Triassic in age (Figs. 2-3, 2-4, 2-6). Upper Triassic strata are mostly terrigenous clastic rocks that commonly contain important coal beds. The Jurassic strata in most places form the uppermost exposed part of the Yangtze succession, but they are missing to the southeast because only older rocks are exposed. The Jurassic strata consist mostly of non-marine terrigenous clastic rocks, commonly red and maroon, and they usually conformably overlie Triassic rocks. Because they are at the top of the exposed section they exhibit variable thickness, but they may reach more than one kilometer thick. In the northwestern, and more rarely in the western, part of the area they are conformably to unconformably overlain by non-marine Cretaceous terrigenous clastic strata that are commonly red and maroon. In turn these strata are locally overlain by lower Cenozoic red clastic strata. Over much of the unit, lower Cenozoic strata consist of small, irregular areas of a few hundred meters of brown, tan, and red non-marine conglomerate, sandstone, mudstone, and local units of lacustrine limestone (E on Plate 1) that rest unconformably on deformed rocks of different ages (particularly in the SE part of the fold belts; Plates 1 and 3), and they are commonly folded and faulted, but not as strongly as the underlying rocks. Most of these strata do not contain diagnostic fossils and are assigned a Paleogene age on the basis of lithological correlation with a few well-dated sections. In the south-central part of the area are several larger areas underlain by well-dated Eocene to Oligocene strata that in places are >1 km thick (e.g., see Fig. 2-5). Outcrops of lower Cenozoic rocks tend to be elongated parallel to the underlying structures where they are involved in post- or synsedimentary faulting and local folding. Non-marine Neogene strata (N) up to a few hundred meters thick are present as tan and light-gray sandstone and mudstone with local conglomerate and lacustrine limestone. These strata occur in small irregular basins or small outcrops locally dated as Miocene, Pliocene, or Quaternary. In some areas, such as that around Kunming, Neogene strata underlie modern lake basins. Many of the Neogene rocks were deposited in basins related to pull-apart basins and other types of young and active fault-related basins (Wang et al., 1998).

The Mesozoic section (Mz, Plate 1) is thickest in the eastern part of the area and thins toward the west, where it oversteps Paleozoic strata and Upper Triassic coarse-grained sandstone and conglomerate and onlaps the basement rocks in parts of the Kungdian high (Plate 4, sections 5, 36; see also Plate 3). In the southeastern part of the South China fold belts the wide belt of thick Triassic strata contains lower and upper units of terrigenous clastic sedimentary rocks and a middle section dominated by limestone. However, to the east of Plate 1 these shallow-marine to locally non-marine strata grade into flysch-like sediments (Fig. 2-5).

## Kungdian High

The Kungdian high is a paleogeographic element (Fig. 2-1) that displays a history of intermittent uplift during Paleozoic and early Mesozoic time that influenced deposition. The high trends generally N-S and exposes large areas of Mesoproterozoic and lower Neoproterozoic rocks unconformably overlain by incomplete sections of uppermost Neoproterozoic and Paleozoic strata that characteristically, but not always, thin westward onto its eastern flank with more numerous unconformities within Paleozoic strata than are present in the South China fold belts in the east (Fig. 2-6; Plates 1, 3, and 4, sections 7, 29, and the eastern ends of sections 32, 33). Much of the western flank of the high is not well exposed, being covered by Mesozoic strata of the Chuxiong basin, but rocks adjacent to the Transitional Unit in the west expose part of its western flank (see below). At various times during the Paleozoic and Triassic time the Kungdian high can be identified to have extended into the southern Longmen Shan area, forming the broad, northwest-plunging Danba antiform (DA; Fig. 2-1 and Plate 2; Chen and Chen, 1987; Burchfiel et al., 1995).

At the crest of the high, basement rocks, including the Neoproterozoic sedimentary and low-grade metamorphic rocks (Za and Z2) of the Yangtze Unit section, are overstepped by Mesozoic strata, commonly Upper Triassic conglomerate and coarse-grained sandstone (Fig. 2-5). Activity within the Kungdian high is sometimes taken to have begun in late Proterozoic time, when thick sections of upper Neoproterozoic strata (Za) were deposited in N-S–trending grabens (see Wang and Li, 2003a; Li et al., 2003; Chen and Chen, 1987). However, this activity is perhaps more logically assigned to the widespread extensional tectonism of Proterozoic age that extended across much of the Yangtze Unit. Because these rocks are mostly exposed today along the Kungdian high, the high has been often regarded as having been part of the Panxi rift, a rift zone that is thought to have influenced sedimentation and structure from Precambrian to Cenozoic time (see, e.g., Chen and Chen, 1987; Liu et al., 2001). We do not adhere to this interpretation, because the Kungdian high displays a broader pattern of uplift during Paleozoic to early Mesozoic time, and was shortened at times in the Mesozoic and Cenozoic. It does exhibit evidence for generally N-S normal faulting during parts of its evolution, which influenced sedimentation during Paleozoic to Cenozoic time (see below).

Along the Kungdian high, and in underlying parts of the Chuxiong basin, the Emei Shan basalt is missing because of erosion prior to the overlap by Upper Triassic strata. This basalt also is missing north of a NW-trending line to the north in Plate 2 (Fig. 2-1; see below). The Emei Shan basalt extends west and north beyond the Yangtze Unit and reaches thicknesses of >5000 m in the Transitional Unit (see below). Mafic dikes and mafic-ultramafic and related plutons were intruded into the Precambrian rocks and remnants of the Paleozoic rocks, including remnants of the Emei Shan basalt, along the eastern margin of the Kungdian high, indicating that Emei Shan magmatic activity was present even though the characteristic basalt is missing. These plutons lie within the north-south–trending belts of Precambrian rocks in the northern part of the Kungdian high, labeled $\delta^3_4$ and $v^3_4$ (see Plate 1), and were studied recently by Shellnutt and Jahn (2010), Zhang et al. (2009), and Shellnutt et al. (2008). The intrusive bodies range from layered mafic-ultramafic intrusions to syenodiorite and A-type granitic rocks. The ages of some of the plutons range from 263 + 3 Ma to 259 + 3 Ma and are coeval with the main body of the Emei Shan, which yields ages of ca. 260 Ma. The interest in these plutons is that they are the host to important deposits of Fe-Ti-V oxide deposits. There may be more Emei Shan–related intrusive bodies that are covered by younger deposits, such as the thick Mesozoic strata of the Chuxiong basin. The Emei Shan basalt is exposed over a large part of the Yangtze and Transitional Units as well as within the southern parts of the Yidun and Songan Ganze Units, where it was extruded into a marine environment. A more extended discussion of the Emei Shan is given in the section on the Transitional Unit, below where the thickest section of Emei Shan basalt is exposed.

The eastern margin of the Kungdian high trends approximately north-south, but it is hard to determine precisely because it is an irregular transition zone with thicker and thinner parts of the Paleozoic section preserved along its margin. Near its eastern margin, and within the eastern part of the high, there is evidence of complex and repeated high-angle faulting that marks the transition from the South China fold belts into the Kungdian high, as shown by the unconformities of different ages that overlie the basement and remnants of lower Paleozoic strata (Plate 3). For example, in the area 40 km NW of Kunming along the N-S–striking active Yimen left-lateral fault (Wang et al., 1998), stratigraphic units from late Neoproterozoic to Mesozoic age show facies and thickness changes, indicating that faulting was active at many different times during sedimentation. There is a thick section of lower Neoproterozoic clastic rocks (Za) east of the fault that are missing west of the fault, but the uppermost Neoprotorozoic dolomite (Z2) is present on both sides of the fault. East of the fault, Upper Triassic strata rest on Permian rocks, whereas west of the fault they lie on uppermost Neoproterozoic (Z2) and Proterozoic metamorphic rocks (pt). These and other N-S–striking faults continued their activity into the Mesozoic, where they affected sedimentation along the northern margin of the Chuxiong basin (for another example, see also Fig. 2-7) and into the Cenozoic, where they controlled many of the N-S–trending exposures of older rocks. The major changes in Paleozoic stratigraphy, and particularly the structurally high position of the basement rocks along the Kungdian high, affected the younger structural development and contributed to different structural styles between the South China fold belts and the Kungdian high (see below).

Although the Kungdian high is most easily distinguished east of the Chuxiong basin as a feature where Upper Triassic strata rest on many different units of Proterozoic and Paleozoic age, its western side is more difficult to determine because it is covered by thick Mesozoic strata of the basin. The axis of the high can be approximately traced around the margins of the Chuxiong basin and in erosional windows below Mesozoic rocks in the northeastern and southeastern parts of the basin (see Plate 1 and Fig. 2-1). Incomplete sections of mainly upper Paleozoic rocks are present northwest of the Chuxiong basin near the boundary with the Transitional Unit, suggesting a generally irregular NNE trend to the northwest margin of the high, but the completeness of the Paleozoic section in this area changes across generally north-striking faults, suggesting that pre–Late Triassic faults have been important also along and west of the Kungdian high. The boundary between the Transitional and Chuxiong units is marked by thrust faults that have telescoped the Paleozoic and Mesozoic sections so that the western margin of the Kungdian high cannot be well located. To the south the western margin of the Kungdian High lies somewhere beneath the western part of the Chuxiong basin. In the western part of the basin, 10 km from its western bounding fault, a complexly thrusted anticline brought up Paleozoic rocks that lie unconformably below the Upper Triassic strata of the basin (Plate 1). Below the Upper Triassic is a section of Middle Devonian limestone unconformably overlain by Lower Permian limestone and Permian Emei Shan basalt. In the footwall of the western bounding thrust fault of the Chuxiong basin a section of Carboniferous and Permian limestone and Emei Shan basalt lie unconformably below Upper Triassic strata of the basin. These two exposures of Paleozoic rocks probably belong to the western flank of the Kungdian high, and the crest of the high should lie somewhere to the east. There is a prominent N-S–trending belt of Precambrian basement rocks exposed by later faulting that lies within the north-central part of the basin and is unconformably overlain by Triassic strata, clearly showing that the axis of the Kungdian high was covered by the thick Mesozoic strata of the Chuxiong basin. This belt of Precambrian rocks is covered by basin sediments to the south, but at the southernmost end of the basin, adjacent to the Ailao Shan metamorphic belt, are several small erosional windows that expose Precambrian basement rocks unconformably overlain by Middle Triassic strata, indicating that these exposures must have lain on or near the axis of the Kungdian high prior to Middle or

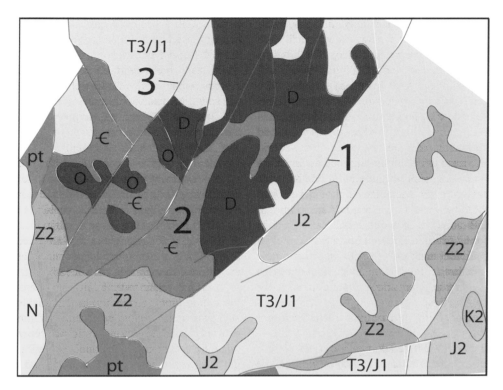

Figure 2-7. A representative geological map, showing the long history of recurrent faulting along the Kungdian high. (Location is x in Fig. 2-2.) Across fault 1, Cambrian and Devonian strata are missing in the erosional window to the east, where Upper Triassic–Lower Jurassic strata rest unconformably on uppermost Proterozoic Sinian strata. Movement on this fault is post-Devonian, pre–Late Triassic. Fault 1 was reactivated in Mesozoic time, as it displaces the Upper Triassic strata; however, note that the Mesozoic age displacement is east side down, counter to what would be the displacement to remove the Paleozoic rocks on the east side. Paleozoic displacement is shown by fault 2, where Ordovician strata lie conformably above Cambrian rocks to the west but are missing below Devonian strata east of the fault. Pre–Late Triassic displacement is shown on fault 3, where Devonian strata are present east of the fault but are missing on its west side. These types of displacements, as well as others, are shown on many of the faults along the Kungdian high. Some of the faults are presently active, indicating that the high has had a long history of faulting, one of the characteristics of the Kungdian high. It must be noted that in many places within the Yangtze platform, faults with Paleozoic displacement are also known (Plate 4, sections 3 and 4) where different Paleozoic-age strata are missing across faults. Such faults, however, are interpreted to be more abundant along the Kungdian high, some with movements interpreted to be Precambrian in age.

Late Triassic time. At this time the high became an area of rapid subsidence as >10 km of post–Middle Triassic strata was deposited along its crest and to the west in the area of the Chuxiong basin (Fig. 2-8; see below).

The Kungdian high can be followed to the north of the map area into the southern part of the Longmen Shan area, where it forms the Cenozoic Danba antiform in Plate 2 (Burchfiel et al., 1995; see Fig. 2-1, DA). Along the axis of the high in the northern part of Plate 1 is a large N-S–elongated plutonic complex that intrudes rocks as young as the Permian Emei Shan basalt and is unconformably overlain by Upper Triassic coarse-grained sedimentary rocks (Plate 1 and Plate 4, section 5). This Triassic plutonic complex contains rocks that range from diabase to granite and its position within the Kungdian high of the Yangtze Unit is anomalous and hard to explain. Its interpretation will be discussed below.

The northern part of the Kungdian high is narrower than to the south and has three unusual features that are difficult to explain. First, while it was a positive area at several times during the Paleozoic it was the site of significant subsidence during Jurassic and Cretaceous time and is overlain obliquely by the NNE-trending foredeep for the Longmen Shan thrust belt lying to the west (see below; Fig. 2-9). Following foredeep deposition, the high was uplifted in Cenozoic time along a trend that was apparently parallel to its earlier Paleozoic and pre–Upper Triassic trend, forming the broad NW-plunging Danba antiform

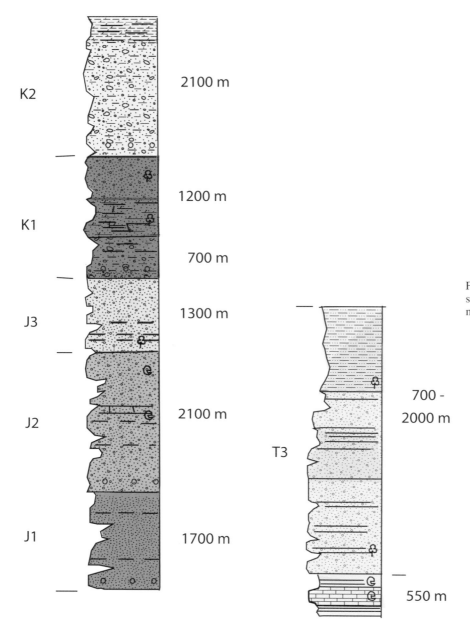

Figure 2-8. Stratigraphic section of Mesozoic strata from the Chuxiong basin (see Fig. 2-2, no. 5, for location).

at its northwest end (Plate 2; see Burchfiel et al., 1995; Fig. 2-1). Why this tectonic element appears to have such a long history of intermittent uplift, and yet was the locus of major superposed subsidence that trends obliquely across it, remains unexplained. Second, during Cenozoic time the NW-plunging Danba antiform was formed obliquely across the Mesozoic NNE–trending foredeep basin, again expressing the older positive character of the high. A major river system was incised along the axis of the antiform, but it is currently part of two rivers, the Dadu and Anninghe Rivers, as the northern part of the river was captured by an east-flowing river near the town of Shimian (Clark et al., 2004). Only the southern, beheaded part of the river flows south in the Anninghe Valley. Third, the active Xianshuihe-Xiaojiang fault system strikes southeast from within tectonic elements to the northwest and locally follows the axis of the Kungdian high in the north but continues obliquely across the crest of the high to the south, where the eastern splays merge into a system of faults as they pass into the South China fold belts, but the western splays continue south to follow the high (Fig. 2-10). Part of the reason for the position of the western faults is that they appear to

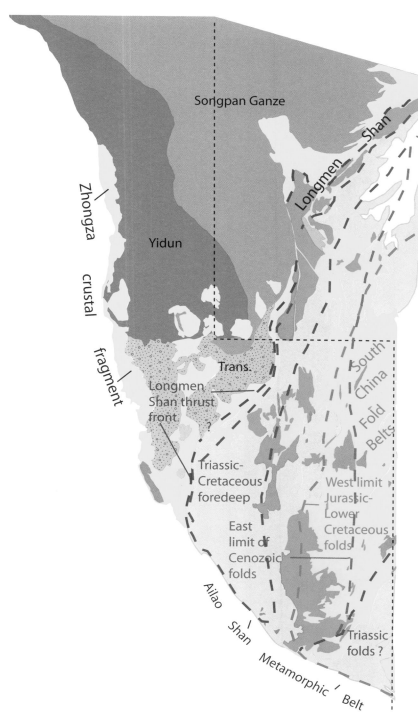

Figure 2-9. Time-space relations between some of the different tectonic elements of the Yangtze platform mentioned in the text. Major elements highlighted are the Triassic-Cretaceous foredeep in the western Yangtze platform (purple dashed line), possible Triassic folds in the SE part of the Yangtze platform (blue), the western limit of Jurassic and Lower Cretaceous folds of the South China fold belts (red), and the eastern limit of Cenozoic folds across the western Yangtze platform (brown dashed line). The Longmen Shan thrust front that figures prominently in the discussion of early Mesozoic and Cenozoic deformation along the boundary between the Yangtze and Transitional Units is shown as a green dashed line.

have reactivated older faults that were related to the development of the high in Proterozoic, Paleozoic, and Mesozoic time (see discussion of the Yimen fault above), but the eastern splays do not appear to have older precursors. Part of the courses of the Dadu and Anninghe Rivers follow the Xianshuihe-Xiaojiang faults and help to explain at least partly the unusual position of these rivers with respect to the Kungdian high. The northern part of the Dadu River, however, does not follow obvious faults related to this fault system.

**Chuxiong Basin**

The Chuxiong basin lies in the southwestern part of the Yangtze Unit (Fig. 2-1) and contains >10 km of Middle to Late

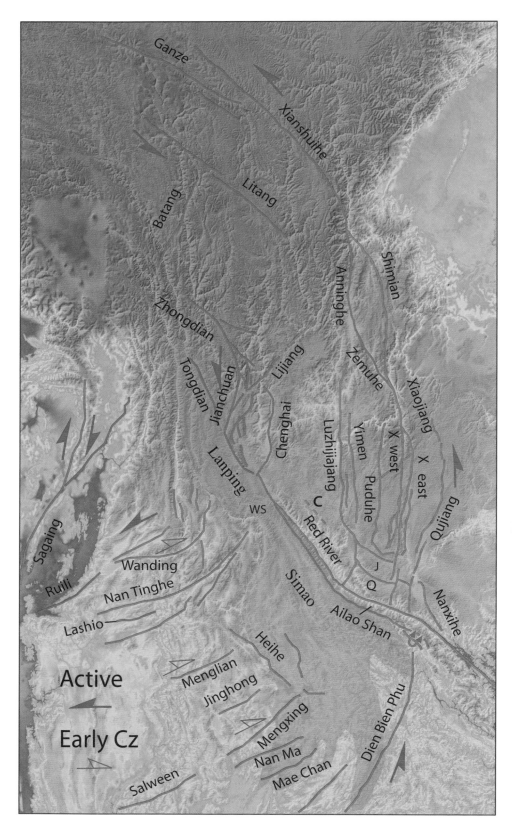

Figure 2-10. Major Cenozoic strike-slip faults of the Yangtze Unit and adjacent regions mentioned in the text. X west and X east refer to two branches of the Xiaojiang fault zone. The Lanping-Simao tectonic unit is highlighted for location on tectonic map and Plate 1. Some faults southwest of the Lanping-Simao unit display early Cenozoic displacement that is right lateral (shown by open-headed arrows) and later that were reactivated by left-lateral displacement (closed-headed arrows), and many of these faults are active. Northeast of the Red River fault, three faults, the Chuxiong (C), Qujiang (Q), and Jinshui (J), show a similar reversal in displacement during the Cenozoic. WS—Wuliang Shan.

Triassic to early Cenozoic strata that are superposed on the crest and western part of the Kungdian high. Tectonically the basin belongs to the Yangtze Unit because pre-Mesozoic rocks that lie beneath it are part of the Yangtze Unit succession (see above). The basin formed as the southern part of a NNE–trending Mesozoic foredeep basin across the northern part of the Kungdian high and continued to the north into the Sichuan basin, where it lies along the eastern margin of the Longmen Shan Mesozoic thrust belt (Burchfiel et al., 1995; Chen and Chen, 1987; Fig. 2-9; Plate 2) where the Mesozoic strata cover a broader area and are generally thicker. The Longmen Shan thrust belt marks the boundary between the Yangtze and Transitional Units to the west; however, the Mesozoic part of the thrust belt in the southern part of Plate 1 is difficult to identify and will be discussed in more detail below. In the northern part of the basin, late Mesozoic and Cenozoic faults and folds generally trend N-S obliquely across the basin margin and divide it and the area north of the basin into discontinuous areas of thick sedimentary rocks separated by uplifts exposing Precambrian basement rocks. The southern part of the basin was not subjected to as severe a disruption and forms a large area of Mesozoic and early Cenozoic strata (Plate 1; Fig. 2-1).

The Chuxing basin contains a section of dominantly non-marine terrigenous clastic strata that range from Middle Triassic to Paleogene in age (Fig. 2-9), and some estimates of the thickness of just the Jurassic to Paleogene strata are >13 km (Yano et al., 1994). The section begins in some places with Middle Triassic limestone, but in most places to the north and east it begins with Upper Triassic non-marine basal sandstone and conglomerate that unconformably overlie Proterozoic basement rocks and eroded remnants of Yangtze Neoproterozoic and Paleozoic strata (Plate 1; Plate 4, sections 6, 7, 29, 32, 33). The basal unit is overlain by Upper Triassic strata that are partly marine in the lower part but become non-marine in the upper part with coal beds. The Upper Triassic strata continue upward by a nearly continuous section of non-marine sandstone, mudstone, shale, and conglomerate, with some lacustrine limestone that ranges from late Late Triassic to Late Cretaceous in age (Fig. 2-9). The Upper Triassic strata are largely pale brown and tan but become red and maroon upward, the dominant color in the Jurassic and Cretaceous strata. Upper Cretaceous rocks were in places deposited paraconformably, marked by conglomerate, on Lower Cretaceous rocks in the northeast part of the basin that locally transgress onto older Mesozoic rocks and in a few places lie unconformably on Proterozoic basement rocks. To the west, however, the unconformities die out within the Cretaceous section (Plates 1 and 3). Lower Cenozoic strata are present only in the northern part of the basin (E on Plate 1) and may not be considered part of the basin succession.

The eastern margin of the Chuxiong basin is marked by thick, mostly Upper Triassic strata that rest directly on Proterozoic basement, or thin, discontinuous remnants of Paleozoic strata on the Kungdian high or on its NW flank (Plate 4, sections 6, 7, 29, 32, 33). Along the basin's northern boundary, Upper Triassic strata lie unconformably on uppermost Neoproterozoic sedimentary rocks (Z2) or thin and discontinuous remnants of Paleozoic rocks that belong to the northwest flank of the Kungdian high. However, the western and southwestern boundaries of the Chuxiong basin are tectonic and are marked on the west by the older Chenghai thrust fault, later reactivated as a normal fault, and to the southwest by the early Cenozoic Red Ailao Shan shear zone and are cut by the late Cenozoic Red River brittle fault (Plate 1). The discontinuity of the Mesozoic outcrops of the northern continuation of the basin is due to post-Oligocene deformation that uplifted basement rocks along general N-S–trending structures (Plate 1; see below). Thus the Chuxiong basin is only the southern remnant of the thick Mesozoic and locally early Cenozoic deposits that formed the NNE–trending foredeep of the Longmen Shan thrust belt of latest Triassic and Early Jurassic age that lay to the west of these deposits (Burchfiel et al., 1995). The Chuxiong basin was the site of important foredeep subsidence during Mesozoic time; however, to the south it abuts the Cenozoic Ailao Shan shear zone, and its southward continuation is not clear and remains questionable (see below).

## STRUCTURE OF THE YANGTZE UNIT

The structure of the Yangtze Unit is complex, having been affected by tectonic activity from Proterozoic to the present. Although the older events are discussed only briefly, tectonic events from the Mesozoic to the present are our major concern. Near the end of the Paleozoic the Yangtze Unit was bounded by oceanic areas to the north, now in the Qinling orogen to the east of southeastern China, and possibly to the west of the Transitional Unit. Its boundaries to the west and south are not clear and require further discussion. The northern boundary in the Qinling Mountains was the site of Triassic collision between the Yangtze Unit (mostly referred to as the *South China block* or *craton* in the literature) and the North China continental crust (Şengör, 1984; Mattauer et al., 1985; Meng and Zhang, 1999; Yin and Nie, 1993), which had little effect upon that part of the Yangtze Unit within the area of Plate 1 but strongly affected the northern part of the Yangtze Unit along the north and northwest parts of the Sichuan basin (Plate 2). Triassic to Late Cretaceous post-collisional convergence along the eastern margin of a combined Yangtze and Cathyasian continental crust did have a major effect on structure within the area of Plate 1 and will be discussed below. Convergence during the latest Triassic and Early Jurassic accreted the Songpan Ganze Unit to the Yangtze Unit along the Longmen Shan Mountains (Plate 2) and southward into the area of Plate 1, but the southward continuation of this boundary across Plate 1 to the Ailao Shan shear zone in the south is complex and will be discussed in more detail below. Deformation along the western boundary of the Yangtze Unit had important effects within the Yangtze Unit and will be discussed within the context of relations between the Transitional and Yangtze Units below. The southern boundary of the Yangtze Unit is along the

Ailao Shan shear zone, a Cenozoic feature that greatly modified the Yangtze Unit and will also be discussed in detail below.

**Precambrian and Paleozoic Deformations**

Several events of middle Proterozoic and early Neoproterozoic deformation, magmatism, and greenschist- to amphibolite-grade metamorphism are recorded within strongly folded schists and gneisses that form the basement of the Yangtze Unit that are best exposed along the axis of the Kungdian high (Wang et al., 2007; Sun et al., 2009; Zhou et al., 2009). Early Neoproterozoic deformation at ca. 950–850 Ma has been well documented in many places. Strong folding and schistosity formation accompanied the deformation with abundant intrusions of large and small plutons that range from gabbro to granite, with felsic rocks being the most abundant. The plutons are best exposed along the axis of the Kungdian high (Plate 1; Li et al., 2003). The early Neoproterozoic deformation was followed by Neoproterozoic extension that began at ca. 820 Ma and lasted until at least 690 Ma. Four periods of rift basin formation can be recognized (see Wang and Li, 2003a, 2003b) during which deposition of coarse-grained sedimentary and locally volcanic rocks of Neoproterozic age occurred (Za in Plates 1 and 2). In many places, coarse-grained sedimentary rocks contain clasts of the underlying or nearby basement rocks, and many of these deposits consist of cobble to boulder conglomerate. The distribution of the stratified rocks, their compositions, and their relations to faults suggest that they were formed mainly by extensional tectonism with rapid changes in thickness and facies owing to their deposition in synsedimentary grabens and half grabens. These rocks are mostly exposed in irregularly shaped uplifts that form N-S–trending belts along the Kungdian high and along the eastern and northern flanks of the Chuxiong basin (Plates 1 and 2).

Uppermost Neoproterozoic strata (Z2 in Plates 1 and 2) transgressed across all the older rock units and structures and began a long period of deposition that continued through the Paleozoic, when the Yangtze Unit was generally stable. There are local areas of uplift and erosion, and at times the Kungdian high was partially elevated (see above). These areas of uplift are marked by paraconformities and low-angle unconformities with the Paleozoic section (Plate 3, Z2 unconformity). During Late Permian time there was extensive extrusion of the Emei Shan basalt (Plate 1, Pβ) that in places west of the Yangtze Unit reaches 5 km thick. The basalt was extruded across most of the Yangtze Unit shown in Plate 1 and also farther west (see below), an event that is related by some workers (e.g., see Xu and Chung, 2001) to Late Permian plume activity; however, this interpretation and others will be considered below. The broad uplift of the Kungdian high occurred several times during the Paleozoic and into early Late Triassic time (discussed above) so that Upper Triassic strata were deposited directly on Proterozoic basement, and remnants of Paleozoic strata and uppermost Neoproterozoic dolomite (Z2) over large and irregular areas (Plate 3, T3 unconformity). It was following deposition of the Upper Triassic rocks when the deformational events most relevant to this study began.

**Mesozoic and Cenozoic Deformations of the Yangtze Unit**

Most of the deformational features of the Yangtze Unit were formed in Mesozoic and Cenozoic time. Formation of folds and faults was temporally and spatially complex, and dating of much of the structure is difficult, because nearly all the Mesozoic and Cenozoic strata are non-marine, their ages are poorly known, and they lack volcanics that would be useful for determining their age. In many places, structures can be assigned only to broad time periods, making it difficult to establish contemporaniety of structures. The structures can be arbitrarily separated into the South China fold belts that lie mainly east of the Kungdian high, N-S structures that lie within the northwestern part of the Yangtze Unit and tend to follow the Kungdian high, and structures in the Chuxing basin in the western part of the area. The forces responsible for the deformations have come from plate boundary activity at both the east and west margins of the Yangtze Unit as well as during Cenozoic time from post–India-Eurasia accretion and associated intracontinental deformation.

*South China Fold Belts*

The style of the structure in the South China fold belts in Plate 1 (Fig. 2-1) has not been well studied. In the east the fold belts consist of broad synclines separating narrower anticlines; the fold axes are curvilinear but trend generally northeast. Toward the west the folds become more closely spaced and more irregular in trend as they merge into the Kungdian high, where the Paleozoic section was thinner. The folds and thrust faults occur in four northeast-trending belts that are dominated in outcrop from south to north by rocks of (1) Neoproterozoic and Paleozoic, (2) Triassic, (3) Paleozoic, and (4) Paleozoic–Mesozoic age (see Plate 1).

The style of the folds and faults in the South China fold belts is uncertain. The NE-trending structures in the southern two belts involve only sedimentary rocks, weakly metamorphosed in their southernmost part. In the southernmost belt the structurally deeper rocks are foliated, and in the deepest structural levels, isoclinal mesoscopic folds are present with shallow-dipping axial planes. At higher structural levels, however, the folds are more upright. No basement rocks are present in outcrop. Farther north within the wide belt of Triassic rocks the folds are generally without cleavage and are more upright, with a tendency to be overturned to the northwest (Plate 4, section 3). Faults dip both to the NW and SE, and most can be interpreted as thrust faults related to the folds. Thus this part of the fold belt could be regarded as a thin-skinned thrust belt that involves rocks that are as old as latest Neoproterozoic and which are weakly metamorphosed to the south. The southern belt of Neoproterozoic and lower Paleozoic rocks ends to the west against the NW-striking Nanxihe fault, with mostly Triassic rocks to the west that are continuous with the belt of folded Triassic rocks. These two southern folded belts are intruded by poorly dated plutons of Jurassic to Cenozoic age (Plate 1).

North of the belt of Triassic strata, the folds and faults that involve mainly Paleozoic rocks also involve both Proterozoic basement rocks (pt) and upper Neoproterozoic strata (Za). The involvement of these older rocks becomes increasingly greater as the folds are traced to the southwest, where structures in the belt of Triassic rocks curve more to the west. In this broad, NE-trending belt are several exposures where Devonian strata rest unconformably (Plate 4, section 3) on basement rocks (pt), and these structures were prefigured by pre-Devonian structure that locally elevated the basement before Mesozoic deformation. Many of the folds are tight, but they are separated by broader, less folded areas (Plate 4, section 28).

Farther north to the northern extent of the South China fold belts, the folds and faults involve extensive exposures of upper Paleozoic and Mesozoic rocks as young as Jurassic. The folds are commonly separated by broad areas of relatively unfolded rocks. The folds still have a dominant northeast trend, but they show an increasing amount of dome and basin geometry to the north and west. Locally the folds and associated thrust faults also involve the lower upper Neoproterozoic rocks: both the sedimentary and volcanic-rich strata (Za) and more locally metamorphic basement (pt). Some folds are asymmetric, overturned both to the SE and NW, but many are upright broad and open folds (Plate 4, sections 4, 5). Some folds show very clear steep to vertical limbs on one side and long, gently dipping limbs on the other side, a geometry characteristic of fault-propagation folds. The steep dips in these rocks occur along anticlines, whereas many of the intervening synclines are broad with subhorizontal strata. Many of the folds can be interpreted to be thin skinned, but they must involve at least the upper part of the basement at depth. This northern belt of folds and faults can be continued into eastern Sichuan (Plate 2: now Chongqing Province), where they are dated as middle Cretaceous (Wang et al., 2010). However, in this area there is little evidence that they involve basement rocks, but they involve older rocks when traced to the southwest into the area of Plate 1 (Fig. 2-11: Note that on Plate 4 in many of the cross sections the base of the Neoproterozoic strata has been projected to depth and clearly indicates an extensive involvement of basement rocks even though the fold-thrust belt is shown to be thin skinned.).

Toward the SW the NE-trending belts of folds and faults enter into the eastern part of the Kungdian high, where they involve increasingly more basement rocks (Plate 1; Plate 4, sections 2, 35, 34). This does not preclude that the structures still have a décollement style structure at depth, but that the thrust faults detach more widely within the metamorphosed foliated basement. Thus the South China fold belts within the area of Plate 1 are interpreted to be the northwestern part of a northwest-vergent foreland fold-thrust belt whose basal décollement lies generally or wholly within basement rocks. The change in fold character to the west is because it merges into the Kungdian high, where the folds appear to die out in an area of thinner sedimentary section and shallower basement. But before ending to the west the folds and thrust faults become more irregular with more north-south trends, have lesser magnitudes of shortening, and appear to die out to the west. Thus there must be a broad, generally northward-trending left-lateral shear within the western margin of the South China fold belts that has not been recognized. The shear may be accommodated by bending or faults; however, this area has been affected by younger, generally north-south–trending structures of Cenozoic age (see below), and the age of any given fault or fold in this area is difficult to determine. The open S-shaped curve in the trend of the South China fold belts in their southwesternmost part (see Plate 1) may also be related to this general left-lateral shear. The fold-thrust belt east of the area of Plate 1 has been interpreted to be basement involved in its older, more easterly part (Yan et al., 2003a, 2003b; Wang et al., 2010) but with a ramping up to a décollement that lies within the sedimentary section to the northwest (Fig. 2-11). Whereas some of the folds and thrust faults shown in Plate 1 may have been detached within the sedimentary section, others clearly involve pre–uppermost Neoproterozoic strata. Also part of the complexity of the fold geometry in the northern part of the Yangtze Unit within Plates 1 and 2 is the result of interference by younger N-S–trending structures that are part of the fold belt of Cenozoic age associated with the Longmen Shan deformation to the north and west that also involves basement rocks (Burchfiel et al., 1995).

*Age of the structures in the South China fold belts.* An understanding of the age of the structures in the South China fold belts (Plate 1) comes from farther east, where the continuation of the structures form the complex fold-thrust belt that is present across most of the eastern half of the Yangtze Unit (Fig. 2-11). In southeastern China the fold-thrust belt can be shown to have developed from east to west, beginning in the east in Late Triassic–Early Jurassic time and ending in middle Cretaceous time. Some authors have suggested that the folds were developed progressively during that time period (Yan et al., 2003a, 2003b; Wang et al., 2010). However, recent work by Xu et al. (2011) has suggested that there were at least two discrete events of shortening, one in the Late Triassic to Middle Jurassic and a second in the Early Cretaceous, separated by a period of extension. Following the younger folding event, extensional tectonism began in the Late Cretaceous and has dominated the tectonism in southeastern China until the present. The area in southeastern China, where the structures have been best studied, still requires more regional mapping to determine the extent and ages of the dated structures. In particular, these better studied areas are far removed from the South China fold belts as shown in Plates 1 and 2, and as shown in Figure 2-11 the structures are not obviously continuous between the two areas; thus care must be taken in making interpretations in the area shown in Plates 1 and 2. The structures in southeastern China involve basement rocks, and the rocks are foliated, locally metamorphosed and intruded by plutons of Mesozoic age, but the structures in the west have less internal strain and are interpreted to be underlain by a décollement in the lower Paleozoic strata (see Wang et al., 2010). In the discussion above it is clear that even the westernmost structures seen in Plate 1 involve basement

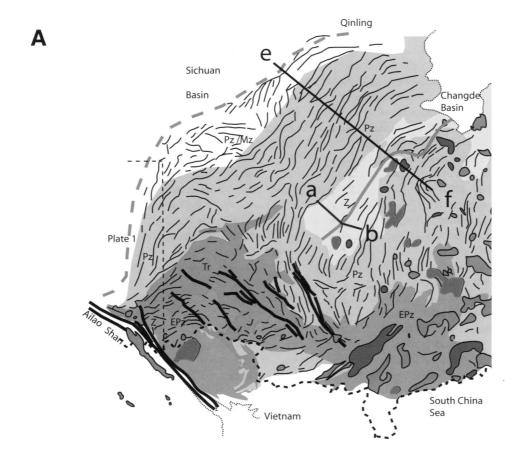

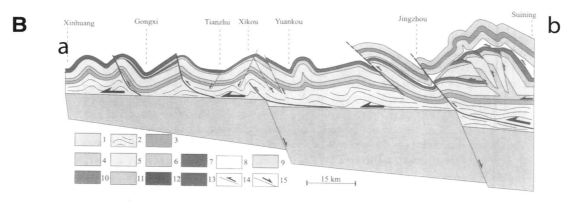

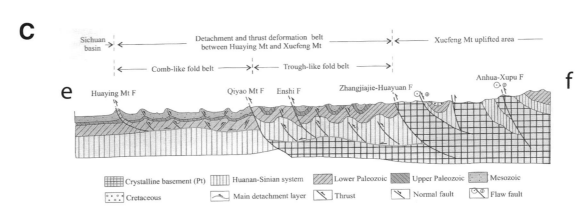

rocks. The western folds involve Early Cretaceous strata and are unconformably overlain by Upper Cretaceous strata, yielding an upper limit to the folding. This fold-thrust belt is interpreted to have formed in a retro-arc setting west of an Andean margin developed along the eastern side of the Yangtze Unit where it was underthrust by oceanic plates (Hsü et al., 1980; Hsü et al., 1988; Yan and Wang, 2000). The magmatic belt in southeastern China ranges from Late Triassic to Cretaceous and Cenozoic in age, but by Late Cretaceous time the shortening in the retro-arc fold belt ended, and the magmatic belt was within a region of extension that has continued to the present.

Within Plate 1 there is little evidence for the age of the South China fold-thrust belt except that related structures are unconformably overlain by Paleogene strata locally (Plate 3). The structures in the southern two belts have generally been regarded as Late Triassic or Early Jurassic in age, based only on the observation that they contain no rocks younger than Late Triassic. These two belts could correlate with the inner and older belts to the east, but they are difficult to trace directly into them, because the structures in the two areas form convex north and west arcs separated by a prominent zone of NW-striking faults (Fig. 2-11). Thus a simple correlation cannot be made. Also there are differences between the two areas of structures; the southern two belts in Plate 1 cannot be shown to involve basement rocks, although they may do so at depth, and they do not apparently involve the complex duplex structures or basement-involved structures known in the east (Fig. 2-11). The southern two belts, shown in Plate 1, do contain plutons, whereas the belts farther north do not, a similarity to the fold-thrust belt farther east, but the abundance of plutons is much less. A major problem within Plate 1 is that the southern part of the fold-thrust belt, along with most of the southern part of the Yangtze Unit, was structurally truncated at a later time (see below), and most of the western part of the Andean arc is missing.

The northern two belts of the South China fold belts in Plate 1 contain rocks as young as Late Jurassic, and in one place Early Cretaceous ($Mz_2$), and thus the structures are younger than the two belts to the south, so it is not unreasonable that the fold-thrust belt (Plate 1) may have developed over the same time period as the fold-thrust belt to the east, but only the northern part of the folds and thrusts can be traced directly into one another (Fig. 2-11). The oldest rocks that unconformably overlie the South China fold belts shown in Plate 1 are Paleogene. If the fold-thrust belt in eastern China is a retro-arc belt, the inner units of this belt do not continue far enough to the southwest to be positioned south of the South China belts in Plate 1. In Plate 1 they are truncated on the south side along the Ailao Shan Unit, but what tectonic elements were there during the folding, and where they have been displaced following deformation, remain major problems that will be explored in Chapter 16.

### Younger N-S Structures in the Western Yangtze Unit

In the northwestern part of the Yangtze Unit, and also within the northern Kungdian high, folds and associated thrust faults of the South China fold belts trend N-S (Plates 1, 2, and Plate 4, sections 5, 5a) and are continuous with N-S structures from the Longmen Shan area to the north (Plate 2). Within the area of Plate 1 the structures involve rocks as young as Late Cretaceous ($Mz_2$) and rarely Paleogene (E), but when followed to the north, these folds and faults lie within the southern parts of the Longmen Shan and the Sichuan basin, where they involve Eocene and lower Oligocene strata and possibly Miocene strata, and many of them are active (Plate 2; Burchfiel et al., 1995). The involvement of Paleogene, Neogene, and Quaternary strata indicates that they are late Cenozoic in age and can be traced southward into the area of Plate 1, where there are few constraints on their age. They constitute yet another, and younger, set of fold-fault structures within the Yangtze Unit and demonstrate the need for the plural designation of the fold-fault structures in the South China fold belts.

The N-S set of structures interact with the older NE-trending structures to the east to form a dome and basin style of interference pattern well developed in the northern part of the area of Plate 1, becoming increasingly more well developed to the north (Plate 2; Burchfiel et al., 1995). The dome and basin character continues south into the western part of the South China fold belts and also into the Kungdian high, where the involvement of Precambrian rocks is extensive (Plate 1). In the south, cross folding and refolding of NE and N-S trends are clearly developed, but whether this pattern is related to the same dynamic system that folded and thrusted the rocks in the southern Longmen Shan is unclear, because the thrust faults that form the eastern margin

Figure 2-11. (A) Generalized map of the South China fold belts in the eastern part of the Yangtze Unit east of Plate 1. Modified from the Geological Map of China. Fold trends are shown with solid lines. Units distinguished are yellow—Neoproterozoic (Z); dark green—mostly early Paleozoic (EPz); light green—mostly upper Paleozoic (Pz); purple—Triassic (Tr); white—Paleozoic-Mesozoic (Pz/Mz) units. Plutons are shown as purple—early Paleozoic; red—Jurassic-Cretaceous. Heavy red line is western limit of Permo-Triassic deformation. Dashed red line is the western limit of the South China fold belt and also marks the western limit of Early Cretaceous folding. Heavy black lines are some major faults. Note that the structures in the eastern part of the belt appear to form two convex west arcs separated by major faults. Only the westernmost folds can be traced directly into Plates 1 and 2 (eastern limit of Plate 1 is shown by black dashed line). (B) Cross section a–b. Taken from Wang et al. (2010). Solid gray (1)—Precambrian basement; gray with lines (2)—Banxi Group of pre-Sinian Precambrian; late Neoproterozoic units (3–11)—Sinian–Cambrian (12); Permian (13). (C) Cross section e–f; taken from Wang et al. (2010). Note that in these cross sections the detachment of the thin-skinned fold-thrust belt is in Precambrian rocks in the southeastern part of the sections and rises to within first, late Proterozoic strata, and then lower Paleozoic strata to the west in the long section e–f. However, within Plate 2 the westernmost parts of the South China fold belts do not involve basement or early Neoproterozoic rocks, but as the folds continue to the southwest into the area of Plate 1 the folds and thrust faults involve basement and early Neoproterozoic rocks.

of the Longmen Shan thrust belt to the north continue southwestward, west of the Chuxiong basin, veering away to the west from the N-S–trending structures. The N-S–trending structures in this southern area east of the Chuxiong basin locally involved dated Paleogene rocks; thus they may be of the same age as the Longmen Shan belt to the north, but whether they result from the same cause remains unclear (see below).

The N-S–trending folds in the northwestern part of the Yangtze Unit are mainly upright, and thrust faults tend to dip steeply to moderately both to the east and west (Plate 4, sections 30, 35, 36). Basement rocks are extensively exposed in the northern part of the area, and the structures have a style reminiscent of the Colorado-Wyoming Rocky Mountains of the western United States. Folds are cored by basement rocks and cut by thrust faults on one or both limbs (Plate 4, sections 5, 30). Some of the sedimentary rocks in the footwalls have a well-developed, steeply dipping cleavage in a narrow zone adjacent to the faults. In the southern part of this belt of N-S–trending structures the folds and faults become irregularly spaced and interference with the older NNE-trending structures makes for a more chaotic pattern of interference, such as north of Kunming (Plate 4, section 34). Some of the structures within the N-S belt were developed from reactivated older structures of Proterozoic, Paleozoic, and Mesozoic age that affected the basement rocks during the evolution of the Kungdian high. The N-S structures extend west into the Chuxiong basin, where they are treated separately below. Their relation to the Transitional Unit to the west is not straightforward. The boundary in the north between the two tectonic units is marked by east-vergent thrust faults that are the continuation of the Cenozoic Longmen Shan thrust belt, and in many places the N-S folds and faults are parallel to this contact. It would be simple to suggest that the N-S structures are related to the same process of E-W shortening as in the Longmen Shan belt, some localized along older faults belonging to the Kungdian high, but at the north end of the Chuxiong basin the N-S structures are cut obliquely by the arcuate Jinhe-Qinhe thrust fault of the Longmen Shan thrust belt at this latitude (Plate 1). These relations are complex and will be discussed below.

### *Structure of the Chuxiong Basin*

The Chuxiong basin contains a thick section (~10 km) of Mesozoic strata that covers a large region in the southwestern part of the Yangtze Unit (Fig. 2-1). The basin began to develop in the Late Triassic, superposed on the western and central parts of the Kungdian high, which shows evidence of complex faulting of several ages (see above), and the margins of the Chuxiong basin, show similar and perhaps long-lasting Mesozoic deformation. For example, along the northeast side of the basin, folds involve Jurassic rocks that are unconformably overlapped at the base of Upper Cretaceous strata (Plate 3, and shown on Plate 1 where Cretaceous strata [Mz2] above the unconformity terminates folds within Mesozoic rocks [Mz]). N-S–striking faults in places juxtapose Jurassic rocks against Proterozoic basement rocks (Plate 1, Mz against pt), but both units are unconformably overlapped by Upper Cretaceous strata (Mz2 on Plate 1; see also Plate 3). The age of this deformation does not fit any of the other established times of deformation within the western part of the Yangtze Unit. The Upper Cretaceous strata and older rocks are folded together. Many similar relationships are present along the eastern margin of the basin and may be related to the synsedimentary subsidence of the basin (note distribution of unconformities along the east side of the Chuxiong basin in Plate 3).

Folds and faults in the Chuxiong basin display both large- and small-scale arcuate trends. On the large scale the trends of the structures are north-south in the northern part of the basin, and when followed to the south they trend more to the southeast, forming a concave northeast arc. The northwestern and western basin margins are marked by the west-dipping Chenghai thrust fault, which has been reactivated in most places as the active west-dipping Chenghai normal fault that forms the east margin of the N-S–trending Binchuan and Chenghai Quaternary basins. Remnants of the Chenghai thrust fault are still preserved along the basin margin where upper Paleozoic rocks were thrust eastward above Jurassic strata (Plate 1). To the west of the Chenghai thrust fault there are no Jurassic or Cretaceous rocks present above Triassic rocks, a characteristic of the Transitional Unit, and all these Triassic strata we place within a Transitional tectonic unit (see below). Thus the western margin of the basin is formed by the east-vergent Chenghai thrust of Paleogene–Neogene age, which telescoped two different Mesozoic sequences. The N-S–trending folds and thrust faults in the northwestern part of the Chuxiong basin can logically be related to this thrusting event. The folds and thrust faults continue to the north, where they involve Eocene strata; thus the folding in this part of the Chuxiong basin is interpreted to be Eocene to post-Eocene in age, mainly of late early Cenozoic age (see Burchfiel and Wang, 2002). However, the same folds farther south curve to trend southeast, parallel to the Ailao Shan mylonitic belt of early Cenozoic age, a structure interpreted to be a major left-lateral shear zone. Thus the folds in the southwestern part of the basin have a complex origin that requires further examination.

Within the central part of the basin the folds are more closely spaced and more numerous, and they form part of the broad, convex, northeastern arcuate pattern of structures in the western Chuxiong basin. Here they show small-scale bending along a poorly defined NW trend. Within the zone of bending are short local segments of faults that have been interpreted to be related to left-lateral shear (Leloup et al., 1995; Burchfiel and Wang, 2003) along the poorly defined northwest-striking Chuxiong fault. This fault is just one of several left-lateral faults within the Yangtze Unit and tectonic units to the NW that lie 50–70 km north of the Ailao Shan left-lateral shear zone and are related to it (Burchfiel and Wang, 2003; see also discussion below). This left-lateral movement was demonstrated to be of early Cenozoic age (Burchfiel and Wang, 2003).

The abundance of structures within the arcuate belt of faults and folds within the western and central part of the Chuxiong basin becomes less and more widely spaced toward the northeastern part of the basin, and older rocks become involved (Plate 1). This suggests that the folding in the basin is related to the major thrust fault along its west side. Relationships are probably more complicated, however, and contrast with the South China fold belts, where the deformation is related to plate boundary activity along the eastern side of the Yangtze Unit. Within the northeastern part of the basin, basement rocks are involved in the basin structures, and they merge with the N-S–trending structures that extensively involve basement rocks farther to the north and northeast as discussed above. Although it seems reasonable, but not proven, to interpret the structures in the western and central parts of the basin as a décollement-style fold and thrust belt, it is clear that basement was involved in the structures in the north and northeastern parts of the basin. The structural interpretation is further complicated by the fact that north of the Mesozoic rocks of the basin the structures in the basin continue to trend N-S, involving incomplete sections of Paleozoic strata that probably lie along the western flank of the Kungdian high, where basement rocks are involved. This relation may indicate that the basement is involved in all the structures that affect the Chuxiong basin. Further complicating the Cenozoic structures in this area is the fact that these N-S faults and folds are truncated on the north by the arcuate Jinhe-Qinhe thrust fault that bounds the southeastern side of the Transitional Unit (Plate 1), clearly indicating that this thrust fault has a very young displacement (see below).

The tectonic setting for the folding in the Chuxiong basin, and the N-S–trending folds and generally steeply dipping thrust faults in the western Yangtze Unit, are not clear. These relationships must be considered: First, the Chuxiong basin forms the southernmost extent of the NNE-trending foredeep of Mesozoic age for the Longmen Shan thrust belt (see above). The foredeep lies above the major basement-controlled paleogeographic positive Kungdian high, which trends north oblique to the foredeep. Second, the basement-involved N-S folds and thrust faults northeast and north of the Chuxiong basin appear to merge with the folds and faults of the Chuxiong basin, but through an area where the folds and faults are less abundant. Third, it may be most obvious to relate these Paleogene–Neogene structures in the Chuxiong basin and in the N-S belt of structures to the northeast to the same deformational processes that formed the Cenozoic Longmen Shan thrust belt (Plate 2) and its southern extension (the eastern boundary of this thrust belt forms the eastern boundary of the Transitional tectonic unit; see below). The variations in structural styles may be related to the complex paleogeography and older structures within the older rocks of the western Yangtze Unit during Paleogene–Neogene deformation the details of which still need to be established. The style of these structures at depth still remains unresolved. In many places these structures clearly involve basement rocks, but their continuation into areas of thicker sedimentary deposits may be more logically interpreted to be thin-skinned décollement-style folding that involves only cover rocks. Thus it is possible that all the structures overlie subhorizontal thrust faults at depth, but the thrust faults over large areas may lie at more than one stratigraphic level; the lowest structural level, however, is within the basement rocks. Because the north-trending folds and thrust faults that lie in the footwall of the Jinhe-Qinhe thrust fault that marks the boundary of the Transitional Unit and is thought to be the continuation of the Longmen Shan thrust belt, the last movement on this fault must postdate the Paleogene–Neogene folding in the Chuxiong basin. The thrust faults in the Longmen Shan belt to the north are both early Mesozoic and late Cenozoic in age, so their truncation of the Paleogene–Neogene folds in later Cenozoic time along the northern part of the Chuxiong basin is to be expected. The youngest structures to affect the Yangtze Unit are the major strike-slip fault systems (see below), and along these active strike-slip faults is abundant evidence for transfer of strike slip to both extension and shortening. A further modification of some of the structures by transfer motion needs to be considered; e.g., left-lateral shear on the Xianshuihe fault may be partly transferred to the Jinhe-Qinhe thrust, causing its very young displacements.

This generalized scenario, however, is complicated by the fact that all the paleogeography in the Yangtze Unit appears to have been truncated by the Ailao Shan shear zone to the south. Southward all folds and thrust faults in the central Chuxiong basin curve to the SE to become parallel with the Ailao Shan shear zone. In the southeastern part of the Chuxiong basin, and at the southern end of the Kungdian high, the structures trend more E-W (Plate 1) and appear to continue to the east into folds of Mesozoic age in the South China fold belts. Both sets of folds and faults change trend more to the NE just before they are truncated by the Ailao Shan shear zone. Why this happens is not clear, because most evidence suggests that major early Cenozoic movements in this area were left-lateral strike-slip related to the development of the Ailao Shan shear zone opposite the S-shaped change in fold trend within the SW part of the South China fold belts (see below). It may be that this change in the trend of the South China fold belts is related to a broad N-S shear along the western margin of the belts, as their displacement dies out westward toward the Kungdian high (see above). In this area just north of the Ailao Shan shear zone the structures of different ages cross nascent complex interference patterns. The nature of the southern boundary of the Yangtze Unit and its relation to the Ailao Shan shear zone are complex and are considered in more detail in a later section.

The structures that mark the western boundary of the Yangtze Unit are regionally important. The Chenghai thrust along the west side of the Chuxiong basin continues south, where it is hard to locate because of late Cenozoic fault disruption and reactivation, but it appears to be truncated by the Red River fault at its south end. The pre-truncation continuity of these structures and other rocks, such as the thick Emei Shan basalt west of the Chuxiong basin, remains one of the major unresolved problems in the area and will be discussed in more detail with the Transitional Unit below. Also, this boundary may have had locally very young activity, as the eastern thrust of the Longmen Shan thrust belt appears to offset the apparent uniform topographic SE slope of the Tibetan Plateau, suggesting late Cenozoic and even active displacement on at least the northern part of the thrust that marks the eastern boundary of the Transitional Unit (Wang et al., 1998; Clark et al., 2005). This is one of the few places where active shortening deformation appears to affect this uniform southeast slope of the plateau.

### *Other Cenozoic Structures in the Yangtze Unit*

Paleogene strata are involved in folds and faults of both the N-S and NE trends in the South China fold belts, but their tectonic time and space relations are difficult to determine because they not only are involved in folds and thrust faults but also with normal faults where the Paleogene strata occur as basin fill. In the NE-trending folds and faults, Paleogene strata lie unconformably on the older rocks (Plate 3), but they are involved in structures that are parallel to older folded structures and are commonly, but not always, less strongly folded (e.g., Plate 4, sections 2, 3; Plate 1 in general). Neogene strata are locally unconformable on these structures; thus the deformation is Paleogene–early Neogene in age. Folding of this age is widespread in the western part of the Yangtze Unit (see below) and may have extended into the eastern part of the unit as well. However, the Paleogene rocks are only present locally, and where they are not present it is unclear if a fold is of Mesozoic or Cenozoic age and/or a reactivated older structure. Most of the Yangtze Unit was affected by late Cenozoic to active strike-slip faulting, and some of the fold and fault relations are probably related to complexities within this major broad region of left-shear deformation (see below). Thus to assign ages to structures in many parts of the area remains difficult and is an unresolved problem. For example, northwest of the Mengzi basin in the southern part of the area of Plate 1, folds in the Triassic rocks show tight arcuate trends concave to the north. A few of the N-S folds involve Paleogene rocks, but most of the folds do not, and many of them can be traced into the NE-trending folds interpreted to be Mesozoic in age. South of these arcuate folds, Miocene and Pliocene rocks are involved in SW-vergent thrusting. Thus, the arcuate trends can be interpreted in several ways, but the ages of the folds where they do not involve Cenozoic rocks are difficult to establish accurately. The closer the rocks are to the NW-trending Cenozoic structures in the Ailao Shan shear zone (see below), the more complex the problem becomes.

Locally, rocks as young as Miocene or Pliocene are involved in thrusts and folds, but over most of the area these strata are unfolded or weakly folded. Many of these rock units are poorly dated and occur in small outcrops, so the extent of the late Cenozoic deformation is uncertain. Many of these structures can be related to strike-slip faults of Paleogene, Neogene, or Recent age. For example, in the southern part of the area of Plate 1, folds and thrust faults in Paleogene rocks have been related to the Paleogene NW-striking, left-lateral, strike-slip Jianshui and Qujiang faults; this Paleogene deformation has been related to localized shear associated with the Ailao Shan shear zone to the south (Fig. 2-10; Burchfiel and Wang, 2003). Quaternary to Recent activity on these same faults is right-lateral and has produced additional folds and thrust faults at their eastern ends (Burchfiel and Wang, 2003). Locally along the Neogene to active strike-slip faults that traverse the Yangtze Unit and beyond (see below) are folds and thrust faults related to constricting bends in the faults or where the faults transfer horizontal displacement into shortening structures (Wang et al., 1998). Thus the age of structures in any one place is hard to determine, or they may be of multiple ages caused by reactivation of older structures.

### Strike-Slip Faults within the Yangtze Unit

There are two major regional strike-slip fault systems of late Cenozoic to currently active age that displace rocks of the Yangtze Unit: the NW-trending Ailao Shan–Red River and the N-S–striking Xianshuihe-Xiaojiang fault systems (Fig. 2-10). The Dali fault system (also shown in Fig. 2-10) regionally belongs to the Xianshuihe-Xiaojiang system but lies to the west largely within the Transitional Unit and will be discussed with that unit below.

The active Xianshuihe-Xiaojiang fault system trends north-south across the central part of the Yangtze Unit as seen on Plate 1 (Allen et al., 1984; Wang et al., 1998; Fig. 2-10) and continues to the northwest across the Longmen Shan thrust belt into the Songpan Ganze Unit (Plate 2; see below). It is a regional fault system related to the late Cenozoic tectonics of the India-Eurasia intracratonic indentation deformation (see Wang et al., 1998) and is a system of faults with great potential seismic hazard; it will be considered in its regional context in more detail below. Within its northern part the Xianshuihe fault consists of a single narrow fault zone, and where it curves south to cut across the southern part of the Transitional Unit and Longmen Shan thrust belt (Plate 2) to continue into the Yangtze Unit of Plate 1, the fault system splays southward into the Anninghe, Shimian, and Zhemu faults (Fig. 2-10). Farther south it continues to splay and forms as many as seven or more individual faults across an E-W width of >100 km. The easternmost of the faults pass through the Yangtze Unit and do not appear related to older structures. However, the most westerly faults of the system pass through the eastern part of the Kungdian high, where faults of similar trend have an older history of Precambrian(?), Paleozoic, and Mesozoic displacement, and they may have reactivated older faults. In the northern part of the Yangtze Unit, no matter how many individual faults are present, the total displacement is ~60 km, a displacement more accurately measured along its singular trace to the north within the Songpan Ganze Unit (Wang et al., 1998). The displacements can be accurately measured on most of the southern fault splays, because they displace older folds and faults of the NE-trending South China fold belts and structures within the Kungdian high.

Many of the faults have jogs along which numerous pull-apart basins have formed, and several of the lakes within the area are formed in part or wholly by these structures. The pull-apart structures are commonly marked by basins with late Neogene or Quaternary deposition (Fig. 2-10). In other places the main N-S–striking faults splay into E-W–trending shortening structures that elevate large exposures of Precambrian rocks. Two prominent examples are the uplifts of the Xi Shan, and west of Dian Chi Lake near Kunming (Plate 1 and Fig. 2-10). In addition, the

N-S–striking faults locally distort the older structures, forming arcuate trends related to left shear near the faults. Between N-S strands of the Xianshuihe-Xiaojiang fault system are numerous shorter faults that trend generally NE. They display right slip and are faults that accommodate counterclockwise rotation of blocks between the main fault strands (see England and Molnar, 1990). As the blocks rotate, they further distort preexisting structures.

The measured amount of displacement on individual faults becomes less in the southern part of the Xianshuihe-Xiaojiang system, and only two of them reach the Red River; none of them cross the Ailao Shan shear zone (Wang et al., 1998; Schoenbolm et al., 2006a, 2006b; see Plate 1 and Fig. 2-10). However, the total amount of shear across the fault system passes through the Ailao Shan shear zone and the Red River fault zone without displacing them but bending them left laterally by ~60 km, a bend clearly evident in the trend of the Ailao Shan rocks as seen on Plate 1. How the shear becomes absorbed through this region is one of the many interesting structural problems in the area (see Wang et al., 1998; Schoenbohm et al., 2006a, 2006b; Burchfiel et al., 2008a).

One of the southeasternmost faults of the Xianshuihe-Xiaojiang system, the Nanxihe fault, strikes SSE from the Mengzi basin and juxtaposes the anomalous section of latest Neoproterozoic and lower Paleozoic rocks to the east against the Triassic rocks of the South China fold belts to the west (Fig. 2-10 and Plate 1; also see above). This fault must have a large displacement, but where it cuts Triassic strata and forms a left-step through the pull-apart Mengzi basin there is little evidence for large displacement; also it may be partly responsible for the arcuate bending of structures in Triassic rocks of the northwest Mengzi basin. In its southeastern part the Nanxihe fault is marked by a narrow belt of faulted Neogene rocks and thus is a young and active fault. More importantly, the Triassic rocks southwest of the fault surround mylonitic metamorphic rocks similar in character to those in the Ailao Shan at the nose of a NW-plunging antiform in the Day Nui Con Voi. The Triassic rocks have been interpreted to be juxtaposed with the mylonitic core rocks in the Day Nui Con Voi by a normal–strike-slip fault of early Cenozoic age that has been folded into the antiform. The core rocks were interpreted to be reworked basement for the Yangtze Unit in this area (Burchfiel et al., 2008b). This complicated structure contrasts greatly with the Neoproterozoic and Paleozoic rocks to the east of the fault and is related to the regional problem of the early Cenozoic Ailao Shan shear zone and will be discussed in more detail below with the Ailao Shan Unit.

With the exception of the active and a few older strike-slip faults, little attention has been paid to many of the strike-slip faults within the Yangtze Unit. Three NW-striking Paleogene-age strike-slip faults in the southern part of the Yangtze Unit, the Jianshui, Qujiang, and Chuxiong faults, were described by Burchfiel and Wang (2003; Fig. 2-10). Their Paleogene displacements are left lateral, but each has been reactivated in Quaternary time to right lateral. The Chuxiong fault that disrupts the fold trends within the central part of the Chuxiong basin was mentioned above. The Jianshui and Qujiang faults had ~13–17 km and ~6–15 km of left-slip displacement, respectively, in early Cenozoic time, and both were reactivated as right-lateral faults in Quaternary time and are presently active (Wang et al., 1998). On strike with these faults and northwest of the Chuxiong basin, the NW-striking Zhongdian fault (Fig. 2-10) offsets units in the Jinsha accretionary complex and the Jinsha River by ~18 km left laterally, an offset of mostly early Cenozoic age. Like the other faults in this zone, it has also been reactivated in Quaternary time as a right-lateral fault. These faults suggest that the region 50–70 km north of the Ailao Shan shear zone has been affected by left-lateral shear in early Cenozoic time and may be at least partly responsible for the arcuate trends of folds and faults in the southern Chuxiong basin. Their relation to the Ailao Shan shear zone and active Red River fault zone is discussed below.

*Other Late Cenozoic Deformation within the Yangtze Unit*

North of the Xianshuihe fault is a second and different late Cenozoic tectonic regime. The eastern margin of the Tibetan Plateau is marked by the steep topographic front of the Longmen Shan northwest of the Sichuan basin. Along the southern and central parts of the range are uplifted antiforms cored by Precambrian basement rocks that fold the early Mesozoic thrust sheets. The steep mountain front turns northward through the Min Shan and the Longmen Shan, loses relief, and is cut by northeast-striking faults with increasing right-slip displacement (Plate 2; Burchfiel et al., 1995). The basement cored antiforms are underlain by steep, west-dipping thrust faults that are part of a system of thrust faults that connect with subhorizontal thrust faults to the east that underlie the Sichuan basin, with their eastern front surfacing in the Longquan anticline, which forms the eastern margin of Quaternary deposition in the Chengdu Plain (Plate 2). This anticline curves northward and obliquely intersects the Longmen Shan where the Quaternary deposition ends (Plate 2). Southward, the Longquan anticline trends south and continues into the area where the late Cenozoic structures of the Longmen Shan cross the northeast-trending structures of the western South China fold belts. Between the Longquan anticline and the Longmen Shan mountain front, which continues to rise to the south, several north-south–trending anticlines rise from the Chengdu Plain and continue south, where they also interfere with the northeast-trending folds, forming a dome and basin pattern. The structures continue to increase in amplitude southward, where they interfere with northwest-trending structures; both sets of structures appear active and involve basement rocks in complex geometries. To the north the structures appear to be underlain by a décollement within the lower part of the Yangtze stratigraphic section, and thus the detachment below the folds drops down into the basement to the south. Westward the décollement below the Chengdu Plain appears to merge into the steep thrust fault beneath the basement-cored antiforms in the Longmen Shan. It was the steep thrust fault below the northern antiform, the Pengguan massif, on which the devastating 2008 Wenchuan M 8 earthquake was triggered (Burchfiel et al., 2008c; Zhang et al., 2010a).

All the structures in this area have topographic expression and are considered active. Their time of initiation remains uncertain. The sedimentary section along the eastern side of the Longmen Shan continues from the Cretaceous into the Oligocene, and these rocks are involved in the structures. Unconformably overlying are poorly dated Neogene rocks that are also folded, and it is interpreted that these Neogene strata, which contain abundant conglomerate derived from the Longmen Shan, signal the beginning of the late Cenozoic deformation. Studies of the Longmen Shan and the Tibet Plateau to the west have suggested that the uplift began ca. 12 Ma, consistent with the late Cenozoic, but dates are poor for the beginning of the Longmen Shan deformation (Kirby et al., 2002).

South of the Chengdu Plain the structures continue north-south, where they parallel the north-south structures interpreted to be Paleogene or early Cenozoic. Here it is hard to determine the age of individual structures and how far south the late Cenozoic deformation extends. In this area south of the Chengdu Plain and north of the Xianshuihe fault system the elevation reaches plateau height and also preserves a low-relief topographic surface similar to that of the plateau to the west. All the structures in this area are the result of shortening and include the Danba antiform; they contrast with the structures and topographic surface south of the Xianshuihe fault, which show little sign of shortening during their uplift, indicating a major change in tectonism across the fault, a difference that will be discussed in more detail in the Synthesis chapter at the end of this volume.

### *Magmatic Rocks within the Yangtze Platform*

The most widespread Phanerozoic magmatic rocks within the Yangtze platform are the Emei Shan basalt of Late Permian age (ca. 260 Ma). These rocks are mapped separately on Plates 1 and 2 and occur all across the southern part of the Yangtze Unit and along the northern margin of the Chuxiong basin. These rocks are missing along the Kungdian high adjacent to the Chuxiong basin, and below the eastern, northwestern, and southern parts of the Chuxiong basin, because of pre–Late Triassic erosion, but they were probably present because of their uniform widespread distribution and the presence of pre–Late Triassic dikes and related plutons of similar composition within the rocks of the Kungdian high (labeled $\nu_4^3$, $\delta_4^3$; see above). Their greatest thickness (at least 5 km) occurs in the southern part of the Transitional Unit directly west of the Chuxiong basin. These rocks have been related to the activity of a Late Permian mantle plume (Xu and Chung, 2001; Xu et al., 2004) beneath this part of China. This interpretation will be discussed in more detail in the section on the Transitional Unit below.

Precambrian plutons are present within the Kungdian high, where they are exposed in the N-S belt of Precambrian rocks (pt) along with plutons related to the Emei Shan mafic magmatism at ca. 260 Ma (to be discussed below) within the uplifts in the northern Chuxiong basin. Within the exposed Kungdian high are plutons of minor magmatic pulses at ca. 252 and 242 ± 2 Ma that consist of felsic plutons and mafic dikes with compositions similar to but younger than the Emei Shan basalt. These younger plutons are considered to be separate magmatic events from the Emei Shan eruptions by Shellnutt et al. (2008).

Phanerozoic plutonic rocks are rare within the Yangtze Unit and occur in only several places. In the southeastern part of the area are two large plutons that intrude deformed rocks; one is within the northern part of the Paleozoic rocks shown in the southeasternmost part of Plate 1 (at the SE end of Plate 4, section 2) that is shown on Chinese maps to be either Jurassic or Cretaceous, but no data are given to support this age assignment. A second plutonic complex intrudes folded Triassic strata west of the Mengzi basin. The complex contains multiple intrusions, including early granite bodies that have yielded a poorly documented Early Cretaceous age (ca. 123 Ma) and younger alkalic bodies (labeled $\varepsilon_6$) that are regarded on a regional basis as Paleogene. These plutons are undeformed and would place at least an upper age limit on the deformation in the rocks they intrude, but the dating is too poor to be very useful. The alkalic bodies may be part of a broad and extensive belt of alkalic plutons that are present throughout much of SE Tibet and extend to the SE into Vietnam (Chung et al., 2005). Numerous similar small alkalic plutons are abundant within the western part of the Chuxiong basin (labeled ε6). These rocks are present in many of the tectonic units in Plate 1 and will be considered in more detail below. In the southern part of the Kungdian high east of the Chuxiong basin are several both large and small plutons that intrude Precambrian basement rocks and locally Mesozoic strata labeled on Chinese maps as Jurassic (labeled $\gamma_5^2$), but no data are given on how this age was assigned.

Perhaps the most interesting plutonic complex is in the Yangtze Unit within the northernmost exposures of the Precambrian rocks marking the Kungdian high on Plate 1 and extending north into the area of Plate 2. It forms a N-S belt of plutons at least 150 km long and 20 km wide along the Anninghe River Valley (Plates 1 and 2). They intrude rocks as young as latest Paleozoic and are unconformably overlain by Upper Triassic non-marine strata (Plate 1; Plate 4, section 5). The plutons are part of an episode of Triassic granitic magmatism that is anomalous with respect to its age and location. Plutonic rocks of this age are abundant and well known west of the Yangtze Unit in the Songpan Ganze and Yidun tectonic units, but none is close to the Yangtze Unit. Within this belt of Triassic plutons are more mafic plutons of late Paleozoic age (labeled $\nu_4^3$, $\delta_4^3$) and also a smaller body of Cenozoic, more felsic, rock (labeled $\varepsilon_6$). The Triassic plutons are difficult to interpret (see, e.g., Şengör, 1984; Şengör et al., 1993). Their origin remains a major unsolved problem. The older mafic plutons may have been feeders for the Emei Shan extrusive rocks or a part of the younger Paleozoic magmatism unrelated to the Emei Shan event.

With the exception of the broad belt of alkalic plutons mentioned above, Cenozoic plutonic rocks are rare within the Yangtze Unit. The large pluton that underlies the high massif of Gonga Shan (7756 m) west of the Xianshuihe fault lies within the easternmost area assigned to the Transitional Unit, but it intrudes

Paleozoic rocks of the Yangtze type in two places and thus is considered here to lie close to the boundary between the two tectonic units (Plate 2) and is an exception. It was studied by Roger et al. (1995), who reported a U/Pb age of 12.8 ± 1.4 Ma and an Rb/Sr age of 11.6 ± 0.4 Ma. They interpreted the magmatism as synkinematic with respect to the Xianshuihe fault, indicating that the fault was at least that old. We support this interpretation but suggest that the intrusion of the pluton was related to a compressional bend in the Xianshuihe fault (see discussion in the Synthesis chapter). It is interesting, and maybe coincidental, that these plutons of three different ages all are intruded within the same general N-S belt within the Kungdian high.

## MAJOR UNRESOLVED PROBLEMS ASSOCIATED WITH THE YANGTZE UNIT

### Southern Margin of the Yangtze Unit

The Yangtze Unit in its broadest sense forms a major part of SE China and is often referred to as the South China block or Yangtze platform, but because of its lack of rigidity to deformation we do not use these names. Its passive margins of latest Precambrian and Paleozoic age are known to the north in the Qinling Mountains, to the northwest in the Longmen Shan, and to the east in eastern China; its southern margin, however, remains unclear because it has been removed by younger, large-scale deformation. Along the southern extent of the Chuxiong basin, the Kungdian high, and the South China fold belts with typical Yangtze stratigraphy and basement rocks, all end abruptly against the NW-trending Ailao Shan metamorphic and mylonitic belt of rocks. This boundary was truncated in early Cenozoic time by major strike-slip faulting along the Ailao Shan range (Leloup et al., 1995), and perhaps also by older deformation in latest Paleozoic time, as suggested by Metcalfe (2006); these suggestions need a more detailed discussion (below).

One important factor in understanding the southern boundary of the Yangtze Unit is to explain the folding in the South China fold belts. As discussed above, the folding within this belt has been related to Mesozoic retro-arc shortening at the southeastern margin of the Yangtze Unit. However, these folds extend to the southwest into the area of Plate 1, but the magmatic activity and metamorphism related to this plate boundary activity are only very weakly developed south of these fold belts shown in Plate 1 but may be present in northern Vietnam. Yet the geologic relations between northern Vietnam and southeastern China remain unclear (see below). As seen on Plate 1 the late Cenozoic Ailao Shan metamorphic belt lies to the south of the unit, and the original Mesozoic relations at this boundary have been disrupted, a problem to be considered in the Synthesis chapter at the end of this volume.

In the southeasternmost part of Plate 1 can be seen the thick succession of Neoproterozoic to Ordovician strata, which are unlike similar-age rocks within the Yangtze sections to the north. Their thickness and facies suggest that they were deposited in a region of greater and more rapid subsidence, perhaps in a separate basin such as the Nanhua basin of eastern China (Fig. 2-11; see above), or related to a passive margin. The Devonian strata unconformably overlie the Ordovician strata, and their deposition and that of the overlying strata into the Upper Permian are similar to the remainder of the Yangtze Unit sequence. Within Plate 1 the anomalous lower part of the section is truncated by the Nanxihe fault and the Day Nui Con Voi shear zone, which forms a NW-plunging antiform at the southeast corner of the area of Plate 1 that continues into Vietnam. In northern Vietnam the thick section of Neoproterozoic and lower Paleozoic strata have been intruded by plutons and metamorphosed similar to the rocks making up the more easterly part of the broad South China fold belt (Fig. 2-11; Maluski et al., 2001), but how these rocks are connected or displaced remains unclear. The nature of the southern margin of the Yangtze Unit of Plate 1 will be discussed below with the Ailao Shan Unit and in the Synthesis at the end of the report. It is fair to say that at present this problem has not been resolved.

### Depositional Environments in the Eastern Part of the Yangtze Unit

The eastern part of the Yangtze Unit section within Plate 1 contains a Paleozoic sequence that shows considerable stability during shallow water deposition. There are unconformities within the section (see Plate 3), but most of these are low-angle paraconformities or disconformities where parts of the sections are missing across faults, indicating faulting during deposition. These are shown in the cross sections on Plate 4 (e.g., sections 2, 3, 4, 5) where Devonian strata commonly rest unconformably on older units. The origin of the probable Paleozoic faulting that caused these unconformable relations has not been discussed in any detail. Commonly there is a low-angle unconformable relationship between the Emei Shan basalt and the rock above and below, indicating that the eruption of this vast quantity of mafic rocks affected the depositional environment far to the east of its center of eruption.

The thick succession of Triassic limestone in the southeastern part of Plate 1 grades eastward beyond Plate 1 into a terrigenous flysch-like succession. How this fits into a regional picture of early Mesozoic paleogeography remains unresolved. Also, this area of Triassic limestone has no younger Mesozoic rocks above it, and it has been suggested that it was folded during Late Triassic time following deposition. Only Paleogene strata rest unconformably above the folds in the Triassic strata; thus the deformation is poorly dated and remains a problem for the evolution of the regional stratigraphy and structure (see above).

### Kungdian High

The dynamics remain unclear for the north-south–trending Kungdian paleogeographic high, which appears to have an evolution that began in at least Neoproterozoic time and continued at different times through Paleozoic to Cenozoic time. The high is well expressed in Plate 1 as an irregular belt of Precambrian outcrops from the Ailao Shan in the south to the Danba antiform in the north (Fig. 2-1), but this map pattern is largely a Cenozoic feature because the northeast-trending Mesozoic foredeep related to the early Mesozoic Longmen Shan thrust belt was continuous obliquely across its northern part. Its most distinguishing feature is that Upper Triassic strata are unconformable on Precambrian rocks marked by the broad area of the T3 unconformity along its axis (Plate 3).

Along the high are thick but incomplete sections of the Neoproterozoic clastic sediments (Za) that have a narrow lateral distribution and were deposited during rifting, a feature that remains to be demonstrated in detail. These rift deposits are unconformably overlain by the latest Neoproterozoic Sinian dolomite that characterizes the basal part of the Yangtze deposition. Along the flanks of the broad N-S trend of the Kungdian high, Paleozoic strata thin and develop unconformities at several levels as they approach the high from the east and less clearly from the west. The Paleozoic strata, including the widespread Emei Shan basalt, are missing from the crest of the Kungdian high where it is overstepped by Upper Triassic terrigenous clastic rocks. Thus the evidence supports the high as having been positive at several times during Paleozoic and early Mesozoic time. Prior to the overlap by the Upper Triassic strata, the axis of the high was the site of an anomalous Triassic plutonism whose origin remains uncertain.

The positive nature of the high was reversed during early Mesozoic time with deposition of a thick NNE–trending succession of Upper Triassic to Cretaceous strata interpreted to be a foredeep for the southward continuation of the early Mesozoic Longmen Shan thrust belt (see below). The foredeep section thickens to the NW across the N-S trend of the Kungdian high and overlies much of the high in the eastern part of the Chuxiong basin. The positive nature of the high reasserted itself in Cenozoic time during the N-S folding and thrusting that developed within most of the western part of the Yangtze Unit and other tectonic units to the west, causing it to be expressed as the N-S–trending belt of Precambrian rocks in the central part of the Yangtze Unit (Plates 1 and 2). Why the Kungdian high has this long history of behavior relative to the crust both to the east and west remains unknown. It is also associated in the south with anomalous crustal and mantle characteristics (see below) at times related to the formation of the Emei Shan plume (see below), but this would not explain its older and younger history.

**Time-Space Relations of the Deformation within the Yangtze Unit**

The timing of deformation of the different structural elements within the Yangtze Unit is not clear, because rocks of the right age to define the time of deformation are either not present or poorly dated. We here focus mainly on the Mesozoic and Cenozoic deformations as of greatest importance for our studies, because they relate to the accretion and intracontinental deformation related to the formation of the present-day crust and lithosphere in the region of Plates 1 and 2.

The oldest Mesozoic deformations may be recorded in three places. They began in the early Late Triassic with uplift of the Kungdian high, which may have been accompanied by the intrusion of felsic and intermediate plutons along the axis of the high. Following uplift along the high, erosion removed rocks down to the Precambrian. This was followed shortly by the formation of the east-vergent Longmen Shan thrust belt to the north, where it can be dated as latest Triassic and Early Jurassic (Plate 2) and which continued south along the boundary with the Transitional Unit (Plate 1; Fig. 2-11; see below) to the west, where its age is less well constrained by only the Mesozoic foredeep. The southern continuation of the Longmen Shan thrust belt may have been west of the Yangtze Unit, but the continuation of the structures formed at this time from the Longmen Shan southward remains a problem because they have been overprinted by strong Cenozoic deformation. The eastern part of the west-dipping, east-vergent Mesozoic thrust belt would have lain west of the Yangtze Unit, but how far west is unknown, for the structures have moved farther east during Cenozoic thrusting. The evidence for this interpretation will be presented with the Transitional Unit below, because that is where all the structures are presently exposed. Within the Yangtze Unit the evidence for this thrusting event is the thick Upper Triassic to Cretaceous strata of the foredeep (Fig. 2-11).

The third area of Upper Triassic or perhaps earliest Jurassic deformation lies in the southeastern part of the area of Plate 1, where the NE-trending folds in the southern two belts of the South China fold belts occurred. The age of this folding is poorly constrained and it has been assumed on the basis of the interpretation that because no rocks younger than Triassic occur in these belts, deformation may have followed the youngest deposition in the area. Correlation of these structures to the east, where they are better dated in eastern South China, is unclear and needs considerable work to unravel the southward continuation of the Southeastern China fold-thrust belt into the area of Plate 1 (Fig. 2-11). These NW-vergent structures contrast with the SW-vergent structures in the Longmen Shan thrust belt to the west, and these deformations were related to convergent activity at two different plate boundaries on either side of the Yangtze Unit.

The age of deformation of the remainder of the northern two belts in the NE-trending South China fold belts is interpreted to have occurred from Middle Jurassic into Early Cretaceous time on the basis of their correlation with the better studied fold-thrust belt in eastern South China (see above; Yan et al., 2003a, 2003b; Wang et al., 2010). The structures in these two northern belts appear to be traceable eastward into the western part of the fold-thrust belt in eastern South China, where they are better dated, but the generally irregular traces of the folds and faults remain to be explained in detail.

There is also minor Jurassic and Cretaceous deformation, but of a very different style, around the north and east margins of the Chuxiong basin. Unconformities at the base of Middle Jurassic and Upper Cretaceous strata show where these strata locally lie on older rocks (see Plate 3). The deformation appears to be related to ~N-S–striking structures associated with the eastern margin of the basin as it continued to subside irregularly, and some of these faults appear to be reactivated older faults related to deformation along the Kungdian high.

Paleogene–Neogene deformation is widespread in the western and central parts of the Yangtze Unit within the area of Plate 1. The structures are mainly N-S–striking folds and faults in the Chuxiong basin across the Kungdian high and superposed on the westernmost extent of the South China fold belts (Plate 1). The N-S Cenozoic structures can be traced northward into the area of interference deformation (southern part of Plate 2 and northern part of Plate 1) to the western boundary of the Yangtze Unit and into the Transitional Unit farther west. The area of interference with the older structures produces a dome and basin pattern, and more commonly causes the older folds to have curvilinear patterns. However, part of the curvilinear patterns is also younger and related to the late Cenozoic north-south strike-slip Xianshuihe-Xiaojiang fault system and deformation within the Longmen Shan region. Thus the problem in this area is determining how much deformation is related to each event.

Traced southward from the region where interference of the N-S trends of the folds and faults dominate, the folds and faults in the Chuxiong basin curve to the southeast, and in the southern part of the basin even to east-west. At the southern end of the Kungdian high, all the structures are mainly east-west trending. The problem in this southern part of the Yangtze Unit is that in our interpretation the deformation along the major Ailao Shan metamorphic and mylonitic belt has been both transpressional and transtensional at different times, and how much of the deformation to its north is a result of deformation within a broader zone north of and related to the Ailao Shan deformation, and how much has been inherited from older deformation in the South China fold belts remains unknown. Some NW-striking faults that are present within 50–75 km north of the Ailao Shan belt exhibit Paleogene left-lateral and younger right-lateral strike-slip displacements similar to those recorded in the Ailao Shan (Burchfiel and Wang, 2003). Thus it seems clear that the Ailao Shan deformation affected this southern part of the Yangtze Unit, but to what degree remains unknown.

Another area of Paleogene and early late Cenozoic deformation occurred in the very southeastern part of the Yangtze Unit, where a NW-plunging antiformal structure of Triassic strata has a core of mylonitic metamorphic rocks of the Day Nui Con Voi and is similar in character to those rocks in the Ailao Shan Unit to its southwest (Plate 1). The mylonitic core rocks consist variably of mylonitized granitic and leucogranitic rocks and metasedimentary rocks reaching sillimanite grade. A narrow carapace of metasedimentary rocks of sillimanite grade, correlated with a Triassic protolith, abruptly decrease to lower grade and are surrounded by unmetamorphosed Triassic strata at the hinge of the fold. The core rocks have been correlated with the Ailao Shan metamorphic rocks, but are the direct continuation of the Day Nui Con Voi mylonitic belt in Vietnam (see, e.g., Jolivet et al., 2001) and are separated from the Ailao Shan rocks to the south by a narrow belt of unmetamorphosed Triassic strata. Thus they are not the direct continuation of the Ailao Shan. The relations here were discussed by Burchfiel et al. (2008b), who have interpreted the core rocks to be partly the basement for the Yangtze Unit and which represent the lower crust in this area that became very hot and mobile during Paleogene and early Neogene time in a transpressional structural environment along the southern margin of the Yangtze Unit. How this hot lower crust developed, and what its relationship is to the Ailao Shan metamorphic and mylonitic rocks and to major extrusional tectonics that occurred along and south of the Ailao Shan Unit in this area, will be discussed in more detail in the Synthesis chapter (Chapter 16), but it still remains a major problem.

**Deeper Crustal Structure within the Yangtze Unit**

As the cross sections of the Yangtze Unit show, the rocks are variably folded, with different trends having formed at different times (Plates 1 and 4; see above) and having formed several different fold and thrust belts. In many places basement rocks are involved in the folds. The cross sections in Plate 4 are taken directly, but somewhat modified, from the 1:200,000-scale Chinese map sheets. Most of the faults are shown as very steep to vertical, cutting the folds, but with these attitudes they will not accommodate the shortening within these fold belts. It must be remembered that these maps were produced either prior to or during the early development of foreland fold-thrust belt geometries we now take to be common interpretations and thus not as a reflection on the geologists who produced these cross sections.

Our current interpretations of the subsurface structure of the Yangtze Unit are only suggestions as to how the deeper structure of the fold belts might look, as it would be a massive undertaking to redo the fieldwork necessary to make more accurate and balanced cross sections in these multiply deformed rocks. We show on Plate 4 our suggested and simplified reinterpretations of the cross sections superposed on the sections taken from the geological maps.

For the South China fold belts, most of the folds and thrust faults can be shown to have involved only Neoproterozoic to Triassic or Jurassic strata except locally, such as the large uplift of basement rocks in the middle part of the belt that exposes only Paleozoic strata (Plate 1 and Plate 4, section 3). This area of basement rocks appears to have been uplifted in pre-Devonian time and may have been a large, but local, uplift of basement that was incorporated into a thin-skinned thrust belt. Other, smaller areas of basement involvement are present, and with additional areas of involvement of earlier Neoproterozoic rocks (Za) and smaller areas of basement rocks involved in the structures to the north, suggest that basement involvement may be widespread and that the thin-skinned thrust belt may have detached extensively or wholly within the Precambrian basement rocks, as shown in our interpretations on Plate 4.

The general vergence of the South China fold belts is not directly obvious, but there may be a dominance of northwest-vergent structures, although both NW- and SE-directed structures are present. Furthermore, in a more regional context as discussed above, this is part of a very broad foreland fold-thrust belt that migrated in time from southeast to northwest, interpreted to be a large-scale northwest-vergent system. A correlation, albeit difficult in part, that this fold-thrust belt is part of a broad backarc Andean belt that trends NE-SW across eastern South China is difficult to demonstrate within the area of Plate 1, because much of its southern and eastern parts are missing, and any restorations of the southern truncated margin of the Yangtze Unit must consider the tectonic setting of the entire South China fold belts.

In tracing the South China fold-belt structures to the southwest within Plate 1 they can be seen to increasingly involve more basement rocks at the surface where they begin to merge with the Kungdian high, and the stratigraphic sections thin drastically or Paleozoic units are missing; in some places the structures involve only the Mesozoic strata (see, e.g., Plate 4, sections 2, 5, 34, and numerous places within Plate 1). This poses an additional problem, however, because this is where the South China fold belts appear to die out to the west within the Kungdian high. If this is true, there should be some structures that accommodate shortening to the east and the lack of shortening to the west. There are numerous, generally N-S–trending faults in this area that may form this accommodation, but they have never been interpreted in that way. Thus the depth of detachment within basement rocks in the South China fold belts and the accommodation for the structures dying out to the west remain major unresolved problems in the eastern part of the Yangtze Unit.

The younger Paleogene to Neogene N-S–trending fold-thrust belt structures in the western part of the Yangtze Unit pose a similar problem. They appear to be related to forces generated west of the Yangtze Unit and represent the eastern part of a generally east-vergent fold-thrust system. In the north the structures clearly involve basement rocks, and in places to the south in the eastern part of the Chuxiong basin basement rocks are also involved (see Plate 4, sections 30, 35, 36). However, the Chuxiong basin contains possibly as much as 10 km of Mesozoic strata that might suggest a basal thin-skinned detachment near the base of the Mesozoic section, but this remains speculative, as structures within the Chuxiong basin, when traced northward, clearly involve basement rocks; thus it is possible that this is true for the entire Chuxiong basin. Finally, the problems for the deeper structure of the South China fold belts are also problems for the younger fold belt in the western Yangtze Unit.

# CHAPTER 3

# *Transitional Unit*

The Transitional Unit is a new unit we identify because it contains Mesozoic rocks, largely Triassic strata (T and Tt, Plates 1 and 2 [available on DVD accompanying this volume and in the GSA Data Repository[1]]), that are transitional between the non-marine Mesozoic rocks of the section in the Yangtze Unit and Mesozoic flysch formations in the Songpan Ganze and Yidun units that are missing between the Yangtze Unit and the Songpan Ganze in the Longmen Shan area (Plate 2). The Triassic rocks overlie Paleozoic strata similar to the Yangtze Unit, with some significant differences, but Paleozoic strata in the Songpan Ganze and Yidun units are either only locally exposed or entirely missing beneath the Mesozoic flysch and flysch with volcanic rocks, respectively, in these two units. The boundaries of the Transitional Unit are marked by a thrust fault zone of two different ages and of great complexity along their contact in the southeast with the Yangtze Unit (see, e.g., Bureau of Geology and Mineral Resources of Sichuan Province, 1991), and both are fault bounded and gradational in the southwest with the Songpan Ganze and Yidun Units, respectively.

In the Longmen Shan region, early Mesozoic thrust faults placed rocks of the Songpan Ganze eastward above rocks of the Yangtze Unit, telescoping and eliminating the transitional rocks that lay between them (modified boundary shown by the white line in Plate 2; Burchfiel et al., 1995). Southward from the Longmen Shan, rocks transitional between these tectonic units appear to be present within the southern part of Plate 2 south of the Xianshuihe fault and continue into the area of Plate 1 in a southward broadening zone. The Triassic strata overlie Paleozoic and locally Proterozoic rocks that are correlative with the Yangtze Unit succession and represent their continuation on the west side of the Kungdian high, where they form a disrupted passive margin sequence. In the northeast part of the area of Plate 1 the Paleozoic and Triassic strata between the Songpan Ganze and Yidun units and the Yangtze Unit are bounded by two west- to northwest-dipping faults, the Jinhe-Qinhe and Xiaojinhe thrust faults, on the south and north, respectively. These thrust faults have a complex history, and although the presently mapped thrust faults have Cenozoic displacement, thrust faults in this position may also have large early Mesozoic displacements. The original Mesozoic thrust faults, however, have been greatly disrupted and are only rarely exposed (a discussion of this relationship will be presented in detail below). In a general way this belt of thrust faults represents the southern continuation of the Longmen Shan thrust belt, formed in both Mesozoic and Cenozoic time, but here the transition from the Songpan Ganze deep-water Triassic flysch (Mzf, Plates 1 and 2) to the shallow-water Yangtze Unit strata is more completely preserved. The Transitional Unit can be traced to the SW to where it is truncated by major NW-trending fault zones that bound the southwest side of the Yangtze and Transitional Units (Fig. 1-1B).

## STRATIGRAPHY OF THE TRANSITIONAL UNIT

Within Plates 1 and 2 the eastern part of the Transitional Unit lies between the Jinhe-Qinhe and Xiaojinhe thrust faults, and contains strata that are missing in the Longmen Shan area shown in Plate 2 (Burchfiel et al., 1995). The contrast in stratigraphy across the Jinhe-Qinhe thrust fault shows considerable foreshortening that may have taken place in both early Mesozoic and Cenozoic times (see below). West of the Anninghe River, in the footwall of the thrust fault, is a thick succession of non-marine Upper Triassic strata, ~15–20 km wide, that unconformably overlies Yangtze basement rocks intruded by Triassic plutonic rocks at the crest and along the west flank of the Kungdian high. These strata thicken to the west and consist of >1000 m of tan, gray to dark-gray sandstone and shale with local coal beds and containing plant fossils of Late Triassic age. They are overlain by >2 km of red sandstone, mudstone, and conglomerate of Early to Late Jurassic age. These Mesozoic rocks form the western part of the foredeep for the southern extension of the Late Triassic and Early Jurassic Longmen Shan thrust belt and belong to the western part of the Yangtze Unit. They are folded and repeated by west-dipping imbricate thrust faults (Plate 4 [see footnote 1], section 6).

The hanging wall of the Jinhe-Qinhe thrust carries uppermost Neoproterozoic (Z2) and Paleozoic rocks with >2–4 km of Permian Emei Shan basalt (Plate 1, Pβ) that are similar to those of the Yangtze Unit (Fig. 3-1; cf. Fig. 2-4). Conformably above the basalt and the overlying Upper Permian strata are >6 km of Triassic rocks that are marine and non-marine terrigenous clastic rocks of Early and Late Triassic age and nearly 2 km of limestone of late Middle Triassic and early Late Triassic age. This section ends in the Upper Triassic with coal-bearing rocks. Like all Mesozoic stratigraphic sections in the Transitional, Songpan Ganze, and Yidun units, it lacks rocks younger than Late Triassic (except for rare local(?) marine Jurassic rocks that have been reported [e.g., Wang et al., 2002; see below]). The next younger rocks are Eocene (E) that locally are >2 km thick in the Yanyuan

---

[1]GSA Data Repository Item 2012297, Plates 1–4, is available at www.geosociety.org/pubs/ft2012.htm, or on request from editing@geosociety.org or Documents Secretary, GSA, P.O. Box 9140, Boulder, CO 80301-9140, USA.

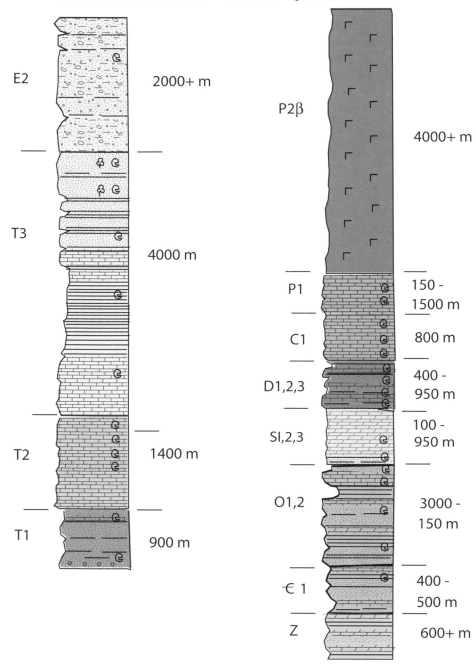

Figure 3-1. Stratigraphic section of rocks in the Transitional Unit north of the Jinhe-Qinhe thrust fault (see Fig. 2-2, no. 6 for location).

basin and surrounding areas, which lie unconformably on folded Triassic (T) and older strata (Plate 4, section 6). The Triassic rocks are much thicker than correlative rocks in the Yangtze Unit east of the Jinhe-Qinhe thrust, indicating much greater subsidence and increasing marine depositional conditions to the north and west. Importantly, to the west and north they show a transition (Tt on Plate 1) to the flysch terrane of the Songpan Ganze and Yidun tectonic units.

Lying tectonically above these transitional rocks is the west-dipping Xiaojinhe thrust fault, whose hanging wall contains Paleozoic rocks overlain by Middle and Upper Triassic rocks (Tt on Plates 1 and 2), which are dominantly fine-grained clastic rocks but include rare limestone. This section contrasts with the rocks from within the Jinhe-Qinhe thrust plate, which contain much less fine clastic strata. The fine-grained Triassic rocks have the characteristics of flysch, similar to Triassic flysch in the

Songpan Ganze Unit, and locally they contain volcanic rocks correlative with rocks in the Yidun Unit to the north and west (Fig. 4-1). The Paleozoic section is mostly complete, but missing (perhaps locally tectonically) are Middle and Upper Cambrian, Upper Ordovician, and Upper Carboniferous strata (Fig. 3-1). The Late Permian Emei Shan basalt is both overlain and underlain by marine limestone and was erupted entirely within a marine environment in contrast to the terrestrial setting within the Yangtze Unit. The transition from a marine to non-marine environment for Emei Shan volcanism occurs with the western part of the Transitional Unit. Locally the lower Paleozoic section is foliated, metamorphosed to low grade, and is more ductilely deformed than rocks to the east. We take the Xiaojinhe thrust fault as the northern boundary for the Transitional Unit, and its hanging wall rocks represent the Yidun and Songpan Ganze Units.

To the south and west the distribution of the transitional rocks is more complex, and how they relate to the major structures in the area is poorly understood, representing one of the major regional problems as shown on Plate 1 (see below). This is particularly the case for the central exposures of the Transitional Unit, so here we skip to the southern part of the Transitional Unit and return to the central part later.

In the southwest the eastern limit of the Transitional Unit is most clearly defined by the west-dipping Chenghai thrust fault, which was largely reactivated as the west-dipping late Cenozoic to active Chenghai normal fault (see below) and is best exposed locally along the east side of the Chenghai and Binchuan basins (Plate 1). At the contact between the Transitional Unit and the Chuxiong basin are remnants of the original thrust fault that in several places contain slivers of Devonian or Permian rocks in its hanging wall. West of the Chenghai thrust fault the Mesozoic section characteristically contains a complete succession of Triassic strata but no Jurassic rocks (Fig. 3-2) in contrast to the Chuxiong basin section, which belongs to the Yangtze Unit where Middle and Upper Triassic strata are unconformable on Paleozoic rocks (Fig. 2-8; see above in discussion of the Kungdian high) and are overlain by a thick and continuous section of Jurassic to Upper Cretaceous non-marine strata. The Paleozoic section in the hanging wall of the Chenghai thrust is typical for the Yangtze platform with two exceptions: the Permian Emei Shan basalt is at least 5 km thick, one of the thickest sections of these rocks known, and the Ordovician rocks, in the southernmost part of the Transitional Unit near Erhai Lake, contain mainly fine-grained, graptolite-bearing clastic rocks with some chert (a point we shall return to below).

The Triassic rocks (T in Plate 1) in the hanging wall of the Chenghai thrust fault we assign to the Transitional Unit, because it contains a complete sequence of Lower to Upper Triassic strata and lacks younger Mesozoic strata. The lower and upper parts

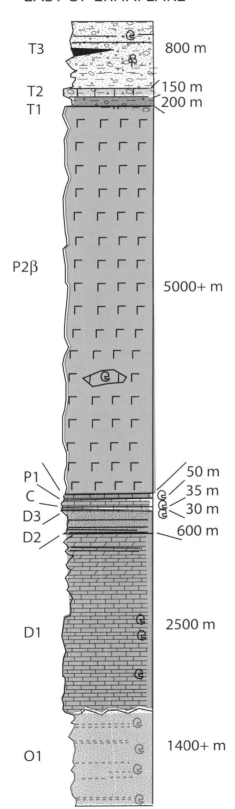

Figure 3-2. Stratigraphic section of rocks in the Transitional Unit east of Erhai Lake. This area contains the thickest section of Emei Shan basalt in all the area of Plate 1 (see Fig. 2-2, no. 7 for location).

of the section are dominated by clastic rocks, and a middle part, largely upper Lower and Middle Triassic, is dominated by limestone. Rocks younger than Late Triassic are lacking, contrasting greatly with the typical Yangtze section of the Chuxiong basin to the east (cf. Fig. 2-8). Toward the northwest the Triassic rocks thicken and consist of >2000 m of interbedded limestone and fine-grained clastic rocks that grade into an almost entirely fine-clastic sequence with some flysch (Tt) farther to the north. The lower part of the Triassic succession is interbedded with basalt that we interpret to be the transition into the lower basalt unit within the Yidun Unit (Figs. 4-1 and 4-2; see below).

The relationship of the Triassic rocks between the northeastern and southwestern areas of the Transitional Unit are difficult to decipher because of a complex and as yet unclear structural picture (see below). Stratigraphically in the northwestern part of the Chuxiong basin, east and north of Chenghai Lake, and lying below a large north-trending exposure of Paleogene strata (Fig. 3-3 and Plate 1), Lower Permian sedimentary rocks are overlain by the Upper Permian Emei Shan basalt that in turn is overlain by Lower Triassic strata. The Triassic section is complete, unlike the characteristic strata of the Chuxiong basin, where Upper Triassic strata generally lie unconformably on Paleozoic rocks. The transition between these two Triassic sections is not exposed. The change in the stratigraphic section occurs in the thrust slice that lies below the Chenghai thrust, a thrust slice that brings up the Paleozoic rocks farther south. To the north and below the Paleogene strata the Triassic section becomes complete, but no Jurassic strata are present. Based on the complete Triassic section, this sequence of rocks we tentatively place within the Transitional Unit; thus this may be one area, the only(?) area, where the transition from Yangtze to Transitional Unit stratigraphy may be exposed. The structure in this area is complex and will be discussed in greater detail below.

To the east along the northern margin of the Chuxiong basin, below the Jinhe-Qinhe thrust fault, the rocks are typical for the west flank of the Kungdian high, where Devonian strata rest unconformably on uppermost Neoproterozoic (Z2) rocks, and Upper Triassic strata of the Chuxiong basin lie unconformably on the Devonian and Permian Emei Shan basalt to the north. Dated Eocene red beds lie unconformably on Upper Triassic strata. In contrast, the rocks in the hanging wall of the Jinhe-Qinhe thrust are typical for the Transitional Unit (Fig. 3-1), with an almost complete section of Upper Neoproterozoic and Paleozoic strata, and thus similar to most other places in which the boundary between the two units has been telescoped, a relationship that can be followed to the northeast into the area of Plate 2 (Burchfiel et al., 1995). The relations between the thrust faults and the different stratigraphic units indicate that the northern continuation of the Jinhe-Qinhe thrust fault north of the Chuxiong basin cuts obliquely across the western part of the Kungdian high Paleozoic strata, and the Transitional Unit's Mesozoic strata become increasingly more telescoped.

Eocene strata are present both in the hanging wall and footwall of the Jinhe-Qinhe and Chenghai thrust faults, but they have a complex and poorly understood relationship to the thrust faults between the northern and southern areas in the Transitional Unit, and how these thrust faults might connect is discussed in detail below.

Paleogene strata (E) are abundant within the Transitional Unit and consist of a wide variety of red, maroon, tan, and yellow sandstone, conglomerate, and mudstone, whose stratigraphy is generally specific to the individual basin in which they were deposited. Everywhere they lie unconformably on deformed Mesozoic and Paleozoic strata, and they occur in numerous small outcrops throughout the area and in several larger areas, such as in the Jinchuan basin, Yanyuan basin, and N-S–trending belts within the middle part of the Transitional Unit (Plate 1). The most thoroughly studied Paleogene strata are those of the large Jinchuan basin, which lies at the western edge of the Transitional Unit (Plate 1). This section contains >4 km of Paleogene rocks that are dated on the basis of fossils in their upper part as Eocene, overlain by unfossiliferous Oligocene(?) strata. These deposits consist of a thick sequence of red mudstone and sandstone in the lower part, unconformably overlain by thick-bedded, tan to white sandstone in the middle part. The top of the section consists of tan to pale-red conglomerate and sandstone. The rocks are not strongly folded except in their western part, where there is abundant conglomerate, and the rocks are locally overturned to the east. The strata in the western part of the basin are faulted against rocks that belong to the Jinsha accretionary complex and to Qiangtang 1 units to the west, but in the southwestern part of the basin Eocene rocks unconformably overlie rocks of both the Jinsha and Lower Triassic rocks of the Qiangtang 1 units (Plate 1; see below). The tectonic relations of these tectonic units and the Transitional Unit at this latitude are some of the most important problems in the region and will be discussed in detail below.

Between the Jinhe-Qinhe thrust to the northeast and the Chenghai thrust to the southwest is a complexly faulted, N-S–trending corridor that contains nearly complete sections of uppermost Neoproterozoic (Z in Fig. 3-3 and Z2 on Plate 1) and Paleozoic rocks with a Yangtze type stratigraphy. Within this corridor the Mesozoic rocks are Chuxiong red beds in the south unconformably overlain by a wide N-S–trending belt of folded Paleogene strata, which to the north unconformably overlie only Triassic rocks that abruptly become to the west and north more clastic and locally contain Lower Triassic basalt similar to rocks in the Yidun Unit. Within the western part of this corridor are several west-dipping thrust faults that foreshorten the stratigraphic sections. This Triassic section appears to show a gradation from the Yangtze Unit section of the Chuxiong basin into the Transitional Unit, both unconformably overlain by thick Paleogene strata. The corridor is bounded by rocks that are the continuation of the Chenghai and Jinhe-Qinhe thrust faults on the west and east, respectively, and could be interpreted as a half window in a continuous thrust sheet, but the relations are much more structurally complex and will be discussed below.

The general interpretation of the stratigraphy in the Transitional unit indicates that the Paleozoic strata of the Yangtze

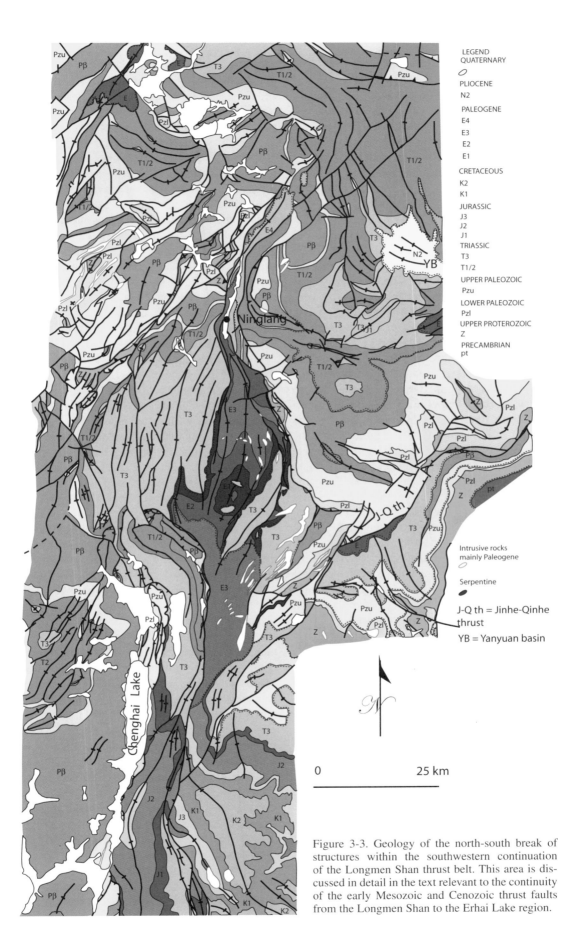

Figure 3-3. Geology of the north-south break of structures within the southwestern continuation of the Longmen Shan thrust belt. This area is discussed in detail in the text relevant to the continuity of the early Mesozoic and Cenozoic thrust faults from the Longmen Shan to the Erhai Lake region.

Unit originally continued to the north below the Triassic strata of the Transitional Unit, which was deposited in a more rapidly subsiding transitional area into the deeper water Triassic flysch sections in the Songpan Ganze and Yidun units to the north. Both the stratigraphic and structural development of this margin in latest Paleozoic and early Mesozoic time is an important feature of the regional tectonics and will be discussed in greater detail below.

## STRUCTURE OF THE TRANSITIONAL UNIT

The thrust faults that define the boundaries of the Transitional Unit in the northeast and southwest are reasonably clear. To the northeast the Transitional Unit is bounded by the Jinhe-Qinhe thrust below and the Xiaojinhe thrust above, and these two thrust faults foreshortened the original stratigraphic transition between the Yangtze Unit to the east and the Songpan Ganze and Yidun Units to the north. These two thrusts have been interpreted as being offset ~60 km in a left-lateral sense by the Cenozoic to Recent Xianshuihe fault zone (Wang et al., 1998). They continue northward into the area of Plate 2, eventually linking with the Longmen Shan thrust faults along the west side of the Sichuan basin (Fig. 2-9; Plate 2; Burchfiel, et al., 1995). To the southwest the Chenghai thrust foreshortens the section between the Chuxiong basin and the Transitional Unit, suggesting that the thrust faults that define the eastern boundary of the unit were stratigraphically controlled in regions of rapid changes in thickness and facies related to increased subsidence in latest Paleozoic and early Mesozoic time. Whereas the Xiaojinhe thrust defines the northern boundary in the northeast, the northern boundary to the west appears to retain its transitional character into the Yidun Unit to the north without major foreshortening.

Outstanding uncertainties regarding the structure of the Transitional Unit include ages of deformation, particularly for the major thrust faults, the continuity of the major thrust faults along the eastern boundary, and how they relate to more regional tectonic events. The age of the eastward thrusting in the Longmen Shan area to the north is well documented to be latest Triassic and Early Jurassic (Plate 2; Indosinian event of Chinese geologists; Burchfiel et al., 1995). The thrust faults are interpreted to be traced southward into the northern part of the area of Plate 1, where the Jinhe-Qinhe and Chenghai thrust faults are their continuations. Those structures involve Paleogene strata and thus indicate demonstrable Cenozoic displacement. However, the extent of Cenozoic deformation on possible correlative structures in the Transitional Unit involve Paleogene strata, and thus have Cenozoic displacement, within the Transitional Unit (Plate 1), but the nature and extent of both the Mesozoic and Cenozoic deformations remain uncertain.

The thick Upper Triassic and Jurassic foredeep section of the westernmost Yangtze Unit that occurs in the footwalls of the of the Jinhe-Qinhe and Chenghai thrust faults is the main evidence that early Mesozoic thrusting continued to the south, but thrust faults involve Eocene rocks and thus must have been reactivated in Cenozoic time. The hanging walls of these thrust faults, and the stratigraphic section of the Transitional Unit, lack Jurassic and Cretaceous strata, as do the Mesozoic thrusts in the Longmen Shan, suggesting that the thrusting produced tectonic relief as a source area for the foredeep rocks.

As shown within Plate 2 and within the Transitional Unit in Plate 1, and within all the units to the west of the Mesozoic thrust faults, there is clear evidence for deformation and metamorphism of early Mesozoic age (see below and as summarized in Burchfiel et al., 1995; Roger et al., 1995). Many of the structures in the units north of the Transitional Unit are intruded by plutons of early Mesozoic, mostly Triassic, age, and only rarely do the plutons show internal deformation (Burchfiel et al., 1995; Roger et al., 2004, 2008) that would place an upper limit on the deformation. The structures in the Transitional Unit show abundant evidence of refolding and form a complex array of folded faults, curved fold axes, and dome and basin interference structures. However, what ages to assign to different structures in the Transitional Unit is difficult because of the rarity of rocks of the ages needed to date them or the lack of detailed geochronologic studies. Many of the structures within the Transitional Unit of Plate 1 are continuations of, or parallel to, structures that can be dated in the Yidun and Songpan Ganze Units and thus are regarded as early Mesozoic in age. Throughout the Transitional Unit, as shown in Plate 1, the oldest strata that overlie deformed Paleozoic to Triassic rocks are Paleogene, and intervening strata are lacking, thus the precise ages for the structures remain poorly determined, but it can be inferred that this area was the sediment source for the thick, continuous Mesozoic section deposited in the foredeep to the east.

Few of the Paleogene strata shown on Plate 1 are dated, and many of the sections have no fossils; their ages are assigned on the basis of lithological correlation with fossiliferous sections elsewhere. The Paleogene strata commonly occur in long, N-S–trending linear belts that extend across the general NNE-SSW or NE-SW trends of known major thrust faults, or in local areas that unconformably overlie strongly folded rocks. Commonly the Paleogene rocks are less folded than the underlying rocks, but in many places they are also strongly folded. The trace of the Jinhe-Qinhe thrust fault in the eastern part of the Transitional Unit is convex, and along its southern trace it overrides local Paleogene strata that reach thicknesses of 1–2 km. Their proximity to, and in a footwall position to, the thrust faults, and because the conglomerate consists dominantly of clasts from hanging wall strata, the Paleogene to younger age of the Jinhe-Qinhe thrust is well established.

The pre-Paleogene structures in the hanging wall of the Jinhe-Qinhe thrust fault consist of tightly folded shallow-water Triassic strata (Plate 4, section 6) that form an arc convex to the south. The folds are unconformably overlain by Paleogene strata in the Yanyuan basin and adjacent areas (Plate 1). These folds are truncated on the north by the Xiaojinhe thrust, which was also probably reactivated in early Cenozoic time but did not involve Paleogene strata directly. The folds in the hanging wall of the Jinhe-Qinhe thrust are interpreted to be early Mesozoic in age, but proof is lacking. The Paleogene strata that unconformably overlie them are generally more weakly folded and are commonly folded along axes that are both parallel and oblique to the underlying folds. Within the southern part of the hanging wall are general N-S–trending folds that are truncated by the Jinhe-Qinhe thrust, indicating that they are older than the thrusting, but how much older is unknown. The Jinhe-Qinhe thrust fault has a large but unknown displacement, as it carried upper Neoproterozoic strata at its base above Paleogene strata in the footwall, juxtaposing two different Paleozoic and Mesozoic sequences. The thrust fault can be followed around a faulted but arcuate trace, convex to the south, and to the west it trends N-S for >25–30 km, overriding a long belt of Paleogene rocks in its footwall (Fig. 3-3; Plate 1). There is also a short splay that continues to the southwest and abuts the north-trending belt of Paleogene rocks. Certainly the thrust has a large displacement, perhaps several tens of kilometers, but how much of its displacement is Mesozoic and how much is post-Paleogene, remains unknown. The nature of this contact and the continuation of the thrust to the southwest is a major problem, discussed below.

Before continuing to address the extent of early Mesozoic deformation within the Transitional Unit, the Cenozoic structures need to be discussed and unraveled. At some place along their traces, nearly all the major thrust faults in the Transitional Unit involve Paleogene strata, but it is not obvious from the map that there is a through-going thrust belt of Cenozoic age along the southern boundary of the unit that would connect the Jinhe-Qinhe and Chenghai thrust faults. Here we offer an interpretation of mapped relations to suggest there is a through-going thrust belt, but local areas of ambiguity exist and need to be investigated further. Along the west side of the Jinhe-Qinhe thrust where it forms a sharp bend, the thrust turns north along a N-S trend and overrides a wide, N-S–trending belt of Paleogene strata (Fig. 3-3). The Paleogene strata are strongly folded along N-S axes and form a long, narrow corridor 2–10 km wide. As described above, the south end of this belt of Paleogene strata unconformably overlies Mesozoic rocks of the northernmost part of the Chuxiong basin, and to the north unconformably overlies Triassic strata that belong to the Transitional Unit. Thus the transition from the Yangtze Unit to the Transitional Unit took place in the Mesozoic rocks below this belt of Paleogene strata, and the Cenozoic Jinhe-Qinhe thrust cuts across the transition.

On the west side of this N-S belt of Paleogene strata the structural relations are likewise complex. In the south, Triassic strata and their underlying upper Paleozoic strata of the Transitional Unit are thrust eastward onto the Paleogene rocks, but at the north end of the belt of Paleogene strata the youngest Paleogene strata (E4; Fig. 3-3) rest unconformably on Triassic strata of Transitional Unit character. The west side of the belt of Triassic strata of the Transitional Unit are cut by several west-dipping thrust faults the westernmost of which carried uppermost Neoproterozoic and lower, middle, and upper Paleozoic strata that in places were thrust eastward over, or in other places are in high-angle fault contact with, the Triassic rocks in the N-S corridor of Triassic and Paleogene strata (Fig. 3-3; Plate 4, section 38). When traced south, this westernmost thrust fault can be followed east of Chenghai Lake to the north end of the Chenghai thrust, which forms the western boundary of the foredeep deposits of the Chuxiong basin. The Chenghai thrust and the fault zone that continues it to the north are of Paleogene or younger age. Thus the Chenghai and Jinhe-Qinhe thrust faults here are parallel and lie on both sides of the N-S–trending belt of Triassic and Paleogene strata. It is plausible that the Triassic–Paleogene units between the faults constitute a narrow half window framed by a single major Cenozoic thrust with large-magnitude (75 km or more) displacement.

However, there is a complication for this hypothesis. At the north end of the corridor the Paleogene rocks (E4) are shown to rest unconformably on Triassic rocks that have clear continuity with both the hanging wall of the Jinhe-Qinhe and Chenghai thrusts, but to the south they also lie unconformably on the footwalls of the thrusts; the details of these relationships are shown in Figure 3-3. The fact that the Triassic and Paleozoic rocks of the hanging walls of the two thrusts have continuity in a large convex north arc around the north end of the Paleogene outcrop appears to support their connection. The pre-Triassic rocks in the hanging walls of these thrust faults have a stratigraphy that is characteristic for the Transitional Unit and unlike that in the footwall, which is of Yangtze character, showing the importance of gaining a good understanding of the regional stratigraphy. It could be hypothesized that rocks at the northern end of the Paleogene corridor were folded and faulted after the thrusting, bringing up the Paleogene rocks from the footwall of a major thrust to form a half window, but the relations at the north end of the corridor make it uncertain whether there is a through-going, early to middle Cenozoic thrust fault. This relationship bears not only on the continuity of Cenozoic, but also early Mesozoic, thrust faults, and will be discussed in detail in the section on major problems within the Transitional Unit, below.

Whether there is one or two thrust faults does not change the fact that the directions of thrusting are complex in this area. If the arcuate trend of a continuous thrust is primary, possibly of Mesozoic age, the direction of thrusting to the southeast may have been influenced by preexisting paleogeography. However, if the thrust traces have been distorted by younger N-S folding and faulting, forming an apparent half window, it would suggest the directions of shortening during Cenozoic deformation changed significantly from southeast to east-west. The presence of a continuous thrust slice that parallels the Chenghai thrust and can be followed into the western side of the "apparent window," and the presence of exclusively N-S structures within the window suggests that younger east-west shortening in this area is the correct interpretation. Interpretations presented below also support this younger period of regional Cenozoic east-west shortening.

West across the Transitional Unit from the apparent half window, the structures trend dominantly N-S but with curved traces. There is a second narrower and discontinuous N-S–trending belt of Paleogene strata that lies ~10–20 km west of the ambiguous half window (Plate 1), where Paleogene strata are both in fault contact and unconformable contact on Triassic Transitional Unit Mesozoic and Paleozoic rocks. The belt is marked by folds and thrust faults and has cut across most of the structures adjacent to it, forming a belt of east-west shortening of Paleogene or post-Paleogene age that is younger than most of the structures in the Transitional Unit. The belt's parallelism and similar structural character with the half-window structure indicates they are of similar age, and many of the structures to the west of this belt that do not involve Paleogene strata are also mainly N-S trending and may also be of the same age (see below). At the south end of this belt of Paleogene rocks, south and east of Lijiang, is a well-documented N-S–trending thrust belt (Plate 4, section 37), where a series of west-dipping thrust faults and associated klippen carry Triassic limestone of Transitional Unit facies in both their hanging walls and footwalls. The thrust belt is ~10–15 km wide and can be traced along strike for ~40 km. The thrust faults involve several hundred meters of Paleogene red sandstone and conglomerate that are part of the N-trending belt of Paleogene strata in the longer N-S–trending belt of deformation. The conglomerate contains large clasts up to boulder size of Triassic rocks derived from hanging wall rocks. The amount of displacement on the thrust faults is only a few kilometers, and hanging wall–footwall cutoffs are exposed. The total displacement may amount only to 10–20 km. The northernmost thrust faults in this belt merge into the narrower belt of N-S structures and may continue forming thrust faults along the west side of a discontinuous narrow belt of Paleogene rocks. In the mountains west of Lijiang are N-S–trending folds that may be part of the same belt of deformation (Plate 4, section 38). Like the large N-S–trending belt of structures that separates the Jinhe-Qinhe and Chenghai thrust faults in the half window to the east, this belt of deformation is of Cenozoic age and postdates any Mesozoic thrust faults that formed the eastern margin of the Transitional Unit.

The western part of the Transitional Unit contains numerous N-S–trending folds and thrust faults that continue northward into the Yidun Unit (Plate 1). Even though in most places these folds and faults do not involve Paleogene rocks, farther north there are numerous places where they do locally involve Paleogene strata, and it can be hypothesized that they are part of a regional Cenozoic E-W shortening that continues far to the north through most of the area of Plate 1 (see below).

Even though the southeastern boundary of the Transition Unit is complex, and some points of its structural character are unclear, the boundary is marked along most of its extent by either the Jinhe-Qinhe or the Chenghai thrust faults. The northern boundary of the Transitional Unit in the west, however, does not appear to be marked by a continuous tectonic contact. In the northeast the Xiaojinhe thrust fault is taken as its northern boundary. Hanging-wall sedimentary rocks are dominated by marine clastic Triassic units (Tt) that are flysch or flysch-like and can be assigned to the Songpan Ganze and Yidun Units (see below). Toward the west, volcanic rocks appear within similar clastic rocks and belong to the southernmost part of the Yidun Unit (Figs. 4-1 and 4-2; see below). The Xiaojinhe thrust is difficult to trace to the west, and although thrust faults are present they do not mark the boundary between volcanic-bearing strata of the Yidun Unit and non-volcanic clastic strata in the northern part of the Transitional Unit. It appears that here the two units grade into one another. Thus the transition between the two units is arbitrarily drawn mainly where the volcanic units of the Yidun Unit first appear (Plate 1). Rocks in the southern part of the Yidun Unit overlie Paleozoic rocks of the Yangtze Unit type, but the Emei Shan basalt in this area is entirely marine. Structures within the hanging wall of the Xiaojinhe thrust are arcuate parallel to the thrust contact, suggesting that these structures are related to S-vergent thrusting. The thrust truncates folds within its footwall that we interpret to be early Mesozoic in age, and we suggest that the last movement on the Xiaojinhe thrust to have been of Cenozoic age. However, in the northern part of the Xiaojinhe thrust sheet, within rocks assigned to the Yidun Unit, are numerous examples of superposed folding that created several large dome and basin structures involving high-grade metamorphic rocks that are of particular regional importance. These structures are probably mostly early Mesozoic in age, but they may carry a Cenozoic overprint (see below).

The Triassic rocks that distinguish the Transitional Unit continue to the west, where they overlie a thick sequence of upper (Pzu) and lower Paleozoic (Pzl) rocks that form a 25-km-wide belt along the western margin of the Transitional Unit. These rocks continue north along the west side of the Yidun Unit for nearly 200 km, and they will be discussed with that unit below. They also form a series of discontinuous exposures north and east of Erhai Lake, where they terminate at the southeastern boundary of the Transitional Unit (Plate 1). This entire belt of rocks has a typical Yangtze stratigraphy and are often referred to as the *Zhongza belt* or *subunit* (Fig. 2-1). All these rocks end to the west against a major N-S–striking fault zone that forms the eastern boundary of the Lanping Simao, Qiangtang 1, and Jinsha accretionary units (Fig. 1-1B and Plate 1). The Paleozoic rocks have one important difference from the Yangtze section. Below dated Upper and Middle Cambrian strata is an extensive thick succession of metasedimentary sandstone, pelite, and locally fine conglomerate that has been metamorphosed to generally greenschist grade but locally reaches lower amphibolite grade (shown as Pzl on Plate 1). Parts of the lower Paleozoic section adjacent to them are also metamorphosed. The rocks are well foliated, with a gently dipping antiformal structure that contains abundant mesoscale folds, but are uniform in composition, so it is unclear with the level of present mapping whether there might be large-scale nappe-like structures within this several kilometer–thick succession. Foliation in the metaclastic rocks is folded around NNW–trending axes, and kinematic indicators indicate a west vergence (Studnicki-Gizbert, 2006). Studnicki-Gizbert (2006) obtained muscovite and biotite $^{40}Ar/^{39}Ar$ ages of ca. 35 Ma from these rocks and interpreted them on a regional basis to be related to Paleogene deformation of the fold-thrust belts to their east discussed above. Along the east side of these metamorphic rocks, faults dip eastward, placing younger Paleozoic rocks against the metamorphic rocks and suggesting that they have been unroofed by extensional faulting of Paleogene age, as these faults are unconformably overlapped by Oligocene strata of the large Jinchuan basin to the south (Plate 1). There are no other rocks in the Yangtze succession that resemble these metaclastic strata, so their depositional age, while pre-Cambrian, is unknown, and their significance remains an unresolved problem.

Yangtze-type strata occur in numerous places below the Triassic strata that characterize the Transitional Unit. Rocks as old as the latest Neoproterozoic dolomite are exposed along the Jinhe-Qinhe thrust, and in a generally irregular domal uplift in the middle of the Transitional Unit (Plate 1 and Plate 4, section 38). This domal uplift was formed by superposed folds in a dome-and-basin-type structure that suggests that at least some of the folds may be early Mesozoic in age. Similar but larger domal structures 50 km to the north within the southern Yidun Unit are interpreted to be metamorphic core complexes of early Mesozoic age (Plate 1; see below) and this domal uplift may be of similar age, but it lacks significant metamorphism.

One area of Paleozoic exposures within the Transitional Unit is exceptional. The Yulong Mountains north of Lijiang are underlain by a N-S–trending, doubly plunging antiform that contains metasedimentary rocks in its core (Plate 1, m; Plate 4, section 38). This area was recently studied by Studnicki-Gizbert (2006), who interpreted the metamorphism to be Paleogene and similar in age to the large area of Precambrian metaclastic rocks to the west that lie below the Jinchuan basin strata. He suggested that the metamorphic rocks here are correlative with the lower Paleozoic clastic sequence in the lower part of the Yangtze succession, and he interpreted the antiform to be a second-generation structure that refolded older west-vergent recumbent folds. The uplift of the Yulong Mountains, whose peaks are 5500 m high and reach ~2000 m above the ambient elevation of the plateau in this region, is interpreted to be late Neogene in age and still active. The Yangtze River is incised across the axis of the antiformal uplift and is antecedent to the uplift, having cut the >3000-m-deep Tiger Leaping Gorge (Lacassin et al., 1996). Unlike Lacassin et al. (1996), Studnicki-Gizbert (2006) argued that the uplift of the Yulong Mountains is the result of extensional tectonism. The mountains are surrounded by mapped normal faults that dip away from the mountain core (Plate 4, section 38) and are part of the extensional component of the active Dali strike-slip–transtensional fault system (see below). The uplift of the Yulong Mountains above the regional height of the Tibet Plateau in this region he interpreted to be the result of upward flow of mobile lower crust into a dilatant area of extension at the corner of rotated blocks within the Dali fault system.

Structures in the southwesternmost part of the Transitional Unit are complex and, in our interpretation, are of both early Mesozoic and Cenozoic age. Rocks of the Transitional Unit end near Erhai Lake in a complexly faulted wedge-shaped area, pointing south between the Lanping Samao Unit to the west and the Yangtze Unit to the east. On the east side of the wedge the southernmost continuation of the Chenghai thrust is present, but its southernmost trace is unclear. It can be traced to the south end of the Quaternary Binchuan basin, where the strike of the thrust projects into an area of Paleozoic strata with a highly complex structure that are intruded by numerous small felsic and mafic plutons and are unlike most of the rest of the Transitional Unit. Based on their complexity and rock units, these highly faulted rocks appear to be part of a major structural complex within the hanging wall of the Cenozoic Chenghai thrust. These rocks are juxtaposed against Triassic strata of the Chuxiong basin, and their boundary is mapped as several high-angle faults, although they are partly covered by Quaternary strata of the Binchuan, Xiangyun, and Midu basins. The Quaternary rocks of the Xiangyun basin may partly conceal a northwest-trending left-lateral fault, another of the series of left-lateral faults that lie north and parallel with the Ailao Shan shear zone (see above). On the southwest side of the wedge the Paleozoic rocks end against a N-striking high-angle fault that we interpret to have had both east-side-up normal, and possibly right-lateral, strike-slip displacement, bringing the complexly deformed Transitional Unit into contact with the Lanping Simao tectonic Unit to the west; however, much of the key contact relations are covered by Quaternary rocks of the Midu basin.

The area in and around Erhai Lake contains a complex structure that we interpret as a major folded-thrust complex (Fig. 3-4). Within this wedge-shaped area at the south end of the Transitional Unit are major structures that can be interpreted to be of early Mesozoic age and which may be the southern part of a Mesozoic thrust belt along the east side of the Transitional Unit. Two structures, not previously described, are of particular interest and require special discussion, as they have regional significance. The first is east of Erhai Lake, where the geological maps of this area can be interpreted to outline a major folded thrust. East of Erhai Lake, Ordovician to Permian rocks define a complexly faulted north-plunging antiform (Fig. 3-4). The Paleozoic section on the NE-dipping limb of the antiform is repeated on a NE-dipping fault, largely covered by Quaternary strata, that continues to the east, where it involves Triassic rocks. The map pattern north and east of the fault defines a large east-vergent, NE-plunging anticline-syncline pair with a recumbent limb at the base of the anticline. In the core of the anticline is a Paleozoic granite, mapped as infolded into the fold pair. The Devonian rocks are shown to rest unconformably on the granite. This structural complex can be viewed as an east-vergent thrust with the fold pair in its hanging wall, repeating a very thick section of generally NE-dipping Emei Shan basalt; both structures are refolded by the north-plunging antiform east of Erhai Lake. The thrust cuts up-section eastward in the footwall, and by interpretation of rocks farther east, also cuts up-section in the hanging wall rocks. West of the northern end of Erhai Lake in the Dian Cang Shan range, a Paleozoic granite is also shown to be unconformably overlain by Devonian strata and is shown to intrude what are mapped as Precambrian low- to medium-grade metasedimentary rocks that underlie all the high Dian Cang Shan range (pt in

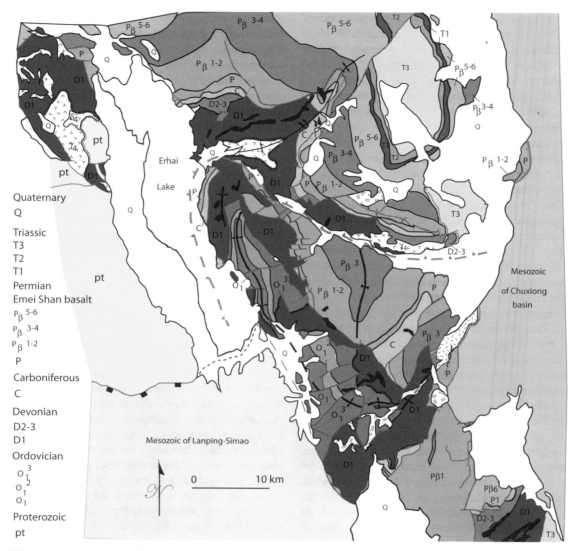

Figure 3-4. An interpretation of the structures in the Erhai Lake region, discussed in detail in the text. This area appears to contain a north-plunging antiform that folds a major east-vergent thrust fault and related structures of early Mesozoic age. The area was redeformed in late Cenozoic time by east-vergent thrust faults and related structures that deformed both the Transitional Unit and the western part of the Yangtze Unit.

Plate 1 and Fig. 3-4). One interpretation is that this granite and the Devonian rocks are the continuation of the folded structure in the hanging wall of the thrust east of the lake. If this is the case, then the metasedimentary rocks of the Dian Cang Shan, at least their eastern half, would also lie in the hanging wall of the thrust, and the thrust would continue south beneath Erhai Lake along the west flank of the antiform. This folded thrust complex, intruded by abundant plutons in its southern part, is unlike any structure in the Chuxing basin to the east, in the tectonic units in the west (see below), or in any other part of the Transitional Unit. In contrast, Leloup et al. (1993) interpreted the relations in the Dian Cang Shan differently. They indicated that there was a major fault between the Devonian and Precambrian strata but admitted that the fault is not well exposed. They did not recognize the folded thrust complex described above.

On the SW side of the Binchuan basin, maps show Upper Triassic strata that are correlative with those of the Chuxing basin to unconformably overlie rocks that belong to both the hanging wall and footwall of the folded thrust, suggesting that this fold-thrust complex is pre-late Late Triassic in age. Such a relation would suggest that the folded thrust may be interpreted to be the southern extent of an early Mesozoic deformational belt lying west of the Yangtze Unit (i.e., the Longmen Shan thrust belt, structures to its west, or continuation of its structures from the Longmen Shan area; Burchfiel et al., 1995). Hanging wall Triassic strata in the folded thrust are continuous from Lower to

Upper Triassic and are of Transitional Unit character (see above). The structural and stratigraphic relations can be interpreted to be the result of Late Triassic eastward thrusting along the eastern boundary of the Transitional Unit that ceased in latest Triassic time, and that the structures formed at that time were overlapped by uppermost Triassic rocks. This is the only place along this belt where a Late Triassic age for the thrusting could be determined south of the Longmen Shan. This complex of structures was carried eastward in the hanging wall of the Chenghai thrust in post-Paleogene time, above the western margin of the Chuxiong basin.

This refolded structure lies within the Transitional Unit, and like some structures discussed above, it suggests that it is of Mesozoic age and was formed somewhere farther west and transported to the east during post-Paleogene thrusting on the Chenghai and Jinhe-Qinhe thrusts. The interpretation of this refolded structure was made from analysis of regional maps only and needs careful field examination because of its magnitude, complexity, and how it compares to other structures in the region. If the above interpretation is correct, the large-scale structure indicates a general eastward direction of movement with a tectonic overlap along the thrust of at least ~30 km north of Erhai Lake, but if the Precambrian rocks west of the lake are also in its hanging wall, the magnitude could be >~50 km, making it one of the largest thrust faults in the region. To the north this thrust fault lies below Quaternary rocks in the Binchuan basin. How it interacts with the Chenghai thrust remains unclear, but hanging wall rocks of this early Mesozoic thrust are also in the hanging wall of the Changhai thrust, so it should continue both to the north and south. At present, however, similar structures remain unidentified, perhaps because refolding by the north-plunging antiform could be of Paleogene age, similar to the north-south–trending structures described above that formed near the western boundary of the Transitional Unit, which were truncated by younger major faults (see below).

Everywhere along the eastern margin of the Jinhe-Qinhe and Chenghai thrust belt the thrust faults locally override Paleogene rocks, and the last movement on these thrust faults was clearly early to late Cenozoic in age. The presence of the continuous Mesozoic foredeep to the east, and the unconformable overlap of complex structures in the Transitional Unit by Paleogene strata within the Jinhe-Qinhe and Xiaojinhe thrust sheets, indicate that the Mesozoic deformational belt was present; however, it appears that the present discontinuity of Mesozoic structures is related to the Cenozoic disruption and reactivation of older structures in the deformational belt.

The structures in the hanging wall of the Jinhe-Qinhe thrust show highly complex patterns, some suggesting folding and refolding, that indicate two periods of deformation (see above), but the ages of the structures cannot be accurately determined. Along the southern part of the hanging wall of the Jinhe-Qinhe thrust sheet, well-developed N-S–trending folds are cut off at right angles by the thrust fault. Also, fold patterns in the hanging wall, farther north and to the west in the northern part of the Chenghai thrust sheet, are strongly curved and in many places are cross folded, forming dome-and-basin-type patterns that locally brought up rocks as old as latest Neoproterozoic (Plate 1). Paleogene strata unconformably overlie these complex folds in many places and are only weakly folded. This suggests that the older structures may be Mesozoic in age. Without intervening strata, however, it is not clear what age this folding might be.

These fold patterns within the thrust sheets of the Transitional Unit are unlike the simpler folds in the rocks of the Yangtze Unit directly to the southeast. Thus we interpret that these folds are part of a Mesozoic deformation event that occurred to the northwest, and that structures formed at that time were first thrust to the southeast in Mesozoic time, in the Late Triassic–Early Jurassic event (or even as other, as yet unidentified, Mesozoic events), and were later thrust again to the east or southeast in Cenozoic time. The Yidun and Songpan Ganze units underwent major deformation in Mesozoic time, and the Transitional Unit, although defined mainly on its Mesozoic stratigraphy, may also be transitional structurally as well (see below) with major early Mesozoic (and perhaps other) tectonic events not recorded in the western part of the Yangtze Unit. Thus the eastern boundary of the Transitional Unit forms an important Mesozoic tectonic boundary within the western part of the Yangtze Unit, with deformation occurring within the South China fold belts in the eastern part of the unit at about the same time, but these two structural units were related to activity along different margins of the Yangtze Unit.

If the present discontinuity of structural trends in the Transitional Unit is the result of reactivation of the older structures, then where do the early Mesozoic and Cenozoic structures continue to the west of Erhai Lake and the Dian Cang Shan? The folding within the Chuxiong basin is Cenozoic in age and is probably related to shortening across the southern part of the thrust belt at the eastern margin of the Transitional Unit (i.e., the Chenghai thrust), but along its southern part the thrust has been largely overprinted (reactivated?) as the N-S–striking Chenghai normal fault, which marks the east side of the Binchuan basin. All the early Mesozoic and Cenozoic structures of the Transitional Unit are cut by a series of late Cenozoic to active NW-striking, left-lateral, strike-slip faults and associated N-S–trending pull-apart basins (see Wang et al., 1998; Studnicki-Gizbert, 2006). NW-striking faults that are present along and parallel to the Ailao Shan in the south continue northward for hundreds of kilometers (Fig. 2-10). These faults lie within several tectonic units, reflect a long complex history, and will be considered in more detail below.

The lower part of the stratigraphic section in the structure at Erhai Lake is of particular interest. The Precambrian rocks in the Dian Cang Shan that may lie in the hanging wall of the hypothesized Mesozoic folded thrust fault, would be the westernmost part of the Transitional Unit in this area and belong to the Yangtze Unit succession (Zhongza subunit, Fig. 1-1B). The Paleozoic section above the Ordovician rocks is similar to that of the Yangtze Unit; however, the Ordovician strata that unconformably underlie Devonian rocks east of Erhai Lake are unlike typical Yangtze strata: They are fine-grained clastic rocks with chert and contain graptolites, but their base is not exposed. The pre-Devonian rocks in the Dian Cang Shan also unconformably underlie Devonian strata, but by reversing thrust movement they would lie several tens of kilometers to the west of the Ordovician rocks that lie in the footwall of the thrust. These two units might be correlative with the rocks in the far southeastern part of the area of Plate 1 that are present in the Ailao Shan near the Vietnam border (see below). These pre-Devonian rocks of the Dian Cang Shan, and the Precambrian rocks that underlie the Jinchuan basin Paleogene strata, might be the protolith for some of the high-grade rocks in the Ailao Shan before they were translated south in early Cenozoic time. The southern part of the Ailao Shan is also characterized by abundant plutonic bodies that range from gabbro to granite, whose ages remain largely uncertain, but they intrude the Paleozoic rocks not unlike the anomalous plutons that intrude the southernmost part of the complex structure near Erhai Lake. These relations may indicate a correlation across at least some of the strike-slip faults and suggest ~400 km of left-lateral separation. A discussion of the regional offsets by strike-slip faults in this area will be considered regionally below.

## IGNEOUS ROCKS OF THE TRANSITIONAL UNIT

The vast extent of Emei Shan continental basalt in the southern part of the area of Plates 1 and 2 has led to the interpretation that these basalts were erupted above a Late Permian plume and are part of the Emei Shan large igneous province (LIP; Shellnutt et al., 2008; Xu and Chung, 2001). Whereas they are extensive within several tectonic units, the area around Erhai Lake contains the thickest section of Emei Shan basalt anywhere in South China. U/Pb SHRIMP (sensitive high-resolution ion microprobe) zircon dates for the Emei Shan LIP have yielded ages of ca. 260 Ma (Zhong and Zhu, 2006; Zhong et al., 2007). The province consists of mainly continental flood basalt that is 5000+ m thick in the area around Erhai Lake in the Transitional Unit, and it thins outward from this area. The volcanic rocks were erupted in a non-marine environment in most of the area, but to the west and north in the Transitional, Yidun, and Songpan Ganze Units the basalt becomes marine and is bounded above and below by marine Permian strata. The dominant rock type is continental flood basalt, but alkali basalt and basalt that plots on both sides of the alkali-tholeiite boundary are also present. Felsic and mafic intrusions in the Panxi area to the east (along the Kungdian high) are related plutonic equivalents and yield similar ages (see below). Xu and Chung (2001) and Xu et al. (2004) contoured the thickness of the Emei Shan basalt to show that the center for a plume eruption was mostly within the Yangtze Unit in the area of the Chuxiong basin. However, they did not take into account possible eastward thrusting in both the early Mesozoic and late Cenozoic along the boundary between the Yangtze and Transitional Units, and their data are limited by their choosing the central area of postulated thickest basalt and the domal shape of the crust during and following eruption in the area covered by Chuxiong basin strata, where most of the basalt has been removed by erosion (Fig. 3-5). Later, Xu and He (2007) presented geophysical evidence that there is a generally circular area of thicker crust up to 64 km thick in the area of the Chuxiong basin that thins to the north and east. P-wave velocities below this central area contain layers of higher than normal velocities in both the upper (6.0–6.6 km/s) and lower (7.1–7.8 km/s) crusts and that the thickness of the higher velocity lower crust is greatest, 20 km, in the postulated central area (Figs. 3-5A and 3-5B). Within the upper mantle below this area, but extending farther east and west, is a lens-shaped region of higher than normal P-wave velocity (8.3–8.6 km/s) that lies at a depth of 80–150 km. Xu and He (2007) presented evidence based on petrogenetic modeling that the higher velocity rocks in the lower crust are probably underplated pyroxenite and olivine pyroxenite cumulates crystallized from picritic melts from the plume. They also drew on evidence presented by He et al. (2003) that tried to show thinning of the Permian Maokou Formation deposited just prior to the eruption of the Emei Shan basalt in the area above the interpreted location of the plume head (Fig. 3-5), a relationship interpreted as the beginning of uplift above a rising plume head. But again, evidence for thinning would lie mostly below the Chuxiong basin strata, so the data are areally incomplete. Xu and He (2007) pointed out that following the eruption the central area was uplifted so that nearly all the post-Precambrian strata of the Yangtze Unit were removed by erosion and that Middle and Upper Triassic strata were the first strata deposited in the area uplifted by the plume head. As they pointed out, this would have been 40+ Ma after the eruption, and deposition of the Upper Triassic strata in this area could be interpreted to have been the result of cooling of the high-velocity material in the crust that became denser, causing subsidence in the area of former uplift. Shellnutt et al. (2008) pointed out that minor magmatic pluses at ca. 252 and 242 ± 2 Ma consist of felsic plutons and mafic dikes with compositions similar to the Emei Shan basalt, which occur within the basement rocks exposed in the eastern part of the Kungdian high, are probably not related to the main eruption of the basalt but are the result of extension that followed eruption, and that the bulk of the Emei Shan basalt was erupted in ca. 1 Ma (Shellnutt et al., 2008).

There is considerable evidence to support an upwelling in the mantle or a possible plume in the area below the western part of the Yangtze Unit from which much of the Emei Shan basalt was erupted, but the greatest volume of material was erupted within the eastern part of the Transitional Unit, where there are no geophysical data, and at the time of its eruption it may have been tens of kilometers farther west. The Emei Shan eruption may also be related to the regional extensional opening of the Songpan Ganze basin (see Burchfiel et al., 1995; Song et al., 2004), a relation that will be discussed below. Whereas there is a general relationship between the geophysical data presented by Xu and He (2007), and a possible location for a central area of eruption, the geometry of the possible plume remains less clear. The area of the postulated plume center is largely obscured by the Mesozoic strata of the Chuxiong basin, and the geology within

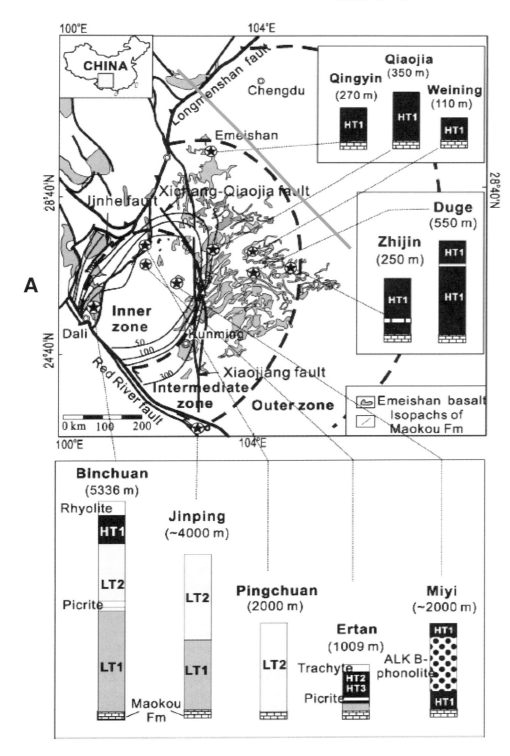

Figure 3-5 *(Continued on following page)*. Relations around the area of a postulated plume as a source for the Emei Shan basalt LIP (from Xu et al., 2004). LIP—large igneous province. (A) General distribution of the Emei Shan basalt and its relations to other features, such as the Longmen Shan thrust belt and Red River fault, are shown. The eruption of the basalt is shown within three zones of generally decreasing thickness: Inner, Intermediate, and Outer zones. The inner zone is in the area of the Chuxiong basin. Dashed lines within the Inner zone depict isopachs that show the thickness of the Maokou Formation, whose deposition just preceded the basalt eruption. The NW-trending green line is the sharp truncation of the Emei Shan basalt to the northeast.

the thickest section of Emei Shan basalt in the Transitional Unit has not been generally considered, so it is difficult to establish the relations between the possible plume and the area of erosion of rocks below the Triassic strata. The general area where pre–Middle or pre–Upper Triassic rocks were removed corresponds to the older, more irregular and persistent uplift along the Kungdian high that trends N-S (Fig. 2-1) and clearly was an area of several uplifts in Paleozoic and early Mesozoic time (Plate 3 [see footnote 1]). The thickest known section of Emei Shan basalt lies in the eastern part of the Transitional Unit, where it

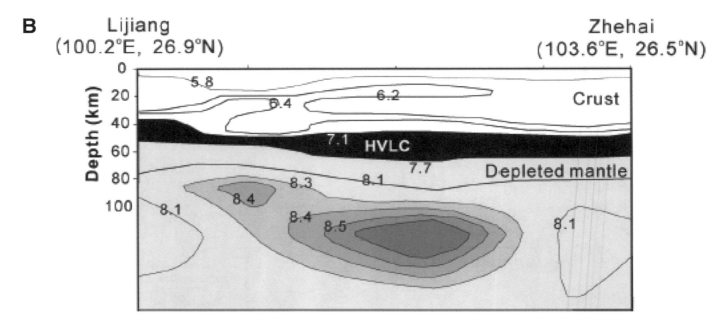

Figure 3-5. *(Continued)*. (B) An approximate E-W tomographic section of P-wave velocities of the lithosphere across the central area of the hypothesized plume. Note the high P-wave velocities in the upper crust, lower crust (HVLC), and upper mantle. Note also that in this figure the thickest section of basalt would be under the Chuxiong basin, where the outcrop is not exposed, and on a regional basis, which was removed before Mesozoic strata were deposited.

is overlain by a section of Triassic rocks that began in the Early Triassic (Fig. 3-2) and lies in the hanging wall of both Mesozoic and Cenozoic east-vergent thrust faults. The effect of these thrust faults on the paleogeography and its possible relation to a plume require, first, confirmation for the existence of the thrust faults, and second, the establishment of their magnitudes, to understand the regional geometry of the basalt. It should be noted that in the figures by Xu et al. (2003) and Xu and He (2007) that show the distribution of pre– and post–Emei Shan rocks, and the circular nature of the crustal thickness in the central area of the postulated plume, also show an abrupt boundary to the west along the Mesozoic and Cenozoic thrust faults (Fig. 3-5). The broad area of subsidence of the Chuxiong basin might be related to the cooling of the denser underplated and intruded mafic and ultramafic rocks in the crust below the basin. However, the Mesozoic strata are part of the foredeep for the Mesozoic thrust belt and have a NNE trend that extends from the postulated plume center northward to the Sichuan basin, and is best related to loading by the thrust faults. The foredeep also crosses obliquely the Kungdian high, and the subsidence and thicknesses of the Mesozoic strata may have more than one origin.

The Emei Shan basalt is not the only magmatic rock of interest within the Transitional Unit. Within the western part of the Transitional Unit are numerous Cenozoic hypabyssal bodies of different kinds of alkalic igneous rocks (see igneous rock labels with subscript 6 on Plate 1; their rock types can be found in the Glossary at the end of this Memoir). These igneous rocks are particularly abundant within the Jianchuan and western Chuxiong basin and are part of a long and wide belt of such shallow crustal plutons that extend from central Tibet to the Vietnam border and intrude the Yidun, Transitional, Yangtze, Lanping Simao, Qiangtang 1, Changning-Menglian, and Lin Cang Units (Fig. 16-21; see also Royden et al., 2008). These igneous rocks intrude rocks as young as Eocene–Oligocene and have yielded ages of ca. 40 to ca. 20 Ma (Chung et al., 2005). Although their origin remains unclear, the alkali-rich magmatic rocks erupted in the eastern plateau from earliest Cenozoic to Pliocene time and had melting temperatures >~1300 °C at depths of 80–100 km (Holbig and Grove, 2008), indicating anomalously high temperatures within the lithosphere of the southeastern plateau and extending into its foreland to the east. The distribution of these rocks plays an important role in the Cenozoic evolution of the region and will be discussed in more detail below.

## STRIKE-SLIP FAULTS WITHIN THE TRANSITIONAL UNIT

The youngest structures to affect the Transitional Unit are late Cenozoic to active strike-slip and related faults of the Dali fault system, a convex-east, arcuate belt of left-lateral strike-slip faults and associated pull-apart basins, now filled with Quaternary and Holocene sediments (Figs. 2-10 and 3-6). These faults were described by Wang et al. (1998) and were recently studied by Studnicki-Gizbert (2006). Some of the pull-apart basins are asymmetric, and the sedimentary fills are tilted to the side with the most active faults. Some of the bounding faults are oblique

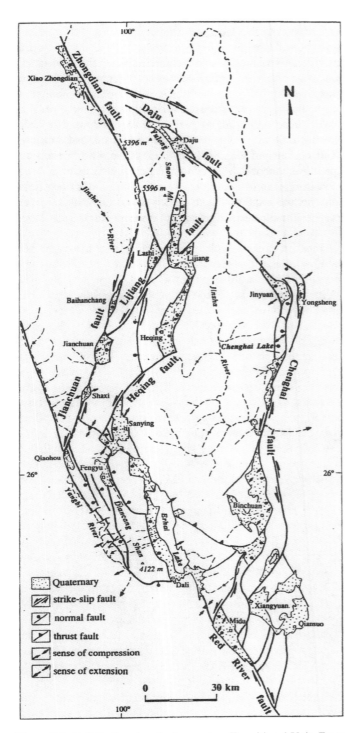

Figure 3-6. Dali fault system in the western Transitional Unit. From Wang et al. (1998).

slip–normal faults and have a very low angle of dip, such as the fault along the east side of the Heqing basin south of Lijiang. At least one of the faults in the Dali fault system has reactivated the early Cenozoic left-lateral Zhongdian fault as a right-lateral fault in the northwesternmost part of the Transitional Unit where it displaces the Jinsha unit (see below; Fig. 2-10), but most of the faults are latest Cenozoic and active and show no evidence for reactivation of older faults. The faults offset rivers, such as the Yangtze, and along their normal- and oblique-slip segments form clear scarps. The left-lateral faults form an integrated system of faults similar to, but west of, the Xianshuihe-Xiaojiang system (see above), and they indicate a major change in the tectonics of the area from regional shortening in the early and middle Cenozoic to strike-slip faulting in late Miocene to Quaternary time within the Transitional Unit. These two strike-slip fault systems accommodate clockwise rotation on crust around the eastern Himalayan syntaxis that began to affect this part of the region as the Indian lithosphere migrated northward during the Cenozoic; earlier effects of similar rotation affected the tectonic units to the west and south of the Transitional Unit (see below).

One area of Quaternary-Holocene sediments of particular interest lies south of the Heqing fault. The stepover of the NNE-SSW–striking fault south of the Heqing basin has its main normal-fault displacement on the west side of the Quaternary-Holocene sediments (Sanying basin), which consists of only a few tens of meters of strata unconformably overlying Triassic and Permian rocks. These young sediments are fan deposits whose source was from Mesozoic rocks west of the fault that are partially ponded, and they unconformably overlie Paleozoic rocks to their east. The strata dip eastward ~5° and are locally incised by streams from the highland to the west. These relations suggest that this is the very beginning of a new basin whose evolution has not proceeded far enough for the basal beds of the basin to have been back-rotated to dip westward into the basin-bounding fault. Thus the process of basin formation in this fault system continues to evolve today.

A N-S–striking western splay of the Dali fault system, the Jinchuan fault, forms the eastern margin of the Jinchuan basin. The fault displays numerous left-lateral stream offsets, indicating that the fault is active. It continues from the Yangtze River south to the NW end of the Dian Cang Shan, where the lower Triassic volcanic-rich strata of the Qiangtang 1 Unit end against it (Fig. 1-1 and Plate 1; see below). The fault ends against a major NW-trending active fault that bounds the northeastern side of the Lanping Simao Unit at this latitude. In fact, all of the faults of the Dali fault system, like those of the Xianshuihe-Xiaojiang system, end before passing into the adjacent Lanping Siamo Unit (Wang et al., 1998; Burchfiel et al., 2008b). How and why these faults end, and what role the Jinchuan fault plays in the termination of the Qiangtang 1 Unit, remain to be resolved (see below).

Many normal faults within the Dali and Xianshuihe-Xiaojiang fault systems formed within releasing bends, but some normal faults throughout the area are not obviously related to strike-slip faulting. Most of them may be related to regional causes. A few show particular relations to thrust faults, such as those along the Chenghai fault, where Paleogene thrust faults were reactivated as young and active normal faults. Other such faults occur along the northernmost part of the Jinhe-Qinhe thrust fault. In its hanging wall are N-S–trending faults that parallel the thrust but consistently show down to the west normal displacement (Plate 1 and Plate 4, section 6). The present-day regional movement of crust in this area is well shown by global positioning system (GPS) results (Zhang et al., 2004; Fig. 16-11). Detailed analysis of the GPS results shows not only the relationship of the strike-slip faults, but also shows divergent motion having taken place on the Dali fault system once the eastward-moving crust passed east of the Eastern Himalayan syntaxis, indicating a general E-W component of extension, a relationship shown geologically by Wang and Burchfiel (2000). The normal faults may have formed in response to the regional extensional strain, but they were sometimes localized by preexisting structures.

## MAJOR UNRESOLVED PROBLEMS WITHIN THE TRANSITIONAL UNIT

From detailed discussion of the stratigraphy and structure within the Transitional Unit, it is clear that there remain numerous first-order problems in resolving not only the geology in this unit but also relations to regionally important tectonic problems. The hypothesized large-scale Mesozoic folded thrust structure near Erhai Lake, discussed above, remains to be confirmed. It would demonstrate the importance of Mesozoic deformation within the Transitional Unit and its relation to the eastern boundary of the unit. The transitional nature of the paleogeography of the unit between the Yangtze Unit and units to the north also needs greater confirmation, because some authors have hypothesized major north-striking faults or suture zones passing through the middle part of the Transitional Unit, connecting the suture between the Yidun and Songpan Ganze Units to the north and the western boundary faults that cut off the Transitional and Yangtze Units in the area of Erhai Lake. However, the geology presented above strongly indicates that rock units form a gradational sequence of strata from the Yangtze Unit into the units to the north and west, but are disrupted by numerous Mesozoic and Cenozoic structures; hence the reason for developing the concept of the Transitional Unit. However, the geology remains poorly known such that the presence of a major structural separation within the Transitional Unit needs to be tested more fully.

Final modification of the Jinhe-Qinhe thrust may be very young. It appears that movement on this thrust offsets the rather uniform southeastward slope of the Tibetan Plateau topographic surface (Wang et al., 1998). This implies late Cenozoic movement. Such thrust displacement may be related to motion on the late Cenozoic to active Xianshuihe-Xiaojiang fault system. Several places along this fault system indicate where left-lateral displacement has been transferred through both shortening and extension to other faults along this system (Wang et al., 1998). Thus the young activity on the Jinhe-Qinhe thrust may be a result of transfer of strike-slip motion to shortening on an older thrust fault. How much this young activity contributes to the present arcuate shape of the thrust fault remains unknown.

## CHAPTER 4

# *Songpan Ganze Unit*

The Songpan Ganze Unit consists mainly of Middle and Upper Triassic flysch that underlies a large triangular shaped area with its apex narrowing to the west in the eastern part of the Tibetan Plateau. This unit was suggested by Weislogel et al. (2010) to be similar to the modern Bengal fan. It lies west of the Longmen Shan, northeast of the Yidun Unit, south of the western extension of the Qinling belt, and it forms the western part of the area of Plate 2 [available on DVD accompanying this volume and in the GSA Data Repository[1]] and the northeastern part of the area of Plate 1 (Fig. 1-1B). Its eastern boundary is along the early Mesozoic thrust belt within the Longmen Shan that continues south into the Xiaojinhe thrust of Plate 1 and has been modified by Cenozoic deformation (Fig. 4-1 and Plate 2). Its southwestern boundary is along a Mesozoic west-dipping subduction zone that separated the Songpan Ganze and the Yidun Units. During part of its deposition the Upper Triassic flysch of the Songpan Ganze and the Yidun Units were part of the same depositional terrane but were separated during Late Triassic subduction of the Songpan Ganze Unit southwestward beneath the Yidun Unit in the northwestern part of the area of Plate 1 (Fig. 1-1B). Westward subduction produced magmatism within the Yidun Unit that is lacking in the Songpan Ganze Unit. Because of the extensive exposure of Triassic flysch, the basement for the Songpan Ganze Unit is only exposed along the eastern part of the unit within and west of the Longmen Shan, and in rare outcrops within the western Songpan Ganze along the Cenozoic Xianshuihe fault.

## STRATIGRAPHY OF THE SONGPAN GANZE UNIT

The Songpan Ganze Unit is characterized by a very thick (>10 km?) section of Middle and Upper Triassic flysch (Fig. 4-2; Mzf on Plates 1 and 2). Exposures of pre-Triassic rocks occur in a northeast-trending belt within and adjacent to the Longmen Shan, where they lie at the base of a major southwest-dipping, southeastward-vergent, early Mesozoic allochthon that carried rocks of the Songpan Ganze Unit above the Yangtze Unit (Fig. 4-1). The pre-Triassic rocks range from thin slivers of basement metamorphic rocks along Mesozoic thrust faults to a section of Neoproterozoic and Paleozoic strata that conformably underlie the Triassic flysch and are similar in stratigraphy to the Yangtze Unit. They are interpreted to be a western, but displaced, part of the Yangtze Unit's western margin. In the northeast is an area that contains a large body of metamorphic rocks, which, with its Neoproterozoic and Paleozoic section, is considered to be a large westward extruded fragment from the Qinling orogen to the east (Plate 2; Burchfiel et al., 1995). In its western part this fragment has a sharp boundary with Triassic flysch to its north, west, and south. All the rocks west of the Longmen Shan are allochthonous and are part of the early Mesozoic Longmen Shan thrust belt. The pre-Triassic rocks in the Longmen Shan occur in folds that plunge west below the Triassic flysch and are rarely exposed farther west.

West of the Longmen Shan and in the southeastern part of the Songpan Ganze Unit, Permian basalt is present within the Paleozoic succession. In the southeastern part of the unit around the domes are two units of basalt that are marine and dated by fossils as Early and Late Permian in age. On Plate 2 the two basalt units are not separate and are shown as Pzuβ because they are not separated on Chinese maps. The Lower Permian basalt is too old to be part of the Emei Shan basalt, but the Upper Permian basalt may be its equivalent, although this remains unconfirmed. In the Paleozoic succession west of the Longmen Shan where it forms the base of the Songpan Ganze succession there is Upper Permian marine basalt dated by fossils. From the map, it appears to cover a broad area but its thickness ranges from 50 to 500+ meters and its broad outcrops are the result of subvertical near isoclinal folding. It is shown on Plate 2 as Pβ* because although it is approximately the correct age to represent the northern extent of the Emei Shan basalt, the correlation has not been confirmed by either age dating or chemistry. These basalt units are important because they suggest that extension took place in the Permian that may have begun the subsidence of the Songpan Ganze basin. The Permian basalt in places is associated with limestone conglomerate that may suggest the beginning of local relief, and the thin Lower Triassic rocks above are mainly fine-grained sedimentary rocks but locally also contain conglomerate. These units may represent the transition to basin formation where the greater subsidence occurred with the deposition of the thick Middle and Upper Triassic flysch so characteristic of the Songpan Ganze succession.

West of these Paleozoic exposures the oldest rocks within the unit are exposed in a narrow zone of upper Paleozoic rocks to which a Permian age is assigned on the basis of fossils in the limestone (Plate 1). The largest outcrop of these rocks occurs in an anticlinal core at the base of thrust faults along the Xianshuihe fault zone (Plate 1). The rocks consist of white limestone, both bedded and massive. In places, limestone, chert, and mafic volcanic and plutonic rocks are interlayered and complexly sheared together. Some of the layering may be primary, but some is clearly

---
[1]GSA Data Repository Item 2012297, Plates 1–4, is available at www.geosociety.org/pubs/ft2012.htm, or on request from editing@geosociety.org or Documents Secretary, GSA, P.O. Box 9140, Boulder, CO 80301-9140, USA.

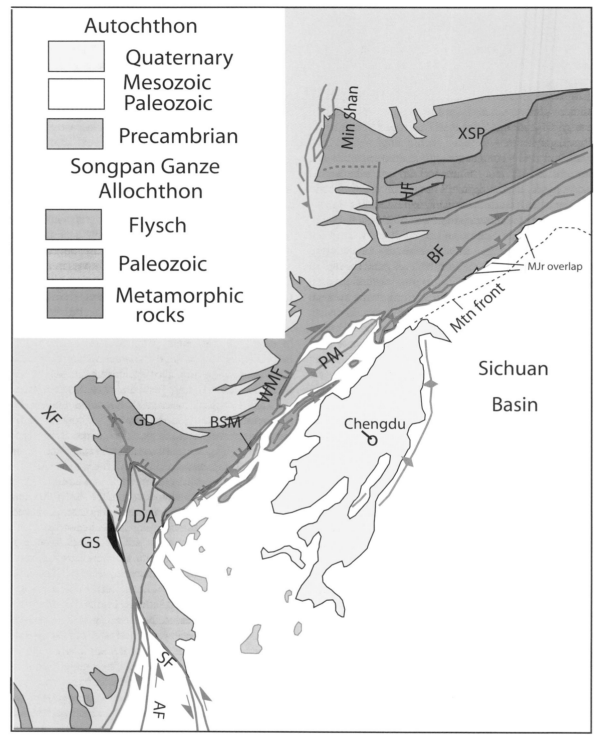

Figure 4-1. Generalized map of the rocks belonging to the Songpan Ganze allochthonous Unit within the Longmen Shan. The eastern boundary is formed by an early Mesozoic thrust belt (red line) that placed rocks of the Songpan Ganze eastward above rocks of the Yangtze Unit. White—Mesozoic and Paleozoic; green—outcrops of Precambrian rocks. Allochthonous rocks consist of Triassic flysch in pale red, and Paleozoic rocks, with some Precambrian rocks, that underlie them in tan. The metamorphic unit from western Qinling is shown in red. Quaternary strata of the Chengdu Plain are shown in yellow. Important Cenozoic structures modify the eastern thrust contacts and also cut across the Songpan Ganze Unit (red lines). Ticked lines are thrust faults, double-dip symbols are normal faults, and arrows indicate strike-slip faults. Important Cenozoic folds that modify the Mesozoic allochthon are shown in blue. AF—Anninghe fault; BF—Beichuan fault; BSM—Baoxing massif; DA—Danba antiform; GD—Gezhong dome; GS—Gonga Shan; HF—Huya fault; MJr—Middle Jurassic overlap unconformity; PM—Pengguang massif; XSP—Xue Shan displaced fragment; SF—Shimian fault; WMF—Wenchuan-Maowen fault; XF—Xianshuihe fault.

# SONGPAN GANZE UNIT

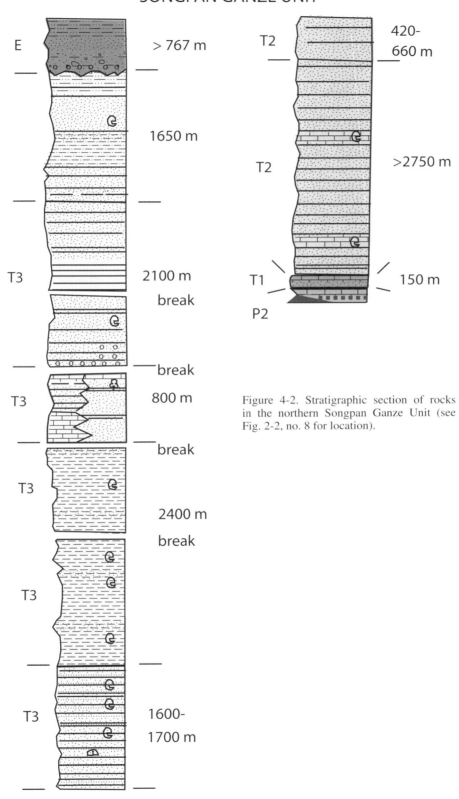

Figure 4-2. Stratigraphic section of rocks in the northern Songpan Ganze Unit (see Fig. 2-2, no. 8 for location).

tectonic; the rocks are foliated, and the limestone has flowed ductilely. In many places the rock units are a chaotically sheared mixture and have the characteristics of mélange, such as along the Xianshuihe fault zone at Dawu (Xu and Kamp, 2000). Marine basalt is present in numerous anticlines along the Xianshuihe fault (Plate 1) and is considered to be probably Permian at the base of the overlying flysch succession. Because the Paleozoic Yangtze type rocks are not exposed within most of the Songpan Ganze Unit, and the Permian basalt is exposed along the Xianshuihe fault, there is considerable speculation about what underlies the thick flysch of the unit—oceanic crust or a thinned continental basement (see below).

Where these rocks are more coherent and in stratigraphic succession with the remainder of the Songpan Ganze flysch sequence, a thin unit (~150 m) of fossiliferous marine Lower Triassic limestone, shale, and fine-grained sandstone crops out at the base (Fig. 4-2). The flysch succession begins in the Middle Triassic and consists of units composed of sandy and shaly flysch sequences with some chert and rare volcanic zones. Both macrofossils and microfossils, mainly conodonts, are present throughout and confirm that the sequence continues into the late Late Triassic. Newly discovered fossils suggest that the sequence may locally extend into the Early Jurassic (e.g., Wang et al., 2002). The sequence is mainly marine, but it grades into a shallow-water sequence near the top, where in places it is brackish water to nonmarine with coal beds. The Jurassic strata discovered by Wang et al. (2002), however, are marine. Volcanic rocks are rare and consist of thin units of mafic, intermediate, and felsic rocks. Most of the volcanic zones are present in the western part of the flysch units near the boundary with magmatic rocks that are abundant in the Yidun Unit. On Chinese maps the flysch has been subdivided into many different mapped units, but this region lies within the high part of the Tibetan Plateau, where much of the area is gently rolling grass-covered terrain with poor outcrops, and sparse incision by rivers to expose significant outcrops; thus the reliability of the mapped units needs to be tested.

The detritus in the main body of the Songpan Ganze has been derived from the Triassic Qinling collision zone to the northeast and its western continuation along the northern margin of the Songpan Ganze marked by a north-dipping subduction zone. A study by Enkelmann et al. (2007) of $^{40}$Ar /$^{39}$Ar ages of detrital white mica, Si in white mica content, and U/Pb ages of detrital zircon concluded that the Middle and Upper Triassic strata in the Songpan Ganze basin (shown on Plate 1 and the southern part of Plate 2), even in its southeastern part, was sourced mainly from the Qinling-Togbai-Hong'an-Dabie collision zone of the South China and North China cratons during Triassic time. A more recent study by Weislogel et al. (2010) supported this general conclusion but showed that the Songpan Ganze terrane consists of three major depocenters of which only the northeastern and central depocenters received detritus mainly from the Qinling-Dabie region, but the southeastern depocenter, southwest of the Xianshuihe fault, also received detritus from the Longmen Shan and Yangtze regions to the east. The sediment transport within the northern and central depocenters indicates that detritus traveled far from its source by southeast flow along the area west of the Longmen Shan. By contrast, sediment in the southern depocenter was less far-traveled. Both Enkelmann et al. (2007) and Weislogel et al. (2010) reported only minor evidence for sources from the Triassic magmatic province in the Yidun Unit to the west. Enkelmann et al. (2007) pointed out that sources from the Yidun magmatic arc adjacent to the Songpan Ganze must have been barred, a conclusion with which we agree, as the geology suggests that a forearc high was between the Yidun and Songpan Ganze Units during part of Late Triassic time (see below). The high is marked by unconformities in Upper Triassic strata along the eastern margin of the Yidun Unit (see Plate 3 [see footnote 1]) that lie along the belt of imbricate thrust faults and mélange west of a subduction zone between the Yidun and Songpan Ganze Units (see below). However, the Songpan Ganze flysch is similar in age and thickness to flysch that constituted most of the rocks within the Yidun Unit of the Songpan Ganze. The source for the Yidun flysch, however, still remains largely unknown and has important implications for the paleogeography during Middle and Late Triassic time.

Paleogene strata unconformably overlie the Triassic rocks in rare small isolated areas within the broad flysch terrane and are locally abundant along thrust faults at or near the boundary with the Yidun Unit (E on Plate 1). They consist of brown, tan, and red conglomerate, sandstone, and mudstone and are a few tens of meters to >700 m thick. Most of the Paleogene rocks are without fossils, and their age is based on lithologic correlation with the few fossiliferous sections present in the area, both in the Songpan Ganze and adjacent Yidun terranes.

Quaternary rocks are present along the river valleys, but are most important in larger areas within pull-apart basins along some of the major strike-slip faults. The Garze basin along the Garze fault, and the Litang basin in the central part of the area, are two of the most prominent regions of Quaternary deposition within strike-slip fault zones.

## STRUCTURE OF THE SONGPAN GANZE UNIT

The main structural features of the Songpan Ganze Unit are broadly linear, NW-trending folds and parallel faults. Folds are mostly upright with near vertical axial surfaces, steep limbs that are both right-side up to locally overturned, and they have a weak to well-developed axial planar cleavage (Plate 4, sections 23, 24, 26 [see footnote 1]). The map shows an abundance of fold axes, but many of them are parasitic folds on larger folds. The faults that parallel the folds are generally steeply inclined and dip both NE and SW and show thrust separations, but kinematic data are generally lacking. Toward the east, north of the Danba antiform, the folds all curve to the north, trend parallel to the Longmen Shan, and contain Paleozoic strata (Plate 2). In this area, some of the fold traces exhibit horseshoe bends, and others show irregular trends, suggesting refolding (Burchfiel et al., 1995). Within the Longmen Shan the base of this folded sequence is marked by a northwest-dipping thrust belt that places the rocks of the Songpan Ganze above the Yangtze Unit in the footwall, indicating that the Songpan Ganze forms a major southeast-vergent allochthon (Fig. 4-1; Plate 2 shows the eastern margin of the allochthon by the white boundary). The basal part of the allochthon has been folded by younger late Cenozoic folds that include the basement-cored antiforms of the Pengguan and Baoxing massifs and the Danba antiform (Fig. 4-1). East of the antiform area, some Songpan Ganze rocks are mapped as klippen. These klippen have been separated from the main allochthon by Cenozoic deformation and subjected to important Cenozoic structural modification (see Burchfiel et al., 1995). South of the Danba antiform, folds in the Songpan Ganze flysch are also arcuate, convex to the south, and have been folded similarly to the fold trends north of the antiform and parallel to the Xiaojinhe thrust, which marks the southern boundary of the unit. However, the Xiaojinhe thrust shows obvious evidence of Cenozoic reactivation.

Within most of the Songpan Ganze west of the Longmen Shan the structures trend northwest, parallel to the boundary with the Yidun Unit as do the structures within the adjacent Yidun Unit. On the Yidun side of this boundary an imbricate thrust belt with mélange involves mostly Triassic rocks (see below) and marks a subduction boundary with the Songpan Ganze Unit. The boundary forms a sharp break between rocks to the west with volcanics, and those to the east that lack volcanics. We interpret the subduction zone to have dipped to the southwest. There are structural complications, however, because Cenozoic deformation along this boundary involves Paleogene strata in narrow belts along northeast-dipping thrust faults, and it appears that the vergence of younger structures at the surface has been reversed (Plate 4, section 23; discussed below).

In the western part of the Songpan Ganze, where there is only flysch exposed, the rocks are well cleaved and metamorphosed to phyllite and low greenschist grade. Two local dome-shaped areas west of the Danba antiform underwent amphibolite-grade metamorphism (shown by dashed isograds on Plates 1 and 2). These two areas have yielded somewhat unreliable K/Ar ages of 192–212 Ma and appear to be related to intrusive bodies at depth. Along the eastern part of the Songpan Ganze Unit are domal areas of metamorphic rocks that are crudely aligned parallel to the boundary with the Yangtze and Transitional Units (Plates 1 and 2). The largest is an area of kyanite/sillimanite grade rocks that occupy a small area that may be remobilized Precambrian basement rocks, the Dong Go dome (m, Plate 2), which has yielded a U/Pb zircon core age of $824 \pm 14$ Ma and a metamorphic rim age of $177 \pm 3$ Ma (Zhou et al., 2002). This area has also yielded a K/Ar age of 109 Ma, an unusual age for rocks within the Songpan Ganze (Burchfiel et al., 1995). Another large area of amphibolite-grade rocks surrounds the circular Xuelongbao dome, cored by Precambrian rocks, directly west of the Pengguan massif (Dirks et al., 1994). Both of these domes have yielded U/Pb zircon ages of $748 \pm 7$ Ma and $824 \pm 24$ Ma, consistent with their being remobilized Precambrian Yangtze basement.

Southwest of the Danba antiform area in the southeasternmost part of the Songpan Ganze are four domal areas that expose Paleozoic strata. The three southwestern domes are cored by metamorphic rocks of unknown affinities (m in Plate 1), but they are surrounded by Lower Paleozoic strata, and thus their metamorphic cores are considered Yangtze Precambrian basement rocks. These domes are characterized by areas of cross folding, hence their domal character. Only the Jianglang dome, the middle of these three domes, has been studied in any detail. The Jianglang dome was studied by Yan et al. (2003b). It has three structural layers: a basal schist of probable Precambrian rocks, a ductilely deformed middle slab of Paleozoic rocks, and an upper unit composed of Triassic flysch of the Songpan Ganze Unit. These three structural units are separated by low-angle faults that Yan et al. refer to as the *lower* and *upper detachment faults*. Rocks of the middle slab were metamorphosed to upper greenschist to amphibolite grade and consist of strata correlated with Ordovician through Permian age and contain basaltic rocks correlated with the Emei Shan basalt (but see discussion above on the Permian basalt). Foliations in the middle and lower units are parallel to subparallel to the detachment surfaces and to stratigraphic units and are axial planar to attenuated folds. The foliations and detachments were folded by late folding or cross folding to form the domal configuration. All the stratigraphic units are thinned, and because the strata can be correlated with the Yangtze succession, they reveal that stratigraphic units are missing at the detachment faults. Five deformations have been recognized. The oldest two affected the basement rocks. The third event affected the middle tectonic unit, forming mylonites and phyllonites within the lower detachment zone and basement rocks, and by brittle deformation within the upper detachment. The first three deformational fabrics were reworked by a fourth foliation that Yan et al. (2003b) related to a compressional event that affected the Triassic flysch. The final event is a brittle/ductile deformation. Their analysis indicates a consistent E to SE kinematic axis that shows shear both to the west and the southeast in different events. All the structures were folded during dome formation. They suggested that the earliest event might have been extensional and related to development of the Emei Shan plume, and that the second event might be related to shortening within the Indosinian Longmen Shan thrust belt. The third event is interpreted to have been a major extensional event that formed the detachment faults and thinned the stratigraphic units. They date this event at $177 \pm 3$ Ma, using U/Pb zircon data of Zhou et al. (2002). The samples dated by Zhou et al., however, came from the metamorphic Gongcai complex in the core of the Dong Go dome, which lies on the Danba antiform 240 km to the northeast. The fourth event is the folding of the Triassic flysch, followed by the fifth event, which they regard as extensional and related to the Cenozoic India-Eurasia collision and its associated deformation. We suggest below that the early events may be related to the opening of the Songpan Ganze basin and to the extrusion of the Emei Shan basalt, as did Zhou et al. (2002). The later two events and detachment formation may be related to the folding of the

Triassic flysch that was detached from its underlying basement rocks, as suggested by Harrowfield and Wilson (2005). The final event may be related to either or both shortening and/or extension, depending upon the interpretation of the Cenozoic development of the Danba antiform.

Unfortunately, little is known of the other domes in the southwestern part of the Songpan Ganze, but they are probably related to several similar large domes to the west, within the Yidun Unit, that have a highly complex internal structure and exhibit the characteristics of metamorphic core complexes (see below).

In marked contrast to the domes that show cores of high-grade metamorphic rocks, the Ge Zhong dome that lies along the Danba antiform, and southeast of the area of amphibolite-grade metamorphism, is cored by Precambrian basement rocks that have yielded a U/Pb zircon age of 864 ± 14 Ma (Zhou et al., 2002) with undeformed near-vertical mafic dikes, indicating that these rocks have not been significantly deformed or metamorphosed since Precambrian time. They are surrounded by greenschist-grade, gently dipping schist of Paleozoic protoliths that form a domal structure around the Precambrian rocks. The contact is marked by a mylonitic carapace that shows down-to-the-northwest shear sense indicators. Unpublished $^{40}Ar/^{39}Ar$ ages for the mylonitic rocks are mid-Cenozoic, indicating young northwest translation of the Songpan Ganze Mesozoic allochthon (Burchfiel et al., 1995). This relationship is similar to the contact between the Songpan Ganze rocks and the Yangtze Precambrian basement rocks to the southeast at the northwest-plunging nose of the Danba antiform. Here the contact is a shear zone that disrupts schists of the Mesozoic allochthon above unmetamorphosed Precambrian basement along a gently northwest-dipping contact that shows northwest-vergent motion of the hanging wall (Burchfiel et al., 1995). These two areas are in contrast to the Gongcai gneissic complex in the core of the Dong Go dome only ~25 km northwest of the Ge Zhong dome, where the basement rocks in the dome were reactivated at kyanite-sillimanite grade during Mesozoic time. These relations suggest important Cenozoic northwest-vergent normal displacement of the Mesozoic allochthon the significance of which remains unclear.

In the northeast part of the region, as shown on Plate 2, there is a large area of metamorphic rocks (Mx on Plate 2), which are overlain by Neoproterozoic and Paleozoic rocks, called the *Xue Shan platform* by Burchfiel et al. (1995; see Fig. 4-1), who considered it a westward displaced fragment from the Qinling orogen to the east. If it is a displaced crustal fragment, then its metamorphic history is tied to the Qinling orogen and is somewhat unrelated to the Songpan Ganze Unit.

The age of the deformation within the Songpan Ganze is well documented within the northern part of the Longmen Shan, where the Pengguan basement-cored massif plunges northeast below the thrust faults that carry the Neoproterozoic, Paleozoic, and Triassic rocks of the Songpan Ganze unit eastward over the Yangtze Unit in its footwall. The eastern parts of the hanging wall rocks are unconformably overlain by Middle Jurassic strata, dating the thrusting as pre-Middle Jurassic (Fig. 4-1). Footwall rocks contain a well-developed thick foredeep of Late Triassic to Early Jurassic age, supporting the thrusting as of that age (Indosinian; Burchfiel et al., 1995). Within the Songpan Ganze terrane to the west the folds and faults within the flysch are intruded by numerous small and large plutons that cut across the structures. These plutons are mostly intermediate in composition, but diabase is also common. The plutons are mostly undeformed and have yielded both Triassic and Jurassic ages that range from 224 ± 4 Ma to 188 ± 2 Ma with one age of 153 ± 3 Ma (Roger et al., 2004, 2010), consistent with a Triassic to Early Jurassic time of deformation.

Rare and poorly dated Paleogene strata unconformably overlie the structures in the Songpan Ganze and are also folded but less tightly than the structures in the underlying Triassic rocks. The extent of Cenozoic folding remains unclear because of the rarity of Paleogene strata. However, compilation of low temperature fission-track and U/Th/He dating of the ages of rocks current at the surface showed that exhumation of deeply buried rocks across much of the Songpan Ganze Unit is uncommon (Clark et al., 2006; Kirby et al., 2002), suggesting that Cenozoic deformation has not been significant, supporting the conclusion of Roger et al. (2010) for a period of general quiescence during Late Jurassic into Cenozoic time. However, it leaves unexplained the thick section of foredeep rocks to the east that continue into the Late Cretaceous.

One of the major tectonic problems of the Songpan Ganze is how the upright near-isoclinal folds project to depth. Harrowfield and Wilson (2005), in their study of the structure of the Songpan Ganze along four profiles that are on both sides of the Danba antiform (Plate 2), and with one profile that ends to the west at Litang, showed that faults generally parallel the fold axes, are steeply dipping, and show abundant kinematic evidence for thrust faulting, but in general do not disrupt the folds. Locally, faults show left-slip kinematics; this is not surprising because some major left-lateral strike-slip faults do cut the unit. Evidence for folded cleavage is locally present but does not appear to be a regional feature. Importantly, Harrowfield and Wilson (2005) showed that the steeply dipping foliation becomes shallower to subhorizontal with depth where the Paleozoic rocks are exposed in the Danba antiform. They interpret this subhorizontal foliation to be related to a décollement at the base of the upright folded section above, decoupling it from deeper rocks below. Fold axes related to the subhorizontal foliation trend NW, and the kinematics determined within the foliation indicate a SW vergence. Steep fold-axis-normal lineations within the upright folds likewise became subhorizontal with depth, but trended NW, interpreted to be a rolling lineation, parallel to the folds within this subhorizontal zone of intense foliation development. The subhorizontal foliation has been warped by the Danba antiform along its northwest trend. The study by Harrowfield and Wilson (2005) is unique in its focus in trying to understand the deeper structure beneath the folded flysch, and this is the only area within the southern Songpan Ganze where a basal décollement has been described. How extensive it is remains speculative, because it is only along the east side the Songpan Ganze Unit that older Paleozoic and Precambrian rocks are exposed where a décollement would be exposed. In our interpretation, however, as well as that of others (e.g., Mattauer et al., 1992; Calassou, 1994; Dirks et al., 1994; Burchfiel et al., 1995; Roger et al., 2004; Harrowfield and Wilson, 2005), the Songpan Ganze Unit probably has a décollement below all the upright folds. The folding is interpreted to be the result of northeast-southwest shortening related to subduction along both its northern and southwestern margins (see below). Along its east side the change in fold trend is interpreted to be related to left-lateral shearing against the northeast-trending margin of the Yangtze Unit during transpressional deformation, forming a major south-facing arc that is well displayed on Plate 2, an interpretation that has been shared by many of the workers cited just above. How this décollement might relate to the northwest-vergent Cenozoic normal-sense deformation in the Ge Zhong dome, and at the nose of the Danba antiform, remains unknown.

**Cenozoic Strike-Slip Faults within the Songpan Ganze Unit**

The youngest structures to affect the Songpan Ganze Unit are strike-slip faults of late Cenozoic age many of which are active. The most prominent of these faults with the largest magnitudes of offsets are the NW-striking Xianshuihe, Garze, and Litang left-lateral strike-slip fault zones (Fig. 2-10). The Xianshuihe fault zone lies entirely within the Songpan Ganze flysch in its western part, but to the SE it cuts across both the Transitional and Yangtze Units (see above; Wang et al., 1998). The Garze fault forms the continuation of the Xianshuihe fault through a left step east of Garze, and most of its trace on Plate 1 lies within the Yidun Unit. The connection between these faults, and how the transfer of motion has taken place, were discussed by Wang and Burchfiel (2000). The eastern end of the Garze fault is marked by the elongated Quaternary pull-apart Garze basin through which the Yalong River has been displaced ~35 km, but this represents only part of the total displacement on this fault, which may be as large as ~100 km (see below). This basin is interesting because the Yalong River has incised deep gorges both north and south of the basin, but where the Yalong flows through the basin it is aggrading. The present headward erosion on the Yalong fault that extends south from southeast corner of the Garze basin suggests that the river had incised a deep valley before it was offset by the Garze fault. The Garze fault zone west of the Garze basin is a well-defined linear strike-slip fault zone that is commonly nearly 500 m wide and consists of numerous fault strands along which are well-developed scarps. It can easily be traced to the west, where it cuts at a low-angle obliquely across structures in the Yidun Unit. At its western end (Plate 1) it forms the boundary between the Yidun and Qiangtang 1 Units, indicating the youthfulness of this major tectonic boundary at this location (see Wang et al., 2008a, 2008b).

The Litang fault zone (Fig. 2-10) consists of several en echelon strands across which the total left-lateral displacement is ~25 km. Several broad, pull-apart basins are present in the releasing bend stepovers, all of which lie within the Yidun Unit (see below), but the eastern end of the fault zone forms the boundary between the Songpan Ganze and Yidun Units (Plate 1) where it turns to the south and has a thrust component.

All these large strike-slip faults strike WNW oblique to the trends of the early Mesozoic structures, a relationship that permits their offsets to be determined. The Xianshuihe fault is the best example. Between Luhuo and Dahuo, cities that lie in small pull-apart basins, the fault offsets a distinctive group of folds and faults. The folds are cored by Permian rocks, and the fold-fault relations are distinct, such that they can be matched on both sides of the Xianshuihe fault (Wang et al., 1998). The total offset is between 42 and 65 km, with the best constrained offsets at ~60 km. The age of initiation of the fault is poorly known, thus an average long term slip rate is poorly determined, but the evidence suggests that the faults initiated between 10 and 4 m.a. ago, yielding an average slip rate for the Xianshuihe fault of 6–14 mm/a. A short-term slip rate of $8 \pm 2$ mm/a was determined by neotectonic studies (Allen et al., 1984; Allen et al., 1991). Global positioning system (GPS) data yield slip rates of ~12 mm/a, but across a broader zone than just the trace of the fault zone (King et al., 1997; Chen et al., 2000; Zhang et al., 2004). Numerous small faults parallel the larger faults, cutting across the older structures and consistently showing left-lateral displacement (Plate 1). Thus left-lateral shear within the area probably occurs across a broader area than just along the major faults. How many of the other NW-trending faults within the Songpan Ganze Unit may prove to have late Cenozoic strike-slip displacement remains unknown.

The segment of the Xianshuihe fault that lies within the Songpan Ganze Unit forms the NE boundary for the crustal fragment that actively rotates clockwise around the eastern Himalayan syntaxis (Wang et al., 1998; King et al., 1997; Chen et al., 2000; Zhang et al., 2004; see above). Within the area of Plate 1 the fault cuts obliquely across preexisting structures, just as it does to the southeast in the Yangtze Unit, and thus is not controlled by preexisting crustal anisotropy. The Garze and Litang fault zones also cut across structures at the boundary of the Yidun and the Songpan Ganze Units and across structures within these units. They appear to be controlled mainly by the regional shear developed by the northward penetration of India into Eurasia at the syntaxis. The Litang fault zone likewise is controlled by a shear gradient, but one smaller than that for the Xianshuihe fault, as shown in the GPS data within the clockwise rotating crust similar to the Dali fault system discussed above (see Zhang et al., 2004).

**MAJOR PROBLEMS FOR THE SONGPAN GANZE UNIT**

Several major tectonic problems for the Songpan Ganze Unit that need to be resolved include the following: What formed the Middle Triassic flysch basin, what is its basement, what caused the folding, what is the magnitude of shortening within the flysch, and what is the origin of the Triassic-Jurassic plutonic rocks that intrude the flysch?

**Formation and Shape of the Songpan Ganze Flysch Basin**

Regionally the Songpan Ganze Unit forms a broad, triangular area with one apex pointing south between the Longmen Shan to the east, and the Yidun Unit to the southwest (Fig. 1-1B). Paleozoic strata that are similar to the Yangtze Unit, but perhaps more strongly faulted, underlie the eastern part of the Songpan Ganze Unit west of the Longmen Shan and also underlie the Transitional Unit and the southern and western part of the Yidun Unit to the southwest (see below; Plates 1 and 2), the south point of the wedge. The subsidence within the Songpan Ganze basin may have begun as early as Permian time, shown by the transition from continental to marine extrusion of Emei Shan basalt that occurred within the Transitional Unit or related to the Early or Late Permian basalt in the southeastern or eastern part of the Songpan Ganze Unit, respectively (see above). The flysch basin extends into the Yidun Unit, where it forms a major part of it (see below). Because of the great amount of shortening represented by folds within the flysch units, the original shape of the basin is unknown. The basin may have been formed by the rifting of the western margin of the Yangtze Unit (South China craton in its larger context) accompanied by counter-clockwise rotation around a pole to the south in the general location of the Transitional Unit, with the rifted fragment being the Yangtze rocks of the Zhongza subunit along the west side of the Yidun Unit (Figs. 1-1B and 2-1). The rifting and the thinning of the crust and lithosphere may have been associated with the massive outpouring of Emei Shan basalt across much of the southern part of the Yangtze, Transitional, and Yidun Units. It is probably no coincidence that the thickest area of Emei Shan basalt is in the area at the southern end or apex of the wedge-shaped Songpan Ganze basin where rifting may have occurred, but where little separation of crust-lithosphere took place. Mattauer et al. (1985, 1992) suggested that the belt of Yangtze rocks along the west side of the Yidun arc (the Zhongza subunit; see Yidun Unit below) was emplaced by right-lateral, strike-slip faulting in early Mesozoic time, displaced from the SW side of the Yangtze Unit. However, there is little evidence of a fault of this magnitude passing through the Transitional and Yangtze Units at the appropriate location. In fact there appears to be continuity of the geology from the Zhongza subunit through the Transitional Unit into the Yangtze Unit (see above). The cause of the rifting and its relation to the origin of the Emei Shan basalt remains unclear. Studies of the Emei Shan basalt in the Transitional Unit certainly support major magmatic upwelling in Late Permian time near the southeast apex of the Songpan Ganze basin that may be related to eruption above a plume, but how eruption and the plume were related to the opening of the Songpan Ganze basin remains unclear. The focus of most studies in this area has been on the magmatism, not its regional relations to structure.

The huge volume of sediments that filled the Songpan Ganze basin has been related to sources mainly to the northeast in the collisional zone between the north and south China cratons in Triassic time in the eastern Qinling mountain belt with possible contributions also from along the western part of this belt that lies north of the Songpan Ganze basin, and in the south to sources in the Yangtze Unit (Weislogel et al., 2006, 2010). The contribution made by the geology shown on Plate 1 is that the Yidun Unit was also dominated by a similar thick flysch succession and may have received detritus directly connected with the Songpan Ganze Unit. This greatly increases the volume of flysch within this region. Regional analysis of paleocurrent and provenance data is still lacking to better constrain the sources and filling history of the Songpan, and particularly the Yidun flysch terranes (see Pullen et al., 2008). The volume of Triassic flysch, however, probably rivals the volume of the modern Bengal fan (Weislogel et al., 2010).

**Basement of the Songpan Ganze Flysch Basin**

The nature of the basement rocks below the Songpan Ganze basin remains a major problem. The oldest rocks exposed beneath the Songpan Ganze flysch are the belt of Precambrian and Paleozoic rocks of the Yangtze type along its eastern margin within and adjacent to the Longmen Shan (Burchfiel et al., 1995; Plate 2). Regional exposures of pre-Triassic rocks in the Songpan Ganze are rare, with narrow belts of an Upper Permian limestone unit lying below a basalt unit that may be the marine equivalent of the Emei Shan or basalts extruded during extensional opening of the basin that crop out along the Xianshuihe fault. In the southern part of the Yidun Unit, below Triassic flysch, are domes that expose strata as old as Precambrian and Paleozoic, some metamorphosed, that are similar to those of the Yangtze Unit (see below; Burchfiel et al., 1995). They also expose both Early and Late Permian basalt. Thus surface geology suggests that the margins of the unit have a basement of Yangtze-type continental crust, but it does little to answer the question of what underlies most of the flysch terrane. Within mélange units along the eastern part of the adjacent Yidun Unit are blocks of Paleozoic rocks as old as Silurian (Plate 1) that were probably incorporated into an accretionary prism during early Mesozoic SW subduction of the Songpan Ganze Unit (see below), which might also suggest that at least some continental crust, albeit perhaps thin, lay below parts of the Songpan Ganze basin.

The thickness of the Songpan Ganze flysch remains uncertain because of deformation, general uniform lithology, and poor outcrops, but if its thickness of ~10+ km is correct, and because of its shallowing upward into a non-marine environment at the top, then isostatic considerations would suggest that the basin is underlain by a thin continental crust. The present thickness of crust below the Songpan Ganze is 60–65 km, but the strata of the basin are thickened by isoclinal folds with generally steep axial planes (see above); thus the thickness of flysch in the basin today reveals little about how much of the crust they may make up.

**Shortening within the Songpan Ganze Basin**

The magnitude of shortening within the Songpan Ganze Unit is difficult to determine, because there are no cross sections from which accurate quantitative measurements can be made (Plate 4, sections 23, 24, 26; see also Harrowfield and Wilson, 2005). The main folds on Chinese maps are determined from the mapping of formational units that at best are hard to distinguish because of poor outcrops and the monotonous nature of the flysch sequence. Large low-angle thrust faults are unknown or unrecognized. All that can be said is that where exposed, the folds are mainly upright, tight to near isoclinal, with vertical axial surfaces; thus the shortening and thickening are large.

One aspect of the shortening within the Songpan Ganze basin is that reconstructions have not addressed the original geometry and extent of the basin. The basin is generally folded along NW-trending fold axes, except in its eastern and southeastern parts, where fold axial surfaces are curved to parallel the basin margins along a transpressional boundary (see above). Taking out the shortening within flysch, the basin must have had a much greater N-S width than the wedge shape that is commonly assumed from its present geometry. In fact, the hypothesis that there was subduction both to the north and south along the present boundaries of the unit would even suggest that some unknown mass of the original basin, both flysch and its basement, may have been subducted. This possibility is further compounded by the fact that the strongly folded flysch within the Yidun Unit probably formed the southern part of the basin during much of Middle and Late Triassic time; this also needs to be retrodeformed as well to know the full extent and geometry of the Songpan Ganze basin (a first, but nonquantitative, attempt is shown in figure 12 of Weislogel et al., 2010).

Evidence along the Songpan Ganze Unit's western boundary with the Yidun Unit indicates that this boundary was a SW-dipping subduction zone in Late Triassic time, and possibly into Early Jurassic time (see below), which is expressed by the Triassic magmatic activity within the Yidun Unit to the southwest (see below). Subduction along this zone was probably responsible for some of the folds and thrusts in the Songpan Ganze shown on Plates 1 and 2 east of its boundary with the Yidun Unit. However, the Songpan Ganze flysch is in the lower plate of this subduction zone, and there is no evidence for collision between the Yidun Unit and a terrane to the north. The folds are rather uniform in character across the entire unit. Subduction also occurred along the northern margin of the Songpan Ganze south of a western Qinling Triassic arc (Şengör and Hsü, 1984; She et al., 2006; Roger et al., 2008); thus subduction occurred on both flanks of the basin. On a regional scale the problem is more complicated and is beyond the focus of this study (see, e.g., Mattauer et al., 1992; Burchfiel et al., 1995; Dirks et al., 1994; Weislogel et al., 2010). What can be said is that because regionally no rocks older than Early or Middle Triassic are exposed, most of the folds in the basin probably lie above a décollement zone at their base (see Mattauer et. al., 1992). How the basement below the décollement accommodates the thin-skinned shortening above remains an unknown problem, but based on the large magnitude of shortening across such a broad area, much of it was shortened and subducted both to the north and to the southwest.

Rocks along the eastern margin of the Songpan Ganze Unit form a convex south arc as seen on Plate 2. Along the eastern side of this arc where the structure trends northeastward the rocks are thrust eastward, forming part of the Late Triassic–Early Jurassic (Indosinan) thrust belt of the Longmen Shan region that extends southward into the area of Plate 1 as the Mesozoic parts of the Xiaojinhe and Jinhe-Qinhe thrust faults (Burchfiel et al., 1995). The arcuate structures within their hanging walls of these two thrust faults form part of this more regional arc (see above). These relations are interpreted to be the result of NE-SW left-lateral transpression between the Songpan Ganze and Yangtze Units to the east in the Longmen Shan as it was subducted to the SW beneath the Yidun Unit (for more discussion, see Mattauer et al., 1992; Burchfiel et al., 1995; Dirks et al., 1994; Harrowfield and Wilson, 2005). This deformation molded the folds into the south-facing wedge that formed the southeast corner of the Songpan Ganze basin and which also affected rocks of the Transitional Unit within the area of Plate 1 (see above). Here the arcuate shape may have been accentuated in late Cenozoic time by transfer of motion from the Xianshuihe fault to shortening and clockwise rotation into the Xiaojinhe and Jinhe-Qinhe thrust faults (see above). Within the Longmen Shan, transitional strata of Triassic age between the Yangtze and Songpan Ganze Units are missing, but in the southern extent of the Longmen Shan thrust belt the hanging wall of the Jinhe-Qinhe thrust fault carries the first indication of the transitional sequence. The two thrust faults diverged southward, and the transitional section became broader, to form the Transitional Unit. This, and the fact that the contact between the Transitional and Yidun Units appears transitional farther south, suggest that the magnitude of the transpressional deformation decreased southward.

**Origin of the Igneous Rocks within the Songpan Ganze**

The origin of the Triassic and Jurassic intrusive rocks in the Songpan Ganze is unclear. They form the wide belt within the eastern part of the Songpan Ganze Unit (Burchfiel et al., 1995; Roger et al., 2004, 2010; Weislogel et al., 2010; Plates 1 and 2). From the discussion of the tectonic setting of the Songpan Ganze above, the belt of plutons did not have a simple origin. In the study by Roger et al. (2010) many of the plutons appear to have crustal sources, but some have a mantle component. Some sources suggest a subduction mechanism related to the transpression boundary to the east; however, the plutonic belt continues south, and some plutons probably related to them intrude the Yangtze Unit where they are overlapped by Upper Triassic strata on the Kungdian high (see above). This relationship was interpreted by Şengör (1984) and Şengör et al. (1993) to be the result of "hidden subduction," where the main subduction of the Songpan Ganze lower lithosphere was to the SE below the Yangtze Unit, and the thrusting in the upper crust was westward-directed above, forming a crustal wedge.

If the Songpan Ganze Unit was subducted both to the north and south, the folding across the 400-km-wide belt remains a problem. Normally in a subduction zone the downgoing plate is not much deformed, except in very wide accretionary prisms. Folds in the Songpan Ganze are generally near-isoclinal upright folds with vertical axial plane cleavage and would require considerable but unknown ~N-S shortening. One possible explanation is that the thick Triassic flysch is underlain by a décollement that extends into both subduction zones and represents broad accretionary thrust belts that meet somewhere in the middle of the unit, similar to that in the active bivergent subduction systems in the Molucca Sea (Silver and Moore, 1978; McCaffrey, 1982). If so, the folds are completely detached at depth, a geometry that is consistent with the fact that none of the folds or thrust faults within the main body of the unit ever expose any rocks older than Middle Triassic, because a large volume of strata and basement was subducted.

## Modifications of the Early Mesozoic Longmen Shan Thrust Belt by Cenozoic Deformation

As the geology and thermochronology show, there appears to have been little tectonic activity within the Songpan Ganze Unit from Late Jurassic to about mid-Cenozoic time. The present land surface is one of extensive low relief, a typical plateau surface, and the thermochronology shows little deep erosion of the unit if there had been major shortening or extensional structures formed (Kirby et al., 2002; Clark et al., 2005). It was during the late Cenozoic that uplift of the eastern part of the plateau took place, perhaps at ca. 10–15 Ma (Kirby et al., 2002; Clark et al., 2005). The Longmen Shan marks the steepest plateau margin slope around Tibet, rising from ~600 m in the Chengdu Plain to nearly 5000 m in the Pengguang massif, and >6000 m at the Si Gou Nong Shan ("four girls") with little evidence for major horizontal shortening (Burchfiel et al., 1995) either at the plateau margin or within most of the plateau to the west. Royden (1997) explained this relation to indicate that the steep topography was the result of eastward flow of lower crustal material, thickening the crust at the bottom and raising its surface with little crustal shortening. The eastward flow was impeded by cold and less ductile crust underlying the Sichuan basin, causing a sharp elevated boundary at the Longmen Shan. Recent geophysical work has tended to support such an interpretation (see Zheng et al., 2010a). The minor shortening that has occurred at the eastern margin of the Longmen Shan has been by folding, thrusting, and uplift of the Precambrian basement in the Pengguang and Baoxing massifs with attendant folding of the early Mesozoic thrust sheet (Fig. 4-1; Burchfiel et al., 1995). The minor amount of shortening documented within the Longmen Shan (28.5 km, Hubbard and Shaw, 2009) is inadequate to develop a ~5-km-high plateau across eastern Tibet.

The elevation of the plateau margin continues to the east from the Longmen Shan and turns north through the Min Shan across the older structural trends in the Songpan Ganze and Qinling orogenic belt (Fig. 4-1). This part of the margin is marked by active faults in the Min Shan and along the Huya fault and related faults (Kirby et al., 2000, 2002). At the same time, uplift within the Longmen Shan was marked by late Cenozoic faults with both right-lateral and normal displacements (Wenchuan-Maowen and Beichaun faults: WMF and BF, respectively, in Fig. 4-1). The Wenchuan-Maowen fault also has a down-to-the-west normal component, displacing the early Mesozoic allochthon. Evidence for down-to-the-west, poorly dated faults of Cenozoic age also is present along the west flank of the Baoxing massif, around the north-plunging nose of the Danba antiform and the Ge Zhong dome (Fig. 4-1).

At about the same time as uplift was occurring along the eastern margin of the plateau the left-lateral Xianshuihe fault began to develop. There is some evidence that some of its left-slip displacement may have been transferred to shortening on the Xiaojinhe and Jin-Qinhe thrust faults, producing steps in the rather uniform slope of the southeastern part of the Tibetan Plateau surface. Also along the west side of the Xianshuihe fault the large Cenozoic Gonga Shan pluton was intruded and was rapidly exposed (12 Ma, Roger et al., 1995; Fig. 4-1). We interpret the exposure of this intrusion to be related to a shortening bend in the Xianshuihe fault, forcing lower crustal melt upward, a feature that characterizes many of the Cenozoic shear zones within the area of Plate 1 (see below). The shortening within the Xianshuihe fault may have had effects that reached eastward into the area of the southern Longmen Shan and Sichuan basin (Burchfiel et al., 1995).

# CHAPTER 5

# *Yidun Unit*

The Yidun Unit is bounded by the Songpan Ganze Unit on the northeast, the Jinsha accretionary complex, sometimes called the *Jinsha suture,* the Qiangtang Unit on the west, and the Transitional Unit to the south (Fig. 1-1B; Plate 1 [available on DVD accompanying this volume and in the GSA Data Repository[1]]). Its southern boundary is arbitrarily defined where Triassic volcanic rocks that characterize this unit end with the accompanying sedimentary rocks grading into the non-volcanic Triassic strata of the Transitional Unit. Like the Transitional Unit, the definition of the Yidun Unit is based on its Mesozoic rocks, which are mainly sedimentary rocks of Triassic age interbedded with volcanic rocks intruded by plutonic rocks that form a Late Triassic magmatic arc (see, e.g., Li et al., 1996). Most of the sedimentary rocks that host the arc rocks are Upper Triassic strata, although in some areas they may be as old as Middle Triassic, and are largely the SW continuation of the Songpan Ganze flysch terrane (see above).

## STRATIGRAPHY OF THE YIDUN UNIT

The stratigraphic succession within the Yidun Unit ranges from Precambrian (Proterozoic?) metamorphic rocks to Upper Triassic strata unconformably overlain in many places by Paleogene sedimentary rocks (Figs. 5-1, 5-2, and 5-3). Almost the entire unit consists of Triassic rocks with pre-Triassic rocks mostly exposed in domes in the southernmost part of the unit and in the Zhongza zone (Chen et al., 1987) along its southwestern side (Fig. 2-1; Plate 1). Permian rocks, mainly basalt but also locally limestone, are present as narrow belts in anticlines and along thrust faults in the southeastern margin of the unit. Locally there are rocks as old as Silurian that occur within blocks and slivers along thrust zones and within mélange mainly along the NE margin of the unit. These rocks suggest that at least locally there may have been Paleozoic continental rocks below the Triassic and that some of these rocks were incorporated into an accretionary prism derived by subduction of thin continental crust that lies below the Songpan Ganze Unit. The main body of the Yidun Unit consists of Triassic rocks, mostly Upper Triassic, and no older rocks are exposed (Plate 1).

The most complete exposures of pre-Triassic rocks occur in two areas: in three structurally complex domes in the southern part of the Yidun Unit where Precambrian and Paleozoic rocks are exposed beneath the Triassic flysch and volcanic-bearing sedimentary rocks, and in the several-hundred-kilometer long Zhongza belt along the west side of the unit (Fig. 2-1; Plate 1). The pre-Triassic rocks in the three domes are metamorphosed, locally reaching amphibolite grade in the western dome (Qiasi dome). In the western and largest of the three domes, Precambrian basement rocks occur in several parasitic anticlines and smaller domes. They consist of a thick sequence (>3 km; structural thickness?) of metasedimentary rocks, interpreted to be Proterozoic in age, unconformably overlain by 30–300 m of sandstone and shale, followed by 250–1000 m of mainly dolomite and limestone (Fig. 5-2; Plate 4 [see footnote 1], sections 10, 11). These rocks are lithologically similar to the uppermost Neoproterozoic (Z2 on Plate 1) in the Yangtze succession. In the western dome the Proterozoic rocks are unconformably overlain by a few meters to >100 m of limestone and sandstone assigned an Early Cambrian age, but without fossils. The only other place where these rocks crop out is within narrow ENE-trending thrust slivers 15 km north of the domes, where they are thrust over Upper Permian limestone and Lower and Middle Triassic clastic rocks, and are unconformably overlain by Upper Triassic strata (Plate 1).

Ordovician strata unconformably overlie the Cambrian rocks in the western dome and form the oldest rocks exposed in the two domes to the east (Figs. 5-1 and 5-2). They consist of fossiliferous sandstone and shale >1000 m thick. Silurian shale >100 m thick lies above. Both the Ordovician and Silurian deposits contain graptolites. Devonian rocks are present only in the western dome and consist of >700 m of sandstone, shale, thin limestone, and local basalt. The Devonian strata unconformably overlie all older rocks units. Limited outcrops of Carboniferous limestone, 150–400 m thick, overlain by Lower Permian limestone, ~600 m thick, are present in the eastern part of the western dome, where they are unconformable or are in fault contact above Silurian rocks. These upper Paleozoic rocks are rapidly cut out along a major unconformity at the base of Upper Permian basalt >1000 m thick and which is the marine equivalent of the Emei Shan basalt. The basalt is present only in the western two domes. The stratigraphy of the upper Paleozoic rocks is similar to that of the Yangtze sequence, but it reflects unstable platform depositional conditions, importantly just prior to the eruption of the Emei Shan basalt. Mapping in these domes shows highly complex structure, and it is possible that some of the complexity in the stratigraphic sequence is structural in origin (see below), but it remains poorly documented. We point out that the regions in which the domes are exposed contain rugged, high relief with poor accessibility, and some Chinese geologists lost their lives mapping in this area.

---
[1]GSA Data Repository Item 2012297, Plates 1–4, is available at www.geosociety.org/pubs/ft2012.htm, or on request from editing@geosociety.org or Documents Secretary, GSA, P.O. Box 9140, Boulder, CO 80301-9140, USA.

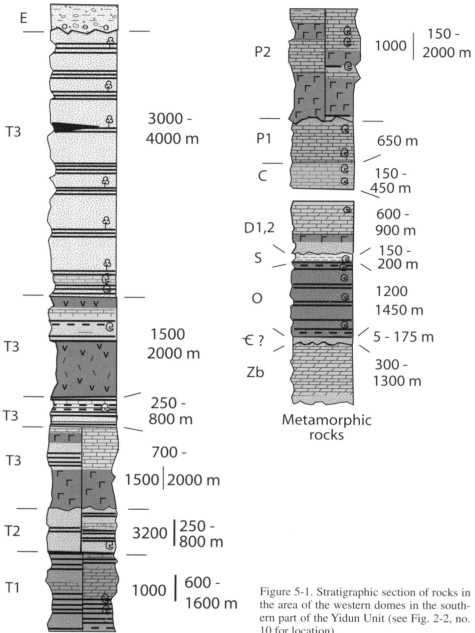

Figure 5-1. Stratigraphic section of rocks in the area of the western domes in the southern part of the Yidun Unit (see Fig. 2-2, no. 10 for location).

In the area around the domes the Triassic sequence begins with fossiliferous Lower Triassic dark shale and sandstone, ~1 km thick, overlain by 1–3 km of Middle Triassic sandstone and shale with some limestone (T1-2 on Plate 1). A major unconformity lies at the base of the Upper Triassic strata, where they transgress over older Triassic rocks in the western dome and onto the oldest rocks in the cores of the eastern two domes. This unconformity is widespread throughout the Yidun Unit and is particularly well developed along its east and west sides (Plate 3 [see footnote 1]). The Upper Triassic sequence begins with >2000 m of basaltic volcanic rocks at the base (Ta$_1$ on Plate 1), interbedded with and overlain by marine fossiliferous limestone that may reach ~1 km thick (Fig. 5-1). Above the limestone is 300–800 m of marine dark-gray sandstone and shale (T3$_1$), overlain by an upper volcanic unit that consists of intermediate to felsic volcanic and volcanogenic sedimentary rocks (Ta$_2$ on Plate 1) interbedded with sandstone, shale, and rare limestone with a total thickness of 1500–2000 m (T3$_2$ on Plate 1). The remainder of the Upper

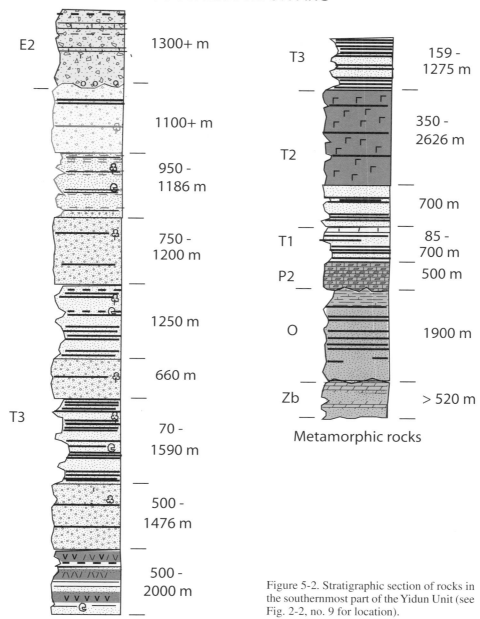

Figure 5-2. Stratigraphic section of rocks in the southernmost part of the Yidun Unit (see Fig. 2-2, no. 9 for location).

Triassic section is >5–6 km thick and grades upward from mainly marine strata overlain by interbedded marine and non-marine strata, containing some coal, and into mainly non-marine strata with coal in its upper part (T3$_2$). This upper part of the Upper Triassic section is typical for the Upper Triassic rocks along the eastern part of the Yidun Unit. At the top they contain more non-marine strata than rocks in the Songpan Ganze Unit, which are mainly marine (even with some Lower Jurassic marine strata [see above]). The lower part of the Upper Triassic sequence is flysch similar to strata in the Songpan Ganze Unit, which may have been deposited in the same broad basin. It is mainly the volcanic rocks that place these rocks within the Yidun Unit. The boundary between the two tectonic units in this southern area is difficult to determine where the volcanic rocks are locally missing.

The section of Triassic rocks in the southern part of the Yidun Unit is typical for most of the remainder of the unit, the bulk of which consists of a thick succession of Lower to Upper Triassic sandstone, shale, and local conglomerate, and limestone with volcanic rocks in two sequences within the Upper Triassic section, and a lower sequence dominated by mafic volcanic rocks and separated by a non-volcanic sequence from an upper sequence dominated by felsic and intermediate volcanic rocks with local basalt. (On Plate 1 the distribution of these two volcanic sections are labeled $Ta_1$ and $Ta_2$, respectively.) In many places the volcanic rocks are missing, but they have lateral equivalents of volcanogenic sedimentary rocks. It is these Triassic volcanic units, along with several very large plutons (see below), that commonly give the name *Yidun magmatic arc* to this unit; however, the stratigraphic sections in the Yidun Unit are dominated by sedimentary rocks, much of flysch character.

Along the western margin of the Yidun Unit the Lower Triassic, and in places Upper Triassic, strata lie both conformably and unconformably on Paleozoic rocks of the Zhongza zone (Plate 1 and Fig. 5-4). This zone of Paleozoic rocks can be followed

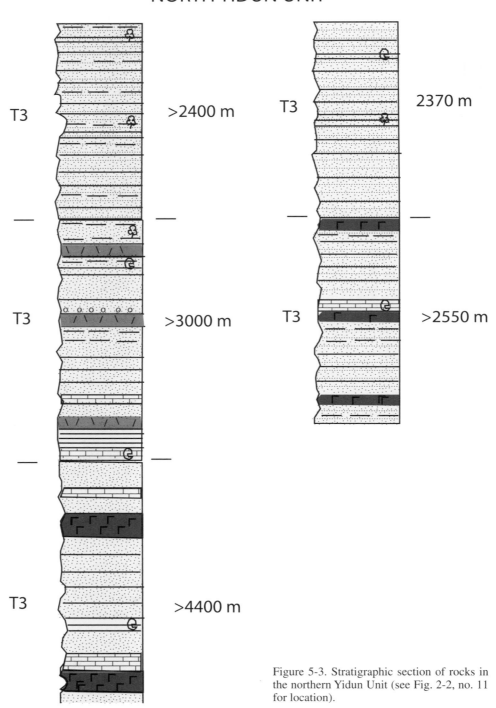

Figure 5-3. Stratigraphic section of rocks in the northern Yidun Unit (see Fig. 2-2, no. 11 for location).

across some major faults from the Yangtze and Transitional Units directly into the base of the Yidun volcanic-bearing section. Cambrian strata make up the basal part of the Zhongza sequence and are locally >3000 m thick, but only the Middle and Upper Cambrian strata contain fossils (as in all these stratigraphic sections, thickness should be regarded as structural thickness in these highly and complexly deformed rocks). Lower Cambrian strata are non-fossiliferous, and their age is based on lithological correlation with similar rocks in the Yangtze Unit to the south. Most of the Cambrian section consists of terrigenous clastic rocks, with limestone forming >1000 m of the Upper Cambrian section, but not its uppermost part. In the northernmost part of the Zhongza belt these rocks are metamorphosed to biotite grade. Ordovician rocks, 1000–1500 m thick, consist mostly of limestone, but some clastic intervals are present at the base and the top, and fossils are abundant throughout. Silurian strata are mainly limestone with units of sandstone and shale, and their total thickness ranges from 1000 to >2000 m. Devonian rocks consist of 700 to >2000 m of fossiliferous limestone overlain by an additional 500 to 1200 m of non-fossiliferous limestone. (Note: Fig. 5-4 is from one area within the Zhongza zone, so some thicknesses and descriptions given in the text do not match exactly those in the figure.) Carboniferous strata consist of ~600 m of fossiliferous limestone. Lower Permian strata are mainly limestone, but in places they contain basalt. Thicknesses given for these rocks range from several hundred meters to more than several thousand meters, and the greater thicknesses are probably structural thicknesses and so are not reliable. Upper Permian rocks contain limestone at the base and >1000 m of basalt near the top, probably equivalent to the Emei Shan basalt, but are not mapped separately, and

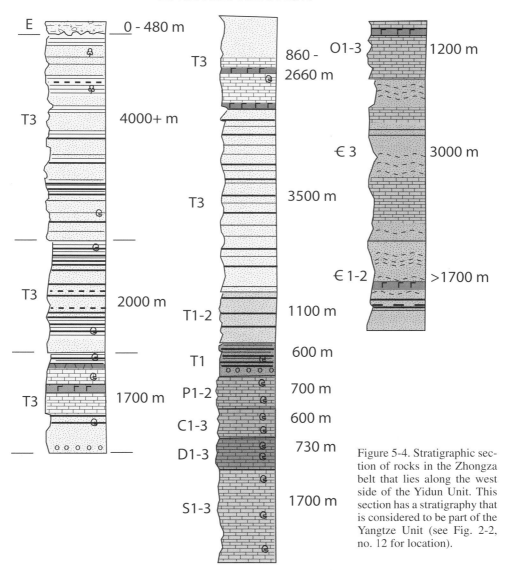

Figure 5-4. Stratigraphic section of rocks in the Zhongza belt that lies along the west side of the Yidun Unit. This section has a stratigraphy that is considered to be part of the Yangtze Unit (see Fig. 2-2, no. 12 for location).

here they were erupted in a marine environment. Basalt is not everywhere present at the top of the Permian section, because the Upper Permian units are commonly missing, and the older Permian strata are unconformably overlain by different parts of the Triassic succession (Plates 1 and 3).

Triassic strata along the west side of the Yidun Unit are >5–6 km thick and are stratigraphically similar to strata along the east side of the unit (see above). Lower Triassic strata are commonly terrigenous clastic rocks, and Middle Triassic strata are also clastic, but some contain thick limestone units (T1-2 on Plate 1). These two parts of the section contain unconformities at their bases, and mostly lie on different parts of the Permian section, but some locally rest on rocks as old as Carboniferous or Devonian (Plate 3). Upper Triassic strata make up >75% of the section and include lower mafic ($Ta_1$) and upper felsic or intermediate ($Ta_2$) volcanic sequences that are separated by non-volcanic clastic strata. The sedimentary parts of the lower mafic sequence are shown as $T3_1$, and the upper sedimentary parts associated with the intermediate and felsic volcanic rocks, and all overlying strata, are shown as $T3_2$ on Plate 1. Although the lower part of the section is marine, the thickest part of the Upper Triassic sequence lies above the upper volcanic unit and is similar to the Triassic sequences described above. Other areas consist of alternating marine and non-marine, terrigenous clastic deposits, becoming more non-marine upward, and commonly contain coal beds in the upper part. The Upper Triassic rocks in places rest unconformably on folded Lower and Middle Triassic rocks, but they rarely lie on rocks as old as lower Paleozoic (Plate 3). The upper part of the Triassic section is estimated to be >7 or 8 km thick; however, this is probably a structural thickness in these strongly deformed and foliated rocks (see below).

In the southeastern part of the Yidun Unit the Triassic volcanic-bearing arc sequence is underlain by rare narrow belts of Permian rocks that crop out below the Triassic rocks in the cores of anticlines and along thrust faults. In the southernmost exposures of these structures the Permian rocks are fossiliferous limestone, but at their northern end a small area of metamorphic rocks, mapped as Proterozoic (m), is present. The Proterozoic(?) rocks occur within an aureole of greenschist-facies metamorphic rocks, and whether they actually constitute Proterozoic basement remains to be demonstrated. Farther north, only Permian basalt with rare limestone is present at the base of the section within this structural zone. This basalt is about the same age as the Emei Shan basalt, and it may be its marine equivalent, but the basalt may form the basement beneath the eastern part of the Yidun Unit and possibly that of the Songpan Ganze Unit (see above). Just west of the northern end of outcrops of Permian rocks a second narrow belt of metamorphic rocks is mapped as Proterozoic basement (m on Plate 1) and is exposed beneath an unconformity at the base of the overlying Upper Triassic basaltic volcanic unit. This belt of rocks also lies within an aureole of greenschist to lower amphibolite-grade metamorphic rocks, and it remains to be determined whether these rocks are really metamorphic basement or metamorphosed younger strata. From surface exposures the Yidun Unit is underlain by the Yangtze platform in the south and west, but the nature of the basement rocks that underlie most of the unit remains unknown.

## STRUCTURE OF THE YIDUN UNIT

The structure within most of the Yidun Unit appears to be simple and consists of NW-trending faults and tight to isoclinal folds with axial planar cleavage that is steeply dipping (Plate 1; Plate 4, sections 16, 25, 39, 40). However, this simplicity is misleading. In a number of transects, the upright folds are second generation, and they fold an earlier, commonly bedding-parallel foliation that is related to isoclinal first-generation folds (see also Reid et al., 2005b). These second-generation folds are outlined on Plate 1 by the contacts between the Lower and Middle Triassic (T1-2, Plate 1) and Upper Triassic units ($T3_1$ and $T3_2$, Plate 1). The nature of this first-generation folding has not been studied, so its geometry and kinematics remain unknown. The folding is generally regarded as having ended in Late Triassic time, as several large Upper Triassic plutons intruded the folded section (see below). However, there has been, at least locally, an important Cenozoic deformation, and until its extent is known, generalizations about the ages of deformation remain uncertain. This makes the internal structure of the Yidun Unit difficult to generalize, but the upright folds are responsible for a severe shortening of the entire sedimentary and volcanic section.

The eastern margin of the Yidun Unit is well defined where Triassic strata, with their two main volcanic units, mostly the lower basaltic unit ($Ta_1$, Plate 1), are juxtaposed with Triassic flysch of the Songpan Ganze (Mzf) across the boundary fault zone (Plate 1). Because sedimentary rocks on both sides of the fault are lithologically similar, it is the presence of Triassic volcanic arc rocks that define the Yidun Unit. The volcanic arc assemblage also contains a major unconformity, commonly marked by conglomerate at the base of the Upper Triassic sequence, which is not present in the Songpan Ganze succession (Plate 3). The boundary fault zone is mapped as steeply dipping thrust faults, but the vergence of the thrust faults is both to the NE and SW along different parts of the boundary (Plate 4, sections 23, 24, 26). In a few places, flysch, belonging to the Songpan Ganze Unit, is shown conformable above Triassic rocks belonging to the arc assemblage, as if the two units locally interfingered before the last phase of faulting. In the northern part of the area of Plate 1 the Songpan Ganze flysch, within 50 km of the arc, contains rare volcanic rocks of arc type, also indicating that arc volcanism may have locally extended eastward beyond the limits of the Yidun Unit as defined here (Plate 1).

In a belt up to 70 km wide west of the northeastern boundary fault zone, abundant blocks of Paleozoic rocks were mapped within the Triassic volcanic-bearing section. These blocks are dominated by Lower Permian fossiliferous limestone (Wp), although fossiliferous Carboniferous limestone (Wc), a long sliver of lower Paleozoic strata (Pzl), basalt, and locally some serpentinite, are also present. Where these blocks occur the sequence is mélange, an interpretation supported by our observations in the northern part of the belt. The serpentinite bodies are of particular interest and are either serpentine or serpentinized gabbro and ultramafic rocks. A number of these blocks occur in the southern part of the Yidun Unit just north of the eastern dome of Paleozoic rocks and along faults and within Upper Triassic rocks near the thrust sliver of Precambrian rocks north of the dome (Plate 1). A few small bodies of serpentinized rocks occur as far south as, and between the domes of, Paleozoic rocks in the southernmost part of the Yidun terrane. The interpretation that these southern fragments are of oceanic lithosphere is hard to justify, as they occur in areas underlain by Yangtze Unit rocks. It is these bodies that some workers have used to extend oceanic crust across the southern part of the Yidun Unit and through the Transition Unit to connect southwestward with the Jinsha accretionary zone, a hypothesis we do not support (see above). These rocks remain, however, to be studied in more detail. All the blocks commonly have tectonic contacts with the Triassic host rocks, but it is not clear how many of these blocks may have been olistostromes deposited within the Triassic strata or occur as tectonically included blocks. This is a typical problem where the rocks have been strongly tectonized, and the original environment in which the blocks were emplaced, is to be determined. These rocks appear as mélange and may have a polygenetic origin.

Along the eastern margin of the Yidun Unit is a belt of numerous north-striking thrust faults that dip both NE and SW (Plate 4, sections 23, 24, 26). The faults along the boundary are both in the Yidun and Songpan Ganze Units and involve numerous narrow outcrops of Paleogene strata (see below), so clearly these faults are not all original faults related to the proposed Triassic subduction of the Songpan Ganze Unit below the Yidun Unit (see below). It is not unreasonable to interpret the eastern boundary of the Yidun Unit as a Triassic accretionary complex that involves both Songpan Ganze flysch and Yidun Triassic strata strongly modified by Cenozoic thrusting. The unconformity in the Upper Triassic rocks is common within this succession and suggests that it formed on an outer arc high in Late Triassic time. Because the volcanic and plutonic rocks of the arc lie southwest of the boundary, we interpret that the subduction zone responsible for the Triassic arc magmatism dipped SW. However, in the southeastern marginal zone of the Yidun sequence, an imbricate sequence of fault slivers brings up Upper Permian limestone lying conformably below Lower Triassic strata. In several places, Upper Triassic strata that contain the basal basaltic volcanic unit ($T3_1$ and $Ta_1$) are shown to lie unconformably on these fault slices (Plate 3), but more generally the same Upper Triassic rocks are mapped to be parallel to the structures. Several of the thrust faults have small serpentinite bodies along them. Whether these thrust faults represent synthetic or antithetic thrusting during the beginning phases of subduction remains to be resolved, and further mapping in these areas is clearly needed. It must be remembered that mapping in these areas was conducted largely before or during the early history of plate tectonics, and many of these areas have not been studied since.

In the middle of the Yidun Unit, west of the large Upper Triassic plutons, are narrow N-S–trending belts with small bodies of serpentinite and serpentinized mafic and ultramafic bodies within Upper Triassic strata (Plate 1). These rocks are most abundant within the middle of the Yidun Unit but also extend into its western part and occur mainly as tectonized lenses with the Triassic strata. They have been interpreted as remnants of oceanic crust that may have developed by Triassic backarc or intra-arc spreading that formed small(?) intra-arc basins floored by oceanic crust (Li et al., 1996). These basins were closed during regional shortening of Late Triassic age, but before the large Triassic plutons were emplaced. Little work has been done on these bodies, so it remains an open question whether they are related to intra-arc spreading or whether they were just tectonically emplaced serpentinized rocks of the lower arc crust.

The western margin of the Yidun Unit is easy to define in some places and very difficult in others, because many different rock units lie along both the western part of the Yidun Unit and within the Jinsha accretionary complex west of it. Along much of the southwestern part of the Yidun Unit the Zhongza belt is a narrow belt of Yangtze-type strata of Paleozoic age (see above). Different parts of the Triassic sedimentary and volcanic rocks of the Yidun Unit lie unconformably on these Paleozoic rocks, usually on lower or middle Paleozoic rocks, making the western boundary easy to define as the western limit of the Zhongza Paleozoic rocks (Plates 1 and 3). Along the northern third of the Yidun Unit the Paleozoic rocks end, and Triassic rocks of the arc, and rocks whose ages are unclear, are present. The strata of uncertain age have been assigned ages from Silurian to Triassic and form what we interpret as a mélange belt (me, Plate 1) that is tectonically juxtaposed with Triassic rocks of the Qiangtang 1 Unit and rocks of the Jinsha accretionary complex to the west, and the boundary is less clear. The fossils that date this belt are mainly from blocks (Ws, Wc, Wt), along with some blocks of serpentinite, within a strongly foliated slaty matrix whose steep foliation becomes more intense from east to west and dips steeply both east and west. This belt we interpret to be a foliated mélange whose initial dip remains unclear, but it is unconformably overlain by Upper Triassic strata of the lower arc unit ($T3_1$, Plate 1). Based on the geology of the Jinsha accretionary complex we suggest that the initial dip was to the west; however, the westward increase in the intensity of foliation and metamorphism might suggest otherwise (see below).

The Paleozoic rocks (Zhongza belt) along the western margin of the Yidun arc range from Cambrian to Permian in age and are strongly folded and cut by numerous generally NW-striking, usually west-dipping faults, so that the rocks occur in large fault-bounded lozenges and slivers. No one fault-bounded sliver contains a complete sequence of Paleozoic rocks; however, parts of sections can be matched between the slivers to form a complete sequence. The anastomosing pattern of faults is suggestive of at least some strike-slip displacement as well as a thrust overprint to the deformation within these rocks. Within many of the fault slivers the rocks are folded with NW-trending axes, and if they are related to the more northerly striking faults it would suggest right-lateral shear, but this is speculative. Rocks as young as Early to Late Triassic lie unconformably on different units within the Zhongza belt, indicating tectonic activity during latest Paleozoic into Triassic time (Plate 3).

The entire Yidun section is strongly folded and was intruded by large, undeformed plutons of Triassic age, so a significant amount of the structure has been assigned a Late Triassic age. The pattern and vergence of the thrust faults along the west side of the Yidun Unit suggest that they may be part of an east-vergent, fold-thrust belt that may be part of a deformation that affects a broader region, including not only the Yidun Unit but the Jinsha accretionary complex and the Qiangtang 1 Unit to the west. The thrust faults that are part of this belt within the Yidun Unit commonly involve Paleogene strata and indicate an important Cenozoic deformation, a conclusion reached for the deformed rocks to their west and east, as well as to the south within the Transitional Unit (see above and below). Thus, there remains uncertainty as to how much of the deformation within the western part of the Yidun Unit is Mesozoic, and how much may be Cenozoic. Some folding may have been older, as Permian rocks, and more rarely Lower Triassic rocks, also unconformably overlie folds (Plates 1 and 3), but these relationships are rare. Even though strongly faulted, the Paleozoic section consistently becomes older to the west, and the Cambrian and Ordovician strata lie in the westernmost part of the Zhongza belt in its central part but occur more rarely along the northern and southern part of this belt, whereas rocks as young as late Paleozoic or even Triassic are at the westernmost boundary north and south of the Zhongza belt. Not all the faults in this belt are thrust faults, as some of the anastomosing faults also have an apparent normal separation down to the east. How much of this is real or part of a complex strike-slip fault system is unknown. Our interpretation of this part of the Yidun Unit is that the section of rocks along the west side has the character of an east-tilted crustal section cut by many east-vergent thrust faults. Some of the tilting occurred in Early and Late Triassic time, but most of it is Late or post–Late Triassic, and some is likely of Cenozoic age. Even the western boundary fault zone of the Yidun Unit in most places dips west and has separation suggesting that it is a thrust fault. This boundary is marked by a system of faults of probably different ages and is often shown as the trace of the Jinsha suture. In our interpretation the rocks to the west of the boundary faults are part of a late Paleozoic and earliest Mesozoic accretionary complex or collage, which may have been formed along the site of closure of an oceanic terrane (see below), but that possibility has been so modified by younger faulting that little of the original suture relationships remain.

In the northern part of the Yidun Unit the western boundary requires considerable interpretation or reinterpretation of existing maps because of mismatches in rock unit designations across map sheet boundaries. For example, at the northern end of the Zhongza belt the eastern part of the Paleozoic section is shown to continue northward across a map boundary by Upper Triassic strata. The western part of the Paleozoic section (lower Paleozoic) is continued to the north by metamorphic rocks shown as Precambrian. What seems clear is that the Paleozoic rocks of the Zhongza belt end, but how they end remains unknown. The Upper Triassic strata that are shown to continue these rocks northward are the unit described above as a highly foliated greenschist facies mélange. The lithological units along the NW margin of the Yidun Unit require serious reinvestigation and remain one of the major problems for Plate 1. We show the ending of the Paleozoic rocks with uncertainty, and the mélange belt that continues to the north (me in Plate 1) must be taken only as our interpretation. The western margin of the Yidun Unit is shown as a fault with volcanic-bearing Triassic rocks in fault contact against the mélange belt, but this relationship also required reinterpretation of existing maps and was not clear from cursory examination in the field. Strata that are clearly part of the Yidun sequence become increasingly more foliated and metamorphosed to the west into the unit we show as mélange, so that the rocks of both the arc sequence and the mélange are highly foliated greenschist and difficult to separate. The north end of the mélange belt is terminated between two faults, and farther north, Triassic strata of the Yidun Unit are in direct fault contact with Triassic strata of very different stratigraphy that belongs to the Qiangtang 1 Unit (see below). We will return to the problem of the western boundary of the Yidun Unit in this part of Plate 1 in the discussion of the Jinsha accretionary complex below.

As viewed on Plate 1, the first-order structural configuration of the Yidun Unit is a large synclinorium with the uppermost Triassic unit ($T3_2$) in the middle, bounded by belts of older Triassic rocks on both sides. Evidence indicates that subduction occurred beneath the Yidun Unit from the east side, but whether and of what age subduction may have occurred on the west side requires considerably more detailed studies. Both sequences of older strata on each side of the Yidun Unit are unconformably overlain by local Paleogene basins, some containing >1000 m of coarse-grained clastic rocks (Plate 1). The strata along the west side are commonly strongly folded and are cut by west-dipping thrust faults (see above). The Paleogene basins on the east side are also strongly folded and are cut by thrust faults, some of which involve the Triassic plutons. Much of the central part of the Yidun Unit is devoid of Paleogene strata, so the extent of Cenozoic deformation within this unit, and the age of its first-order synclinorial structure, remain unknown.

The youngest structures to affect the Yidun Unit are the late Cenozoic and active NW-striking, left-lateral, strike-slip faults and NE-striking right-lateral faults. As is common throughout the area, many of the strike-slip faults are associated with pull-apart basins. The Ganze fault, the western continuation of the Xianshuihe fault in the north, and the Litang fault system in its middle part of the Yidun Unit, are the two main left-lateral fault zones within the Yidun Unit, although there are many smaller parallel faults. The Ganze fault has ~75–100 km of displacement (Wang and Burchfiel, 2000) as measured by the offset of the Ganze pluton. The fault zone contains several small basins parallel to it, but only the Ganze basin appears to be an obvious pull-apart basin. The fault zone is locally up to several kilometers wide, marked by numerous active scarps and stream offsets, and in its eastern part it bounds the north side of the high Chola Shan range, with many peaks >6000 m high. Although Wang and Burchfiel (2000) interpreted this fault zone to be transtensional, the high range to the south and the global positioning system (GPS) velocity field suggests that it is obliquely transpressional. The nature of the Quaternary basins along the fault needs to be reexamined. For example, the narrow, elongate WNW–trending basin along the fault in the northwestern part of the area of Plate 1 is where the fault makes a bend to the west, which should be a releasing bend. The western continuation of the fault forms the boundary (Plate 1) between the Yidun and Qiangtang Units. Little work has been done on this very active fault; thus its nature remains unclear, but a recent study by Wang et al. (2008a, 2008b) of this fault zone shows evidence of its recent activity and complexity, and that it loses displacement westward. The Yushu fault, the western part of the Ganze fault, was the site of a major Ms 7.1 earthquake in 2010 with sinistral displacement (Chen et al., 2010; Lin et al., 2011; Zha et al., 2011).

The Litang fault zone consists of discontinuous left-stepping faults and many smaller parallel faults that cross the entire Yidun Unit along which is a series of Quaternary filled pull-apart basins (Plate 1). The total displacement on this fault zone is ~25 km (Wang and Burchfiel, 2000). The displacement on the western part of this zone appears to end before crossing into the Jinsha accretionary complex (see below). To the east a continuation of a fault along the Litang fault zone forms the eastern boundary of the Yidun Unit, where it cuts obliquely across folds and faults and becomes generally parallel to structures in the Songpan Ganze Unit (Plate 1). Here the fault curves slightly south along its eastern end, dips northeast, and begins to pick up a thrust component and thus is often referred to as the *Litang thrust*. The structures in its hanging wall curve to the southeast and east into the area of Plate 2 (see footnote 1) and form part of the regionally convex, south arcuate structures discussed above. On a regional scale the left-slip faults are part of numerous faults (Fig. 2-10) that are formed within a large area of crust that rotates clockwise around the eastern Himalayan syntaxis and is traversed by several faults that accommodate similar but more local rotations (see above and also Wang and Burchfiel, 2000).

Farther south the active left-lateral Zhongdian fault, the northwesternmost fault in the Dali fault system (Fig. 2-10), displaces the western part of the Yidun Unit and continues into the Jinsha accretionary complex, where rock units are offset ~18 km (see below). It has been interpreted that the Zhongdian fault had two periods of left-lateral displacement, the larger in early Cenozoic time, and the smaller during its presently active history (see above). Like the Litang fault, its current sense of displacement is related to differential clockwise rotation within a larger clockwise rotating crust.

The NE-striking Batang fault is the only major right-lateral fault that displaces the Paleozoic basement rocks in the Zhongza zone. It also displaces rocks of the Jinsha accretionary complex to the west and ends to the east before entering the Triassic volcanic succession. There are several other parallel but smaller right-lateral faults within the western part of the arc, but they are of lesser importance. The right-lateral faults are considered conjugate faults to the more important left-lateral faults.

## PLUTONS IN THE YIDUN UNIT

Several large plutonic bodies and many smaller bodies intrude the sedimentary and volcanic Triassic succession mostly in the eastern part of the Yidun Unit (Plate 1) and are dominantly intermediate in composition, but granite, tonalite, and gabbro are also present. Many of the plutons are compound, consisting of an assemblage of different plutons or zoned plutons indicating differentiated bodies (hence the several labels indicating different compositions, shown without boundaries within the plutons on Plate 1). These plutons in general intrude into folded rocks, cutting sharply across folds with well-developed contact aureoles and locally distorting the folds near the pluton boundaries. The relations to the folds, and the fact that most of the plutons are not deformed or tectonically foliated, indicate that they are mostly late tectonic intrusions; thus they place an upper limit of Late Triassic on much of the folding (see Reid et al., 2005a, 2005b). Most of the dating of these plutons is old, and was done by the K/Ar method, which yielded a range in age from 200 to 160 Ma, so the plutons were interpreted to be Late Triassic and Jurassic. More recent work by Reid et al. (2005a) confirmed a Late Triassic age for the emplacement of two plutons. They reported U-Pb ages of 218 ± 1.5 Ma for the Garze pluton in the north and 215 ± 1.7 Ma for the large Daocheng pluton in the south (Plate 1). Based on their associated $^{40}Ar/^{39}Ar$ studies of these plutons, Reid et al. (2005a) further suggest that they were intruded at shallow crustal levels and at some places cooled to below 300 °C by the beginning of the Jurassic. The dated plutons are younger than the ages for the arc volcanism, consistent with their cross-cutting relations to the folded volcanic sequences. They are also about the same age as most of the plutons dated from the Songpan Ganze Unit, as pointed out by Roger et al. (2004) and Reid et al. (2005a, 2007). Weislogel (2008) adds two new zircon ages of 215 ± 2 and 225 ± 2 Ma from two plutons in the northern part of the unit that agree well with the study by Reid et al. (2005a); however, she came to the conclusion that these ages suggest they were not related to SW-directed subduction of the Songpan Ganze Unit below the Yidun Unit because they are older than the folding in the Songpan Ganze Unit, a suggestion that could be debated.

Reid et al. (2005a) also reported Cretaceous ages for two plutons in the Yidun Unit. The Chuer Shan pluton in the north yielded a U-Pb age of 104.7 ± 1.0 Ma, and the Haizi pluton in the middle of the western part of the unit yielded an age of 94.4 ± 1.2 Ma. They interpreted them to have been intruded into very shallow crustal levels, cooling rapidly. The origin of Cretaceous plutons in this locality is not easy to explain and will be discussed below. There remain many plutons and large parts of the dated larger plutons that are unstudied, and which require study, as many of the plutons are compound intrusions. Before a truly adequate picture of the Yidun magmatism is known, these rocks require greater study, but the work by Reid et al. (2005a, 2005b) was ground breaking.

There are also several intrusive bodies into the arc terrane, shown as alkalic plutons, of Cenozoic age (e.g., $\varepsilon_6$ on Plate 1), but the ages reported for these rocks are old K-Ar ages and are probably not very reliable. Such plutons are not unexpected, as many shallowly intruded plutons of this type are present in the Jinchuan basin within the Transitional Unit and in the Qiangtang 1 Unit to the south and west of the Yidun Unit, respectively (see Fig. 16-21).

## MAJOR UNRESOLVED PROBLEMS IN THE YIDUN UNIT

### Origin of the Magmatism within the Yidun Unit

The occurrence of a mafic to felsic succession of volcanic rocks and large intermediate to felsic plutons of Late Triassic age are interpreted to have developed within a magmatic arc environment, and the unit is often referred to as the *Yidun arc* and is related to subduction processes. This interpretation is reasonable, but there are many parts of this interpretation that remain unresolved. It appears most reasonable from the geology presented above that a southwest-dipping subduction in Late Triassic time was taking place along the eastern side of the Yidun Unit, with the Songpan Ganze Unit on the subducting plate. The eastern side of the Yidun Unit is marked by mélange that may be more widespread than presently shown on Plate 1. However, the arc sequences continue up to a sharp boundary that separates them from non-volcanic units of the Songpan Ganze. The eastern boundary of the large plutons is similar, coming to within just a few kilometers of the boundary. There is little space for any accretionary complex or forearc features to be present. Part of the answer may be that the boundary has been strongly modified by Paleogene deformation, as shown by Paleogene strata in the footwalls of many thrust faults on the east side of the Yidun Unit. Most of these thrust faults dip eastward along the eastern side of the Yidun Unit and thus cannot be remnants of west-dipping subduction (see, e.g., Plate 4, section 23). This younger post-subduction thrusting and associated deformation may have foreshortened the Triassic forearc region, or locally covered it or been responsible for its erosional removal. While southwestward subduction below the Yidun Unit appears most reasonable, it is not without some uncertainties. It is important to note that the Triassic volcanism and plutonism terminated to the south before reaching the Transitional Unit at about the same latitude that the Songpan Ganze basin grades into the Transitional Unit, indicating that the subduction zone ended there.

The western boundary of the Yidun Unit has been interpreted to have been an east-dipping Triassic subduction zone by some workers (e.g., Roger et al., 2010). However, the boundary has been so modified by younger faults that most of the evidence for subduction along it has been obscured. The short belt of mélange along the northern part of the boundary can be interpreted as both related to east- or west-dipping subduction. The most compelling evidence for west-dipping subduction along this boundary comes from the Jinsha accretionary complex and will be discussed with that unit below. It remains possible that subduction was mainly west dipping, but for a short period it was also east dipping.

The Cretaceous plutons within the western part of the Yidun Unit are hard to explain. The evidence for Cretaceous subduction beneath the Yidun Unit is lacking. Evidence for eastward subduction along the western margin of the Yidun Unit is limited, and was not of the correct age to have produced these plutons. The mélange belt along the northwest side of the Yidun Unit (me) described above is not present west of the Cretaceous plutons and is probably a Triassic mélange. Mélange units that might mark subduction zones are present within the Jinsha accretionary complex to the west, but they likewise are too old to be related to Cretaceous plutonism (see below). The general eastward dip of the crustal Zhongza section might suggest eastward underthrusting; however, its formation is probably also too old (see also below). It is possible that there has been so much modification of the western boundary of the Yidun Unit that much of the evidence for eastward subduction has been removed. The problem of the Cretaceous magmatism thus remains unresolved.

### *Extent of Paleogene Deformation*

The Paleogene rocks involved in structures within the Yidun Unit clearly indicate extensive Cenozoic deformation (see above). Abundant narrow, generally north-south–trending belts of Paleogene sandstone, shale, and conglomerate range from a few hundred meters to >2000 m thick, mostly without fossils, that lie within the footwalls of numerous thrust faults within the Yidun Unit (Plate 1). They also occur intermittently along the faults taken to define the eastern boundary of the unit. The abundance of Paleogene strata involved in thrust faults along the eastern part of the Yidun Unit indicate that Cenozoic deformation strongly overprinted the original Late Triassic structures, but to what extent remains unclear. The NE-dipping thrust faults within the eastern part of the unit consistently have Paleogene strata in their footwalls, so if the original subduction system was east vergent (see above), the younger deformation has a reverse sense of vergence on faults in many places. In other parts of the unit, Paleogene strata lie unconformably above folded Triassic strata and are also folded. In places, such as along the eastern side of the large Triassic plutons, Paleogene strata lie unconformably above the plutonic rocks and are folded, indicating that the post-Paleogene deformation was at least locally intense. The work by Yang and Yao (1998) and Zou (1993) on the Garze pluton in the northeastern part of the Yidun Unit shows major northeast-vergent thrusting that involves not only overturned Paleogene strata but also slivers of granite and Silurian rocks that are locally mylonitic. It was pointed out above the evidence for strong Paleogene folding and faulting along the western part of the Yidun Unit. In many places the Paleogene strata contain abundant locally derived coarse debris, suggesting local relief and that many of the deposits were syntectonic. The evidence is clear that Paleogene and post-Paleogene deformation has been important within the Yidun Unit and probably has been greatly underestimated. It is difficult to determine the magnitude of this younger deformation, because the Paleogene strata are only sporadically developed. The age of folding and faulting that involved the Paleogene rocks is also unclear, because the age of these strata is poorly determined. Recently strong Paleogene deformation has become more evident in southern Tibet, such as in the Gonjo basin within the Qiangtang 1 Unit to the west (Studnicki-Gizbert et al., 2008), and in other basins to the northwest by Horton et al. (2002) and Kapp et al. (2005a) as well as in the Transitional and Yangtze Units, as discussed above. Neogene rocks are rare and are only weakly folded, but just south of Litang along the eastern Yidun Unit boundary Neogene strata lie in the footwall of the SW-vergent thrusts, but recall that these faults are probably along constraining bends on Neogene strike-slip faults and may be reflective of overly local deformation.

**Core Complexes in the Southern Part of the Yidun Unit**

In the central part of the Yidun Unit the dominant trend of the folds and faults is north-south, and when traced southward they bend to the east, where three large domes of older rocks are presently framed by Triassic rocks (Plate 1). Similar bending and domes are present, as shown on Plate 2, to the east (see above), and collectively they form a crude belt of domes and metamorphic rocks that extend from the southern Yidun Unit northeastward into the Longmen Shan area (Burchfiel et al., 1995). The structure of the domes suggests a pattern of NW- and ENE–trending cross-folds, yielding a dome-and-basin structure. Within the stratigraphic succession in the larger western domes, sections of pre-Triassic rocks are commonly missing at contacts or along faults. Within the large western dome the metamorphic isograds in its western part, and the biotite-muscovite isograd includes Triassic rocks and rocks mapped as Precambrian basement that reach garnet grade. The structure in the Triassic frame of the dome clearly shows a domal configuration that lies athwart the regional fold trends, and low-angle faults are present within the outer part of the dome, supporting the interpretation that this structure is a metamorphic core complex. Recent mapping of the larger dome suggests that stratigraphic units are missing along low-angle faults. More detailed mapping in the Jianglang dome within Plate 2 (Yan et al., 2003b) shows a complex history of low-angle faults that cut out section in a style that is similar to that of extensional core complexes in the western North American Cordillera (e.g., Lister and Davis, 1989). The Jianglang dome is interpreted to have undergone a first period of extensional deformation that is older than 200 Ma, which Yan et al. (2003b) suggested may be related to the extrusion of the Permian Emei Shan basalt, followed by a compressional event that they relate to early Jurassic shortening along the Longmen Shan thrust belt. A second extensional event followed, dated by U-Pb on zircon by Zhou et al. (2002) at 177 ± 3 Ma. Both extensional events have a general ESE extension direction. Structures formed in these three events are intruded by a poorly dated pluton that yielded dates from 115 to 131 Ma (Bureau of Geology and Mineral Resources of Sichuan Province, 1991). The entire complex of structures has been folded along a north-south trend by younger east-west shortening. The earliest extensional event in the Jianglang dome and an unconformity at or near the base of the unit, correlative with the Emei Shan basalt in all three domes (Plate 1), may suggest a regional extension in Permian time that can be related to the opening of the Songpan Ganze basin.

Unfortunately the age of the structural evolution in the domes is poorly known, and regional and local evidence suggests that the domal structures were formed in early Mesozoic, Cretaceous, and Cenozoic time (Yan et al., 2003b). The domal structure involves Triassic strata of the Yidun sequence and distorts the regional structural trends in the Yidun Unit, so they must be no older than Late Permian or Triassic. Only the northeasternmost dome (Plate 1) shows Paleogene rocks lying unconformably above Ordovician rocks, indicating that the older rocks were exposed by Paleogene time, but whether they, or any of the other core rocks of the domes, were exposed as early as early Mesozoic time remains unproven. Why the domes are present only in the southern part of the Yidun Unit and form a NE trend belt toward the Longmen Shan is unclear.

A second feature within the domes worth speculation is that they may show evidence for faulting and tilting of blocks, perhaps in extension, during late Paleozoic and early Mesozoic time. An unconformity at or near the base of the basalt unit (Pβ) is correlative with the Emei Shan basalt in the two southern domes, and in the eastern of the two domes it rests on lower Paleozoic rocks. As the rocks in the domes are of the Yangtze Unit type, this might suggest that they were displaced prior to eruption of the basalt, perhaps during the breakup of the crust during the beginning of the subsidence to form the deep-water basins of the Songpan Ganze and Yidun Units. For example, in the middle of the large western dome, a NW-striking fault along the east side of the dome juxtaposes Lower Triassic and Upper Permian strata against Silurian and Carboniferous strata. The same Upper Triassic unit lies unconformably on different rock units on opposite sides of the fault. Similar relations are present along other faults in the dome, suggesting pre–Late Triassic faulting. Work by Yan et al. (2003b) suggested that the oldest structures in the Jianglang dome are low-angle to subhorizontal faults of probable extensional origin of Permian or Triassic age. Coupling the high-angle and low-angle faults together, Yan et al. (2003b) suggested that the faulting is of the core-complex type. The Yangtze Paleozoic succession thus can be interpreted to be part of an unstable margin, and the faulting suggests that the instability was related to normal faulting perhaps related to the rifting of the Zhongza block away from the Yangtze Unit in Permian to Early Triassic time to form the Songpan Ganze depositional area. Other faults have mapped relations that suggest younger thrust faulting during dome formation, which may be related to shortening of

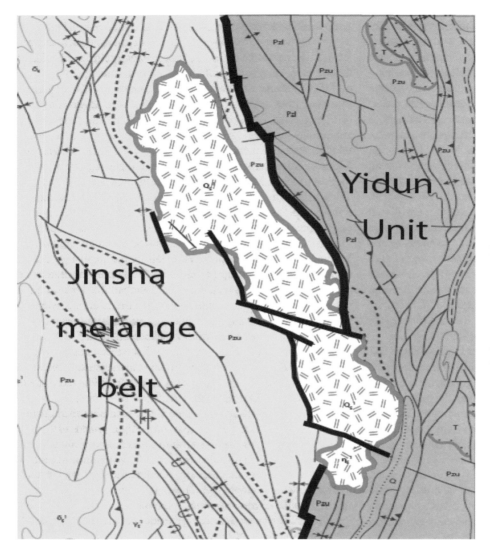

Figure 5-5. Geology of the mafic pluton near Batang that intrudes both the Jinsha accretionary complex and the Yidun Unit (red line is intrusive contact). Isograds related to the pluton emplacement also cross the boundary between the two units (see Plate 1). The pluton is considered to be Triassic in age on the basis of a date from a small internal pluton within the larger body. It is important, because some regional structural interpretations make the boundary between the Jinsha and Yidun rocks a major early Cenozoic strike-slip zone related to major early Cenozoic southeastward extrusion.

Late Triassic and Early Jurassic rocks in the Songpan Ganze and Yidun Units, and in the Longmen Shan. These domes lie within a terrain with significant relief, mountain peaks reaching 6 km high and very poor access, so it may be some time before the nature of their structure is known and mapped well enough for their origin to be determined.

Rifting and basalt eruption in late Paleozoic time may have provided the mafic rocks and their metamorphic equivalents for the serpentinite bodies in the areas around the domes that have been used as evidence for an oceanic basement. These mafic rocks may be only related to rifting of the continental crust in this area. However, the structure, tectonic setting, and age of the deformation in the domes remain major unresolved problems.

**Basement for the Yidun Unit**

Rocks older than Late Permian or Early Triassic are rare within the Yidun Unit except for the Zhongza belt and within the domes in the southern part of the unit. Otherwise only thin, tectonically emplaced slivers of metamorphic rocks of unknown protolith and Paleozoic strata, which could be remnants of a rifted basement, are present. The fact that the Yidun Unit consists mostly of a very thick flysch sequence, not unlike the Songpan Ganze, suggests that they are underlain by a similar crust of similar origin as for the Songpan Ganze Unit. The differences in the Upper Triassic sections is well explained by the development of the volcanic arc within the western part of the flysch terrane. The great thickness of flysch reported for both units suggests a thin continental crust or perhaps even a local oceanic crust. If the Songpan Ganze Unit was subducted beneath the Yidun crust, the blocks of Paleozoic strata in the subduction mélange suggest that at least part of its crust was continental. The present crust in this area is >60 km thick, but how much of it consists of isoclinally folded Triassic flysch remains unknown. In any case the problem of the nature of the basement beneath the Yidun and Songpan Ganze Units is similar.

## Western Boundary of the Yidun Unit

Regionally the interpretation of the western boundary of the Yidun Unit is extremely important, as it has been interpreted to mark one of the major Mesozoic sutures, the Jinsha suture, within the Tibetan Plateau. Structures along the boundary have a long and complex history, but younger structures have clearly modified the original structural relations related to the juxtaposition of the Yidun Unit to units to the west. The boundary with the Yidun Unit has been described above, but its possibility as a suture zone will be treated in more detail in the section on the Jinsha accretionary complex, below. The western boundary fault zone also has been proposed for a candidate for a major early Cenozoic left-lateral, strike-slip zone during which a large body of crust to its southwest was extruded to the southeast, caused by indentation of India into Eurasia (Remplemaz and Tapponnier, 2004), a feature that will be discussed with regional interpretations below. However, we point out that the area south of Batang has a large NNW–elongated mafic pluton that intrudes across the western boundary fault zone (Fig. 5-5). A small pluton intruded into this body has yielded a K-Ar age of 190 Ma, but this age must be validated for the entire pluton, because it would provide an upper limit on the age of movement on the boundary fault zone at this latitude. If the relationship of the pluton to the boundary fault zone and its published age are correct, it indicates that movement on this segment of the fault zone would have ceased in Late Triassic to Early Jurassic time, a point that must be considered in any tectonic synthesis.

CHAPTER 6

# Qiangtang Unit

The Qiangtang Unit lies in the north-central part of the area west of the Yidun Unit (Fig. 1-1b: Plate 1 [available on DVD accompanying this volume and in the GSA Data Repository[1]]). Regionally the Qiangtang Unit covers a vast east-west–trending area in central Tibet lying between the Bangong suture on the south and the Songpan Ganze Unit on the north. In central Tibet it is a compound unit that includes a complex stratigraphy and structure (Zhang et al., 2006a; Kapp et al., 2003, 2007), and within the area of Plate 1 it is a similarly complex unit, both stratigraphically and structurally. We divide the Qiangtang Unit into three subdivisions: Qiangtang I, II, and III units (note the lower-case "u"), based largely on their differences in Mesozoic stratigraphy and their possible separations by major faults (Fig. 1-1B). In numerous erosional windows through the Mesozoic rocks a wide variety of sedimentary, metamorphic, and magmatic rocks are exposed, and their variable character we interpret to be the result of the accretion of arc, continental, and oceanic fragments representing an accretionary collage. The collage is most completely exposed at the eastern side of the Qiangtang I unit and is referred to here as the *Jinsha accretionary complex,* where it serves as a characteristic basement for the Qiangtang Unit and possibly the Lanping-Simao Unit to the south. Thus in our interpretation the Qiangtang Unit is not a discrete tectonic element but is a collage of Paleozoic and lower Mesozoic rocks overlain by a largely Upper Triassic to late Mesozoic sedimentary cover that defines the unit.

## QIANGTANG I UNIT

The Qiangtang I unit covers the largest part of the Qiangtang (sensu lato) and lies west of the Yidun Unit and east of major fault zones that separate it from the Qiangtang II and III units, and to the south from the Lhasa Unit (Fig. 1-1B). We divide the unit into two parts, the Jinsha accretionary complex, which forms its basement, best exposed along its eastern side, and all the rocks to the west that consist of Paleozoic basement exposed in erosional windows and a Mesozoic and Cenozoic sedimentary cover. This division is necessary because rocks of the Jinsha complex are thought to mark a major suture zone, the Jinsha suture, which extends across central Tibet eastward into the area of Plate 1. The Jinsha complex, shortened for brevity, had a complex history, and to evaluate its assignment as a suture zone requires considerable discussion, as it was affected by great structural modification from its inception in the late Paleozoic to the present.

### Jinsha Accretionary Complex

The Jinsha complex consists of numerous different rock assemblages whose relationships are often uncertain. Unlike the other tectonic units of Plate 1, the Jinsha complex consists of mostly Paleozoic rocks that have been strongly modified during Mesozoic and Cenozoic time. It figures prominently in the tectonic evolution of SE Tibet, where it is interpreted to be the site of the former Paleotethyan ocean, and it consists of a collage of rock units assembled during late Paleozoic into Triassic time. It has been the locus of numerous later tectonic deformations, and thus much of its primary history has been overprinted, so it is difficult to unravel evidence of deformation related to its original collage formation from structures related to younger deformation. We describe below the complex in more detail than other areas we interpret as collage, because it is better exposed along the eastern side of the Qiangtang I unit than elsewhere; however, its tectonic assembly remains poorly understood.

The Jinsha complex within the area of Plate 1 extends from near the north end of Plate 1 southward, where it ends at the north end of the Dian Cang Shan (Fig. 1-1B). It consists of a highly varied assemblage of rock types that include upper Paleozoic strata with abundant mafic volcanic rocks and fewer lower Paleozoic strata, invaded by plutons and metamorphosed from greenschist to locally kyanite-sillimanite grade (Pzum and Pzlm, Plate 1). Some belts of mélange with serpentine and ultramafic bodies are recognized as separate rock assemblages, but how widespread they are is uncertain. In some areas the Paleozoic strata are in large coherent fragments up to tens of kilometers long. Rocks are locally metamorphosed to varying degrees and locally to a degree that their protoliths are unknown (m). The metamorphism, structural complexity, difficulty of access, and mostly older mapping have rendered the geology of the Jinsha complex rather poorly known.

The western boundary of the Jinsha complex is arbitrarily taken as the unconformity with overlying Mesozoic rocks of the Qiangtang I unit, but locally lower parts of the Triassic succession are considered to form part of the Jinsha complex. Also, erosional windows within the remainder of the Qiangtang I unit expose pre-Mesozoic rocks that are considered to be of an accretionary complex but are described separately (see below).

Unlike the other units discussed, the Jinsha complex is characterized by mainly Paleozoic rocks that lie below unconformities

---
[1]GSA Data Repository Item 2012297, Plates 1–4, is available at www.geosociety.org/pubs/ft2012.htm, or on request from editing@geosociety.org or Documents Secretary, GSA, P.O. Box 9140, Boulder, CO 80301-9140, USA.

at the base of Lower and/or Middle Triassic rocks, or below the unconformity at the base of the lower Upper Triassic rocks where the older Triassic rocks are missing (Plates 1 and 3 [see footnote 1]). These older Triassic rocks we have included in the Qiangtang I unit; thus the western boundary of the exposed Jinsha complex has a highly irregular western limit, in some cases of Lower and in other cases Middle or even lower Upper Triassic rocks, depending upon their relations with the more uniform deposition of Upper Triassic to Cretaceous strata that form the characteristic Qiangtang I unit succession. In many places the western boundary is a fault, but in others it is an unconformity; however, the general location of the boundary is reasonably clear on Plate 1. The eastern boundary of the Jinsha complex lies along a fault zone that consists of faults of different ages and is often hard to define (see above). The fault zone separates it from the Triassic rocks of the Yidun Unit and Paleozoic rocks of the Zhongza belt (Plate 1 and Fig. 1-1B). What makes the rocks of the Jinsha complex difficult to decipher is that they are variably metamorphosed, strongly deformed, and are intruded by numerous plutons. We discuss the rocks systematically from north to south and include Lower to lower Upper Triassic volcanic-bearing sections, because they form a transition between more chaotic rocks in the Jinsha complex proper to the widespread characteristic stratigraphic succession that forms the bulk of the Qiangtang I unit above.

At the northern end of the area of Plate 1 the Jinsha complex is missing along the WNW–striking, left-lateral, active Ganze fault, which separates the Yidun Unit and the Qiangtang I unit. To the south the Ganze fault terminates a more northerly striking fault that separates the western part of the Yidun Unit from the Qiangtang I unit. Farther south the more northerly striking fault cuts off the north end of a mélange belt (me), whose vergence is unclear, but we interpret but whose western part we interpret to be formed by west-dipping subduction, although the structures and rocks in the eastern part of the mélange may belong to the Yidun Unit to the east because of the apparent eastward gradation into rocks of the Yidun Unit where they are overlain unconformably by upper parts of the Yidun sedimentary succession; the origin of this highly foliated mélange remains unclear (see above). On the older maps this mélange belt is mapped as a unit of Silurian rocks; however, its characteristics are that of a mélange, and the rocks that yield older ages are interpreted as blocks within the mélange. To the south, directly on strike with these rocks, are fossiliferous Upper Triassic strata that contain numerous blocks of serpentinite, metamorphic rocks, basalt, and Silurian sedimentary rocks; thus we interpret these rocks as a part of the mélange that may be as young as Late Triassic. In the field these rocks show progressively greater foliation development and increased metamorphism to greenschist grade to the west, and they also grade eastward into Triassic strata of the Yidun Unit. In its southernmost part the mélange is unconformably overlain by Upper Triassic strata of the Yidun Unit (Plate 1). The age of mélange formation may extend from late Paleozoic into Triassic time, an interpretation that is favored below. In this segment of the Jinsha complex its eastern boundary is within the mélange. The overlap by Upper Triassic strata of the Yidun Unit indicates that closure of the Paleotethyan suture was Late Triassic. As discussed above, the mélange belt strikes south into rocks mapped as Paleozoic at the north end of the Zhongza belt, a relation that is ambiguous (it occurs at a map sheet join) and remains an unresolved problem.

West of the mélange belt, the main part of the Jinsha complex is exposed, lying unconformably below Lower and/or Middle Triassic volcanic-bearing rocks (Mz1 on Plate 1; this is the lowest of two major unconformities in these Triassic rocks; see Plate 3). Unconformities are present between Lower Triassic rocks and underlying rocks, and between Middle Triassic and Lower Triassic and Paleozoic igneous and metamorphic rocks, within which are several different fault-bounded units of Paleozoic and pre–Upper Triassic rocks (Plates 1 and 3). Paleozoic rocks (Pzum and probably m, Plate 1) below the unconformities are generally more metamorphosed, more complexly deformed, and intruded by numerous plutons, as are some of the older Triassic rocks. The lower Triassic strata (Mz1) are more continuous than the underlying Paleozoic rocks and generally form the lowest part of the Mesozoic Qiangtang I succession, but they are more deformed, with abundant volcanic rocks and some pluton intrusions. We interpret these rocks to be part of a transitional sequence that overlaps strata of the more fragmented Jinsha complex and indicates that activity within the Jinsha complex continued into perhaps early Late Triassic time. This assemblage of lower Triassic strata (Mz1) can be followed for ~150 km southward, where they are mostly unconformable on the older rocks (Plates 1 and 3).

Below the Triassic rocks are folded low-grade Lower to Upper Devonian clastic strata that contain felsic volcanic rocks in the Upper Devonian. They are unconformably overlain by Lower Carboniferous limestone that contains mafic and felsic volcanic rocks in its middle part. Both units (Pzum, Plate 1) are intruded by granite, tonalite, and diabase, which lie below the lower unconformity (T1, Plate 3) and are thus late Paleozoic in age. These plutons make up the western part of the pre-Triassic outcrops. Younger plutons intrude the Paleozoic and strata above the lower unconformity as young as early Late Triassic and may be the plutonic equivalents of the Triassic volcanic rocks they intrude. This assemblage of rocks is unconformably overlain by Upper Triassic strata (Mz2a) along the upper unconformity (T3, Plate 3), which appear to cap the youngest Triassic magmatic and structural activity in the Jinsha complex.

Southward for ~120 km the upper Paleozoic rocks are continued by a belt of metamorphic rocks (m, Plate 1) intruded by numerous plutons and assigned a Precambrian age on some maps. However, these schists, gneisses, and marbles are almost certainly Paleozoic in age, but of what age (late Paleozoic?) remains unknown. Within the middle of this belt of metamorphic rocks are large and small bodies of serpentinized mafic and ultramafic rocks that are the main evidence for an oceanic crust in this area. Intruded into the metamorphic rocks is a complex of mostly mafic and intermediate plutons. Many of these plutons also intruded the Lower and Middle Triassic rocks (Mz1, Plate 1) below the upper Triassic (T3) unconformity. The Lower and

Middle Triassic rocks contain abundant mafic and intermediate volcanic rocks, and the intrusive rocks are probably their plutonic equivalents. In many places they are unconformably overlain by the lower Upper Triassic (Mz2a, Plates 1 and 3), largely non-volcanic sedimentary rocks, thus limiting these plutons to an Early or Middle Triassic age. Rare plutons intrude into the Upper Triassic rocks above the upper unconformity.

The metamorphic rocks extend south of the intrusive plutonic complex, where metasedimentary rocks are present with abundant mafic volcanic rocks (Pzum and Pzlm). Most of these rocks are assigned a Permian age, although fossils are rare, and the evidence is considered equivocal. East of the belt of metamorphic rocks is a second belt of clastic rocks that contain large blocks of limestone of possible Silurian age (Ws, Plate 1). These clastic rocks are assigned a Carboniferous–Permian age (Pzum) on poor fossil evidence, and they continue south, where they include rocks assigned Silurian and Ordovician ages (Pzlm). This section is described to have Silurian rocks overlain by Permian strata, and, if true, they contrast greatly with the rocks in the northernmost part of the Zhongza belt (Yidun Unit) to their east (Plate 1). Near the north end of the Carboniferous-Permian rocks along the east side of the Jinsha complex a large klippe of the metamorphic rocks is shown lying above the Carboniferous-Permian rocks. If the mapping is correct, it is one of the few places where a major low-angle thrust has been mapped within the Jinsha complex. This part of the Jinsha complex has fewer plutons but also has local areas where the rocks have been metamorphosed to high grade, locally containing kyanite and sillimanite (see isograds on Plate 1). Also within this belt the Paleozoic rocks, largely Permian(?), contain abundant mafic volcanic rocks in their western part, and to the south contain small bodies of gabbro. The rocks are strongly deformed, with a well-developed foliation, and commonly they are pervasively sheared. Some areas are shown to have mappable units, but other areas contain strongly sheared rocks and may possibly be mélange, but how much of this belt is mélange remains unclear. Along this segment of the Jinsha complex, south of the mélange belt, its eastern boundary is marked by a west-dipping fault between the more metamorphosed and deformed rocks of the Jinsha complex in its hanging wall, and rocks of the Zhongza belt to the east.

We interpret the above described section of the Jinsha complex to have had a basement of upper and lower Paleozoic rocks, at least in its northern part, and it is characterized by arc magmatism of Permian to early Late Triassic age. The Permian part of the arc is manifested by the mafic volcanic rocks associated with Permian sedimentary rocks and plutons that are unconformably overlapped by Lower Triassic sedimentary rocks. The Lower and Middle Triassic volcanic rocks and the plutons that intrude them are interpreted to form the transitional units at the base of the Qiangtang I unit; the complex is overlapped by the unconformity at the base of the Upper Triassic sedimentary rocks of the Qiangtang I unit (T3, Plate 3). Small outliers of lower Triassic rocks to the south correlate with the overlap assemblage, suggesting that much if not all of the Jinsha complex was unconformably overlain by these rocks and that much of the deformation and metamorphism was pre-Triassic, a suggestion that needs confirmation. Part of the mélange that is present at the north end of the exposed Jinsha complex (me, Plate 1) is interpreted to be the accretionary prism in a west-dipping subduction zone and related to the late Paleozoic and to Lower to Middle Triassic magmatism to its west and in the units of the Jinsha complex described above.

Southward, the Lower Triassic rocks (Mz1) of the traceable transitional assemblage along the western side of the Jinsha complex end and become discontinuous, and most of the boundaries with the Paleozoic rocks are mapped as faults. Upper Triassic strata in many places rest unconformably (T3, Plate 3) on the Paleozoic rocks of the Jinsha complex. At Batang there is a large NNW–trending mafic pluton ($\omega_5^1$, Plate 1) that is mapped as intruding both the Jinsha complex and the western part of the Zhongza belt, the only pluton to cross this boundary. A small interior pluton to the larger body has yielded an old determined age of 190 Ma. Isograds of metamorphic rocks related to the pluton reach kyanite grade and also cross the boundary between the two units. If this mapping is correct, it indicates that major deformation within the Jinsha complex and movement on its eastern boundary fault had ceased at this latitude by the end of Triassic time, an important constraint on the regional tectonics of part of SE Tibet (see below). At the latitude of this pluton the upper Paleozoic rocks of the Jinsha complex can be divided into two sections, both metamorphosed to at least greenschist facies, a western section rich in mafic volcanic rocks, and an eastern section largely devoid of mafic rocks. The mafic-poor section forms lenses bounded by faults on both sides. Both sections are locally affected by high-grade metamorphism, containing kyanite and sillimanite. The western section consists of metasedimentary rocks with >2000 m of mafic volcanic rocks with rare limestone interbeds that contain middle Carboniferous fossils. With the paucity of fossils within this zone the ages of many of these mafic volcanic dominated sequences are uncertain. Southward the eastern volcanic poor section wedges out along faults, and from here southward the entire width of the Jinsha complex is occupied by upper Paleozoic rocks rich in mafic volcanic rocks.

In the middle of this belt at 29° N latitude, Upper Triassic rocks with a Qiangtang I stratigraphy (Mz2a) lie unconformably (T3, Plate 3) above the mafic volcanic dominated section (Pzum) along their east side, but they are overthrust by these Paleozoic rocks on the west side. This thrust has Paleogene rocks that lie unconformably on Upper Triassic strata in its footwall, demonstrating that some faulting within the Jinsha complex is Cenozoic, but the extent of Cenozoic deformation is uncertain because of the rarity of Paleogene strata.

In the same area along the eastern side of the Jinsha complex is a large area where a thick section of dark shales and siltstones is overlain by a thick sequence of mafic and felsic volcanic rocks. In the lower sedimentary part they contain lower Triassic fossils (Mz1, Plate 1). The Lower Triassic rocks contain large bodies of mafic intrusive rocks as well as some bodies of serpentinite. The northern and southern boundaries of these rocks are not clear, but at their southernmost boundary (Mz1) they lie unconformably (T1, Plate 3) on the Paleozoic mafic volcanic rocks, and the Paleozoic rocks appear here to be mélange. All these rocks are unconformably overlain by Upper Triassic strata (Mz2a, Plate 1; T3, Plate 3) similar to the larger area to the west that has a Qiangtang I Mesozoic stratigraphy. We have interpreted the Lower Triassic rocks to be the equivalent of the basal lower Triassic part of the Qiangtang I unit (Plate 1), and these outcrops suggest that they originally covered all the Jinsha complex. Both the Lower and Upper Triassic rocks are overridden by thrust faults on both sides: on the east side by Paleozoic rocks of the Yidun Unit (Zhongza) and on the west side by Paleozoic rocks with mafic volcanics of the Jinsha complex. At this latitude it appears that both the Lower and Upper Triassic rocks of the Qiangtang I unit also extended across the entire Jinsha complex and were in fault contact with the Yidun Unit on their east side.

To the south, and lying below the Lower Triassic rocks (Mz1) along the east side of the Jinsha complex, are metamorphosed clastic rocks with abundant mafic volcanic rocks and some chert and limestone (Pzum). Rare fossils indicate that some of the rocks are Permian. They are strongly sheared, contain lenses of serpentinite and one larger serpentinite body along the eastern bounding fault, and blocks of limestone (w, Plate 1), some with Carboniferous fossils. These rocks are partly or wholly mélange. They are intruded by several plutons of diabase to intermediate compositions. The ages of the plutons area shown as late Paleozoic and Triassic, but with no radiometric dates. The western pluton is unconformably overlain by Upper Triassic Qiangtang 1 strata (T3, Plate 3).

The west side of this belt of Upper Triassic strata that rest unconformably on the mafic volcanic–rich upper Paleozoic rocks (Pzum) is in the footwall of a second belt of upper Paleozoic mafic-rich volcanic rocks (Pzum) thrust eastward. Within this belt are several areas of Upper Triassic rocks that are mapped as klippen above the upper Paleozoic rocks; if this is so they are another, but rare, example of low-angle faults present within the Jinsha complex.

Along this central segment of the Jinsha complex the upper Paleozoic rocks with abundant mafic volcanic rocks have a long, N-S–trending belt of Qiangtang I Upper Triassic strata (Mz2a) in the middle, and a second belt on the west side that has thrust faults on both sides. West of the western belt of Mesozoic rocks, the upper Paleozoic rocks (Pzu) are different, as they lack the abundant mafic volcanic component. They have more the characteristics of the upper Paleozoic rocks that are exposed in the erosional windows below the western and northern parts of the Qiangtang I Mesozoic strata (see below). For this reason we place the western boundary of the Jinsha complex between these two Paleozoic sections. The separation is clear, but the contact between these two types of upper Paleozoic rocks is exposed in only one place, where it is an east-dipping fault. This is probably a fundamental boundary between upper Paleozoic arc sequences and a hinterland (western) unit of unknown character and extent.

The belt of Qiangtang I Upper Triassic (Ma2a) strata that separates the Paleozoic strata with abundant volcanic rocks into two parts can be followed south. Here, the Lower and Middle Triassic volcanic-bearing assemblages (Mz1, Plate 1), both in depositional contact below them (T3, Plate 3) and in footwall position to thrust faults above them, become more abundant, and all these Triassic rocks continue southward, where they end at the north end of the Dian Cang Shan (Plate 1). The lower Triassic sequence contains several lenses of serpentine and serpentinized ultramafic rocks, which may suggest that part of this sequence contains mélange. It also contains a N-S–trending belt of >2500 m of abundant intermediate and spilitic volcanic rocks with Late Triassic fossils (T3b, Plate 3) that lies in their eastern part. These spilitic rocks may represent an intra-arc rift similar to the one of the same age(?) that occurs in the western arc sequence of the Lanping-Simao Unit to the southwest (T3b, Plate 3; see below). The Paleozoic volcanic-rich rocks west of these Mesozoic Qiangtang I rocks are highly foliated, metamorphosed to amphibolite grade, and consist of mafic amphibolites whose protoliths are uncertain but are shown on Plate 1 as metamorphosed upper Paleozoic rocks (Pzu). They are bounded by steep thrust faults that dip below them from both sides, and they contain abundant small bodies of shared serpentinite in highly foliated rocks that appear in some places to be mélange. This belt of fault-bounded Paleozoic metamorphic rocks can be followed south, where they pinch out between Lower Triassic rocks on the east and a narrow belt of Jurassic red beds (Mz2b) with a fault on its east side that connects with the Lanping-Simao Unit on the west (Plate 1). The fault on the east side of these Jurassic rocks we interpret as one of the major faults within the region, and it plays an important role in the Cenozoic extrusion tectonics of southeastern Tibet (see below).

The belt of Paleozoic rocks on the east side of the Qiangtang I accretionary complex and east of the belt of Upper Triassic spilitic rocks (T3b) consists of Permian strata in the north, but southward it includes Carboniferous and Devonian rocks that form a north-plunging syncline at their south end (Plate 1). The contact between these Paleozoic rocks and the Triassic rocks to the west is intruded by a complex N-S–trending pluton for ~120 km. The pluton is shown as Triassic ($\gamma_5^1$) but with no geochronological data. This eastern belt of Paleozoic rocks is shown to have a greater abundance of clastic rocks than the same age rocks adjacent to them to the east that belong to the Yidun Unit, and the fault between them is taken as the boundary between the Jinsha complex and the Zhongza belt of the Yidun and Transitional Units. Like the Lower and Middle Triassic rocks to their west, the eastern belt of Paleozoic rocks, and the pluton that intrudes their contact, end against the eastern boundary fault, but ~75 km north of where the Triassic rocks of the Qiangtang I unit end.

At latitude ~28°10′ N the NW-striking Zhongdian fault offsets left laterally all the rocks in the eastern part of the Jinsha complex (Plate 1 and Fig. 2-10). South of the eastern part of the fault within the Triassic rocks (Mz1) of the Jinsha complex there is a well-developed mélange that includes abundant lenses of ophiolitic rocks that have been well described in the literature (e.g., see Mo et al., 1998). These Triassic rocks are shown on Chinese maps as Upper Triassic, but they lie unconformably below the characteristic non-volcanic basal Upper Triassic rocks of the Qiangtang I unit, and we group them with basal transitional rocks into the Jinsha complex of the Qiangtang I unit. Like most of the rocks within the Jinsha complex and parts of the transitional lower Mesozoic Qiangtang I unit, they contain sections that are mixed mélange and more coherent units.

In this southern part of the Jinsha complex, its eastern boundary fault zone is defined by faults east of which the rocks are generally older than those in the Jinsha complex. We take the boundary to be along an east-dipping fault in the north, with lower Paleozoic rocks of the Zhongza sequence in its hanging wall. Farther south it contains rocks as young as Triassic to the east, and the boundary fault is offset left laterally along the active Zhongdian fault. South of the Zhongdian fault, rocks on the east side are Devonian strata, and at the southern end are metamorphic clastic rocks of uncertain age (Pzl on Plate 1, pre-Cambrian[?]; see above) that lie below the lower Cenozoic Jinchuan basin strata. The eastern boundary fault continues south and displaces the Paleogene strata of the Jinchuan basin and thus had early and possibly late Cenozoic movement; however, as mapped, the fault has the same Cenozoic rocks (E, Plate 1) on both sides. The fault can be traced south to the north end of the Dian Cang Shan where it intersects the NE-striking Jinchuan fault of the late Cenozoic Dali fault system. This part of the boundary fault figures prominently in the regional tectonic analysis and will be discussed in more detail below.

*Evolution of the Jinsha Unit and Its Eastern Boundary Faults*

In nearly all regional interpretations a Paleotethyan suture of late Paleozoic or Triassic age is placed along the Jinsha complex (Mo et al., 1998). The evidence presented above suggests that there probably was a suture along this zone on the basis of the presence of mafic and ultramafic rocks, belts of mélange, arc volcanics, the chaotic structure, and the accretionary characteristics of the mapped rock units. However, from the discussion above it is obvious that the suture zone has had a complex evolution and has been strongly modified by younger structural events. The evolution includes a long complex history for the eastern boundary fault of the Jinsha complex with the Yidun Unit.

The age of possible ocean floor within the Jinsha complex is poorly known. Recent studies by Jian et al. (2009a, 2009b) have attempted to address the age of the ocean floor by sensitive high-resolution ion microprobe (SHRIMP) U/Pb zircon dating of a variety of rocks that occur as tectonically included blocks within a sheared matrix that they call *mélange* (Fig. 6-1). The samples dated are gabbro, trondhjemite, anorthosite, and granodiorite, which geochemically are interpreted to be fragments of rocks formed during seafloor intra-oceanic subduction and arc formation. Samples yield a wide range of ages for possible oceanic rocks from latest Devonian to Triassic, but Carboniferous and Permian ages are most common. The ages are from single samples and thus have a large uncertainty, but they do suggest that these rock fragments formed in oceanic settings. The granodiorite samples are from a belt of abundant volcanic rocks and are probably part of the Permian arc magmatism. Thus the Jinsha complex probably had one or more sites of oceanic and arc environments of late Paleozoic and possibly Early Triassic age, but the spatial and temporal relations of these rocks remain unclear. Post-collage structures have disrupted the relations between all the units, making it difficult to unravel their original relations without more detailed mapping.

We propose the following scenario for the formation of the Jinsha accretionary complex. The volcanic rocks of late Paleozoic age and Early to early Late Triassic age were part of a collage of accreted arc units within an east-facing arc system that included abundant plutons of late Paleozoic and early Mesozoic age, many of which may be the deep-seated equivalents of arc volcanic rocks (we here differ with a recent paper by Reid et al., 2005a, 2005b). Plutonic and volcanic rocks of these ages are not present in the Yidun Unit, where the volcanic and plutonic arc rocks are almost entirely Late Triassic in age. The polarity of the arc or arcs within the Jinsha complex remains difficult to determine, because parts of the arc systems appear to be missing or hard to recognize, such as forearc basins, subduction mélange, or high-pressure/low-temperature metamorphic rocks. An accretionary mélange belt may be present in the north, but this mélange has ambiguous relations to the Yidun Unit and the Qiangtang I unit to the east and west, respectively (see above), and it is difficult to follow to the south as a discrete unit. Although evidence for discontinuous mélange belts do exist farther south, these belts do not extend along the entire Jinsha complex nor at its eastern margin. The mélange in the north is bounded by west-dipping faults on both sides, and on its west side it is in direct fault contact with high-grade metamorphic rocks. In its southern part it is unconformably overlain by Upper Triassic strata of the Yidun Unit, which might suggest that the mélange, or part of it, belongs to the Yidun Unit; but because of extensive post-arc faulting, this relationship is not definitive. The mélange belt ends to the north where the Qiangtang I unit and the Yidun Unit are in fault contact along younger faults. From this mélange belt to the south, the Jinsha complex is in fault contact with the Yidun Unit, marked by the juxtaposition of different rock units of the Jinsha complex against a continuous belt of Yangtze-type rocks (Zhongza belt) that do not contain arc type units. At its southern end the Jinsha complex and the Lower and lower Upper Triassic volcanic belt of the Qiangtang I unit ends, and its southern continuation is not obvious. Both the Paleozoic and Triassic strata on both sides of the boundary fault are distinctly different, and it appears that two different terranes have been juxtaposed across this boundary fault, where few rocks define a subduction zone and/or its polarity. Several mélange belts occur to the west, but the nature and tectonic setting of these belts are unknown. They commonly contain tectonically included blocks of mafic and ultramafic rocks

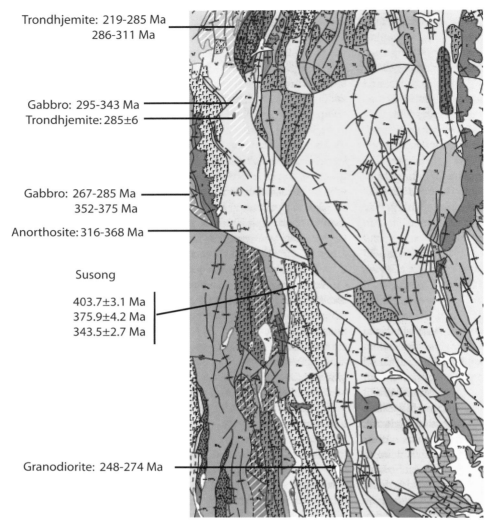

Figure 6-1. Distribution of sensitive high-resolution ion microprobe (SHRIMP) U/Pb zircon ages for blocks of possible oceanic and arc rocks within the Jinsha complex, presented by Jian et al. (2009a, 2009b). The samples yielded a wide range of ages from Late Devonian to Triassic, but the most common ages were Carboniferous and Permian, consistent with the known geology in this complex, which we interpret as a collage of different tectonic units and environments. These data suggest that the oceanic crust in this area is Devonian to Permian, and the oceanic rocks may have been formed in several different environments, separating different units within the collage. Post-collage structures may have disrupted the relations between all the units, making it difficult to unravel their original relations without more detailed mapping.

and serpentine, but whether they were derived from oceanic or arc crust, intra-arc rifts, or blocks incorporated into shear zones remains unknown.

We tentatively suggest that the Jinsha complex was assembled as a collage of fragments within a west-dipping subduction zone, and the complex and its Triassic cover were accreted to the Yidun Unit probably in Late Triassic time. Since then it has been strongly modified by younger deformation, such as that along the eastern boundary faults, where its eastern parts were thrust below the Yidun Unit, forming the east-dipping crustal section at the western margin of the Yidun Unit. At that time the eastern parts of the Jinsha complex disappeared below the Yidun Unit. If the Jinsha complex had been thrust below the Yidun Unit on a long-lived eastward subduction zone, evidence for the subduction within the Yidun Unit should be present, but all the major plutons and most of the volcanic rocks are present west of the boundary fault. The tectonic setting and modification of the Jinsha complex remains one of the major problems within southeastern Tibet.

The most important post-Triassic deformations within the Jinsha complex remain uncertain. Deformation within the Qiangtang I unit may be Late Cretaceous, Paleogene, and early late Cenozoic (see below), and the deformation within the adjacent Yidun Unit involves Paleogene rocks (see above). However, the lack of Upper Cretaceous rocks within the Jinsha complex and the Yidun Unit makes identifying Late Cretaceous structures difficult, and only rarely are Paleogene strata present within the Jinsha complex. Where Paleogene strata do occur they are commonly folded and displaced by faults, many of which have unknown displacements. Clasts within some of the Paleogene strata were locally derived, indicating that the deformation was Paleogene; however, the Paleogene strata are poorly dated, and thus some deformation may have been Neogene. These deformational events may be parts of protracted deformational events responsible for the major folding and faulting within the Qiangtang I unit and the Yidun Unit, and structures formed by these events must extend into the Jinsha complex; however, the data are limited but suggest that early Cenozoic deformation played an important role.

## Modification of the Eastern Boundary Faults of the Jinsha Complex

In the north the eastern boundary fault juxtaposes the Upper Triassic rocks of the Yidun Unit and the Qiangtang I unit, where both units contain volcanic rocks, but their associated strata are very different: flysch on the east, and shallow-marine rocks on the west. There is no direct evidence to explain when this juxtaposition occurred, except that it postdates the structures on both sides. The age of the structures on the east side of the Yidun Unit is interpreted to be Late Triassic, as they are cut by plutons some of which have yielded Late Triassic ages (Plate 1). The age of the structures on the west side in the Qiangtang I unit are poorly constrained. They deformed Upper Triassic rocks, but based on regional relations they could have been formed in latest Cretaceous, early Paleogene, or post-Paleogene time (see below).

Southward the eastern boundary fault strikes into the Jinsha complex, and the major fault displacement appears to be on the west-dipping thrust fault along the west side of the mélange belt (me). This fault parallels faults to its east that displace Paleogene strata, suggesting a post-Paleogene displacement. Farther south there is little information on the age of the fault boundary, as only Paleozoic to Upper Triassic rocks occur on either side of the fault, and post–Late Triassic displacement again juxtaposed different Upper Triassic sections of the Yidun Unit and the Qiangtang I unit.

In its southernmost extent the boundary fault displaces the Paleogene strata of the Jinchuan basin, clearly indicating that this part of the boundary fault has Cenozoic displacement, but rocks belonging to the basin are present on both sides of the fault, suggesting that Cenozoic displacement has not been large. If there was a large pre-Paleogene displacement, then the large mafic pluton and its metamorphic aureole near Batang, which cross the boundary, become very important. The age of this pluton is poorly constrained by an old K-Ar age of 190 Ma for a smaller internal pluton, but if it is Triassic, then large post-Triassic lateral displacement on this segment of the fault is prohibited.

Regionally this fault boundary was suggested by some workers (Leloup et al., 1995) to have a large, early Cenozoic left-lateral displacement and to be related to the extrusion of a large crustal fragment to the southeast that was caused by the indentation of India into Eurasia. There is abundant evidence that many segments of the boundary fault have Cenozoic displacement, but based on evidence presented above, displacements of this age do not appear to have been large. Large displacement could be pre-Paleogene, but the nature and regional importance of this fault boundary remain one of the major problems in this part of Tibet. In our interpretation the large displacements related to Cenozoic extrusion are on faults west of this fault (see below).

Overlying the Jinsha complex, the basement for the Qiangtang I unit along its eastern side, and extending westward to a long linear belt of Paleozoic and metamorphic rocks, the Qiangtang I unit includes all the Mesozoic and Cenozoic strata in the north-central part of the area of Plate 1 (Fig. 1-1B). The Qiangtang I unit extends southward into a series of narrow N-S–trending belts of Mesozoic and Paleozoic rock, where it appears to merge with the Lanping-Simao Unit. However, in our interpretation the two units are separated by a major N-S–striking fault zone, a relation to be discussed in detail below. The unit includes isolated exposures of Paleozoic rocks that occur in northwest-trending linear belts that lie below all the Mesozoic rocks, including the Lower and lower Upper Triassic rocks, with abundant volcanic rocks that lie above and mostly west of the Jinsha complex and all the Mesozoic and Cenozoic strata above them (Fig. 6-2).

The Lower and Middle Triassic rocks form a transitional sequence from the Jinsha complex below into the mainly nonvolcanic rocks of the Upper Triassic to Cretaceous-Paleogene strata of the Qiangtang I unit above. The Lower and Middle Triassic strata are included here in the Qiangtang I unit for reasons given above and are exposed in some of the erosional windows through the Mesozoic strata west of the Jinsha complex discussed above.

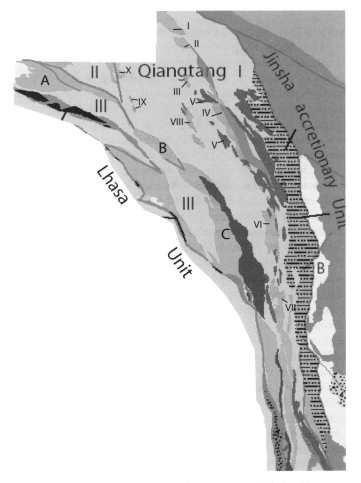

Figure 6-2. Locations of pre-Mesozoic exposures within the Qiangtang I unit. Roman numerals refer to references in the text to Paleozoic strata at these locations. A, B, and C are lens-shaped bodies of pre-Mesozoic rocks that separate Qiangtang I, II, and III units.

These Triassic transitional rocks extend beneath the eastern part of the Qiangtang I unit, but farther west the erosional windows expose Paleozoic and locally Lower Triassic rocks of different character. This suggests that the basement of the Mesozoic Qiangtang I unit may be a complex assemblage of Paleozoic rocks accreted together in pre-Triassic time. The Qiangtang I unit is defined mainly on the basis of its Mesozoic strata.

**Pre-Mesozoic Rocks of the Qiangtang I Unit West of the Jinsha Complex**

Within the broad expanse of Mesozoic strata in the northern part of the Qiangtang I unit are structures that brought up rocks older than the Late Triassic strata. The unconformity at the base of the Upper Triassic strata is present throughout the Qiangtang Unit (T3, Plate 3). The northern two uplifts (I, II in Fig. 6-2) are bounded by both thrust faults and unconformities with the overlying Upper Triassic strata, exposing metamorphic rocks of unknown protolith intruded by plutons shown as Triassic. The plutons intruded the adjacent Upper Triassic strata. In the southern of these two exposures the metamorphic rocks are unconformably overlain by Paleogene strata, with volcanic rocks similar to those at the north end of the Paleogene Gonjo basin. Farther south is another irregular exposure of similar metamorphic rocks entirely surrounded by unconformable and overlying Upper Triassic strata (III, Fig. 6-2).

All along the west side of the Paleogene Gonjo basin, lower and upper Paleozoic rocks are exposed (IV, Fig. 6-2). Fossiliferous Lower Ordovician fine clastics, with chert and some limestone (Pzl), were folded along NNE–trending axes before Lower Devonian limestone (Pzu) was unconformably deposited above them (Plate 1). Above lie more than 3000 m of Devonian to Upper Permian strata dominated by limestone, and with shale and terrigenous clastic rocks at the top (Fig. 6-3). The Paleozoic rocks were weakly deformed and eroded before the basal unit of the Upper Triassic strata was deposited unconformably above them (T3, Plate 3). The Mesozoic strata and their Paleozoic underpinnings were deformed together along NNW–trending axes before the Paleogene rocks of the Gonjo basin were deposited unconformably above (Plates 1 and 3). These Paleozoic rocks are not exposed east of the Gonjo basin, which suggests that the Gonjo basin may overlie a major pre-Paleogene fault. West of the Paleozoic rocks at the west side of the Gonjo basin are several exposures below the Upper Triassic unconformity of Lower and Middle Triassic strata and with some volcanic rocks intruded by plutons (V, Fig. 6-2). These rocks are only present locally below the unconformity at the base of the Upper Triassic, indicating an irregular distribution similar to the lower Triassic rocks that lie in the transitional sequence above the Jinsha complex to the east. The Lower and Middle Triassic rocks can be followed to the south in isolated exposures, where they merge with similar rocks that lie directly on the Paleozoic rocks of the Jinsha complex (Fig. 6-2). This belt of rocks and the exposures to the north indicate that some of the rocks similar to the Jinsha complex and the Lower Triassic transitional assemblage underlie at least the NE one-third of the Qiangtang I unit in the north.

Farther south is a large rectangular exposure of lower and mostly upper Paleozoic rocks (VI, Fig. 6-2). Lower and Middle Ordovician limestone with some sandstone and shale, ~1100 m thick, are the oldest rocks here and are in fault contact with fossiliferous Middle and Upper Devonian limestone (~500 m), overlain by 250 m of Lower Carboniferous limestone (Plate 1). Permian rocks make up most of this exposure and consist of a lower limestone unit overlain by sandstone, shale, limestone, and mafic to felsic volcanic rocks at the top (Fig. 6-3). This belt of Paleozoic rocks, like the ones to the south, is strongly folded and faulted on N-S trends, and both are unconformably overlain (T1, Plate 3) by the Lower Triassic volcanic-rich unit in one place, but more commonly they are unconformably overlain (T3, Plate 3) by the Upper Triassic non-volcanic section. These Paleozoic rocks are much less metamorphosed than similar-age rocks in the Jinsha complex to the east and contain fewer volcanic and plutonic rocks.

South of this large rectangular exposure the Mesozoic rocks of the Qiangtang I unit become separated into N-S belts by more exposures of the underlying Paleozoic rocks. One of these belts of Paleozoic rocks is metamorphosed to amphibolite grade and is well foliated and locally mylonitized (VIII, Fig. 6-2). This belt is associated with narrow belts of mélange, with some blocks of serpentinite and mafic and ultramafic rocks. These belts of Paleozoic and Mesozoic rocks are in the transitional area between the Qiangtang I unit and the Lanping-Simao Unit. How they may be related to the Lanping-Simao Unit farther south is discussed in greater detail below.

In the northwestern part of the Qiangtang 1 Unit are several more exposures of pre–Upper Triassic rocks, but only one exposure contains Lower and Middle Triassic transitional-type rocks locally at the base. About 30 km west of the Gonjo basin at Toma is another exposure of Paleozoic rocks lying unconformably below the Upper Triassic sequence (VIII, Fig. 6-2). At the north end are rare exposures of interlayered felsic volcanic and sedimentary rocks, which are assigned an Early Triassic age; they lie unconformably on Upper Permian rocks (Plate 1; Plate 3, T1). Below the more widespread unconformity at the base of the Upper Triassic sequence is Upper Devonian, Lower and middle Carboniferous, and Lower and Upper Permian rocks. Devonian strata consist of 525 m of limestone with some silty limestone, but the base is not exposed. Fossiliferous Carboniferous limestone is >1200 m thick and is overlain by fossiliferous Lower Permian limestone and silty and sandy limestone ~430 m thick. Upper Permian shale, sandstone, and coal are ~800 m thick and are unconformably overlain by Lower Triassic shale and felsic volcanic rocks. This Paleozoic sequence is remarkably similar to other Paleozoic sections to the south and east, but it contains only rare volcanic rocks at the top of the Permian section, so characteristic of the Jinsha complex. The Paleozoic rocks were folded along N-S axes before deposition of the Upper Triassic strata.

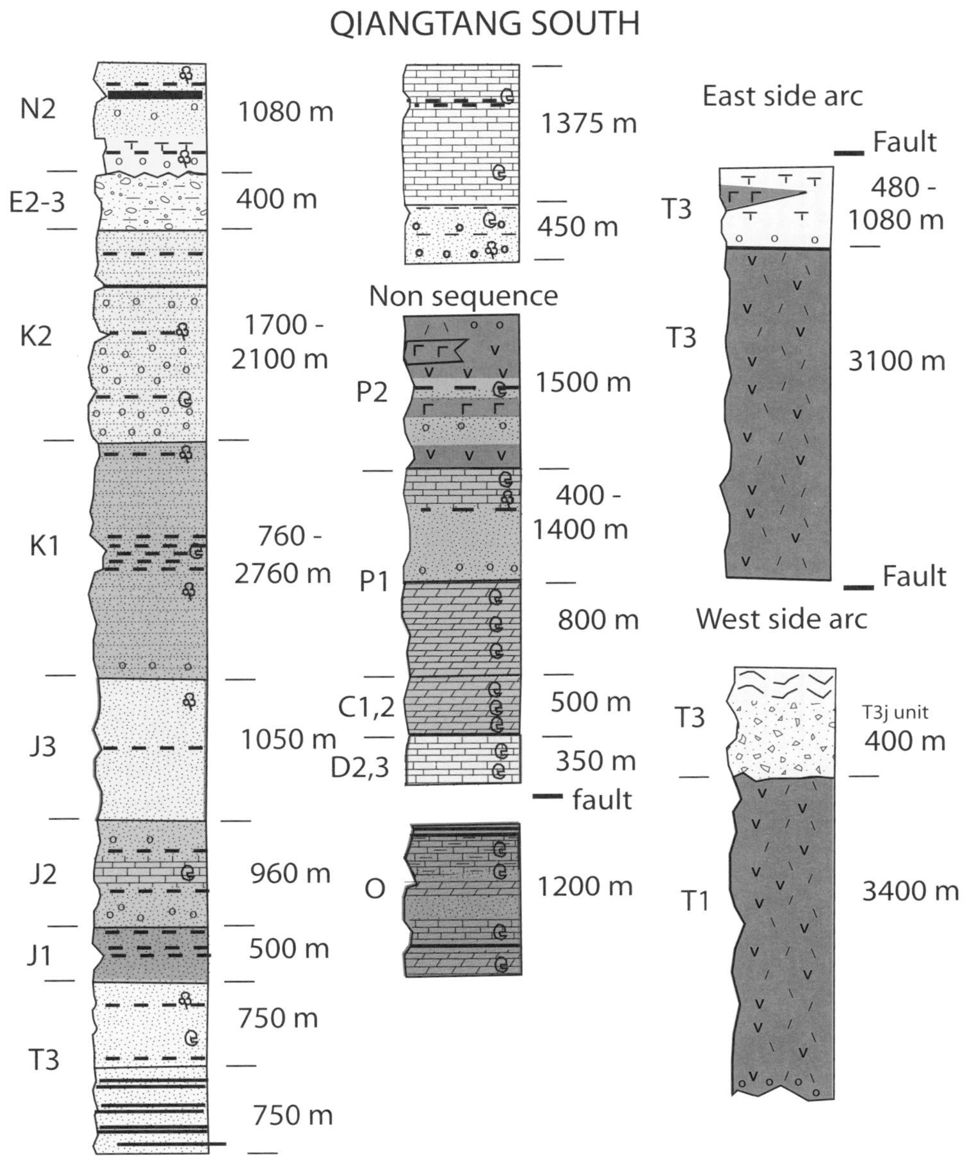

## QIANGTANG SOUTH

Figure 6-3. Stratigraphic section from the southern part of the Qiangtang I unit. (See Fig. 2-2, no. 13 for location.) The east and west side arcs refer to the volcanic rocks that are present on the two sides of the Qiangtang I unit (see cross section 20, Plate 4).

Two small areas of Carboniferous strata are exposed in the western part of the Qiangtang I unit (IX, Fig. 6-2), which consist of clastic and calcareous rocks. They contain only rare felsic volcanic rocks and were strongly folded before deposition of the surrounding unconformably overlying Triassic strata. These Carboniferous strata were refolded with Upper Triassic strata at a later time (Cretaceous or Paleogene; see below).

In the northwestern part of the Qiangtang I unit are exposures of upper Paleozoic rocks dominated by 2000+ m of mainly limestone of Carboniferous and Early Permian age, with rare felsic volcanic rocks; this section becomes more clastic in the upper part of the Permian section (X, Fig. 6-2).

All these sections of Paleozoic basement rocks for the Qiangtang I unit below the Mesozoic sequence are characterized by carbonate rocks, and they do not exhibit the complex geology or abundant magmatic activity in the upper Paleozoic rocks of the Jinsha complex. This suggests that the western limit of the Jinsha complex is characterized by a sheared mélange; by volcanic-sedimentary, metamorphic, and plutonic rocks; and by rare mafic and ultramafic bodies. This area probably lies along a NNW line along the Gonjo basin and east of the large rectangular exposure of Paleozoic rocks to its south (VI, Fig. 6-2). West of this line the Paleozoic strata have a simpler stratigraphy, are dominated by carbonate rocks, and are largely unmetamorphosed, with only rare volcanic rocks. These sections are hard to correlate with one another, but the upper Paleozoic rocks also differ from the dominantly thick clastic sections of age-correlative rocks in the Lhasa Unit to the west. Unfortunately detailed stratigraphic and paleogeographic reconstructions of these Paleozoic rocks has not been done, but their differences suggest a complex assemblage of somewhat disparate rock assemblages that include only rarely Lower Triassic rocks assembled into the basement on which the Upper Triassic strata of the Qiangtang I unit was deposited.

## MESOZOIC AND CENOZOIC STRATIGRAPHY OF THE QIANGTANG I UNIT

The Lower, Middle, and lower Upper Triassic transitional rocks were described briefly with the Jinsha complex above, but because they are the basal unit belonging to Mesozoic rocks of the Qiangtang I unit they are described in more detail here. This unit of Lower and Middle Triassic rocks (Mz1, Plate 1) lies between the unconformity at its base (T1, Plate 3) and an unconformity below lower Upper Triassic (T3, Plate 3) mainly non-volcanic sedimentary rocks above. The rocks between the two unconformities we interpret as a transitional assemblage between depositionally and structurally unstable conditions during formation of the Jinsha complex below, and a broader assemblage of Mesozoic strata deposited in a more stable environment above. It is the upper sequence of Mesozoic rocks that overlie them that are the most characteristic and that define the Qiangtang I unit (Figs. 6-3 and 6-4). The Lower, Middle, and lower Upper Triassic rocks consist of complex sequences of sedimentary rocks interlayered with abundant mafic and intermediate volcanic rocks, and which were often intruded by small plutons that range from diabase to granite. Mapped units within this succession are unconformities between the rock types (Plate 3) that are commonly highly variable in composition laterally. These units are unmetamorphosed except around the plutons, and are generally less strongly folded than the rocks of the Jinsha complex below that may have been locally metamorphosed to upper amphibolite grade.

The lower part of the Triassic sequence is best exposed around the northwest end of the Jinsha complex (Plate 1). Lower Triassic rocks consist of a lower unit of red conglomerate and sandstone with intermediate volcanic rocks and an upper unit of limestone, and both units contain fossils. They rest unconformably (T1, Plate 3) on upper Paleozoic granite that intrudes Devonian and Carboniferous rocks. The Lower Triassic rocks range from a few hundred meters to >1500 m thick. Middle and lower Upper Triassic rocks unconformably (T2 and T3, respectively, Plate 3) overlie both the Lower Triassic rocks and the Jinsha complex. In places existing maps show the contact between the Lower and Middle and the lower Upper Triassic rocks as conformable. Mapped relations suggest that these rocks may have been at least partly deposited in an extensional setting where fault blocks were tilted during deposition. The Middle and lower Upper Triassic rocks consist of limestone interbedded with sandstone, shale, and local conglomerate, and coal. The section contains abundant felsic, intermediate, and mafic volcanic rocks throughout.

Along the west side of the Qiangtang I unit is a thick volcanic-rich succession of >4000 m of felsic and intermediate volcanic rocks with some hypabyssal plutons, tuffs, ignimbrites, and agglomerates (Mz1a) with basalt (Mz1b) and volcanogenic sediments at the top (Mz1b) (Fig. 6-3). Southward, where these rocks are adjacent to the Lanping-Siamo Unit, they contain a basal sequence of fossiliferous sedimentary rocks of Lower and Middle Triassic age. These rocks form a large, NW-trending, lens-shaped outcrop everywhere fault bounded, except at its northern end, where the rocks lie unconformably on Carboniferous rocks (Pzu), on a contact that needs further study. They pinch out structurally to the south but reappear farther south as a narrow belt that can be followed continuously southward to the west side of the Lanping-Simao Unit (Plate 1). Mafic tuff at the top contains fossils of early Late Triassic age. The bulk of this section is assigned a Middle Triassic age. The main volcanic outcrops in this western area are intruded by plutons assigned a Triassic age, which are interpreted to be part of the plutonic roots of this volcanic assemblage. These rocks end abruptly to the north between two faults. We interpret this section of rocks to be separate from the volcanic belt to the east (Fig. 6-3) because they have a very different composition. Their importance lies in their connection to the Lanping-Simao Unit to the south and will be discussed in greater detail with that unit (see below).

Above the regional unconformity that lies at the base of the upper Upper Triassic rocks (T3, Plate 3) is a thick section of strata that range in age from Late Triassic to Cretaceous, with abundant red beds, which characterizes the Qiangtang I unit (Figs. 6-3 and 6-4). In contrast to the older Triassic rocks, they are generally lacking in volcanic rocks, except to the northeast, and they have much less lateral variation in rock types so that a similar sequence can be recognized across nearly all of the Qiangtang I unit (Figs. 6-3 and 6-4). This unit is characterized by a lower threefold sequence with (Mz2a, Plate 1): (1) a basal unit 500–700 m thick of red and yellow sandstone, siltstone, conglomerate, and rare volcanic rocks; fossils indicate a Late Triassic age for these strata, deposited in a mainly non-marine environment; (2) a gray to light-gray fossiliferous marine limestone that ranges from a few hundred meters to locally >1300 m thick; and (3) a unit of black to dark-gray marine mudstone, shale, argillite, and fine-grained sandstone that ranges from 1100 to 1700 m thick. The top of the Triassic section contains sandstone, shale, and mudstone up to 1200 m thick, with fossiliferous non-marine red beds that locally contain coal. Jurassic rocks (Mz2b, Plate 1) are 1500–3700 m thick and consist of red sandstone, mudstone, siltstone, and conglomerate. Fossils are rare, but this section is underlain by uppermost Triassic strata and contain a few fossils of Late Jurassic age at the top. The Jurassic strata are most widespread in the central part of the Qiangtang I unit (Plate 1), where Cretaceous rocks are only locally present. Cretaceous strata (Mz2c, Plate 1) consist of >3000–4000 m of red sandstone, conglomerate, mudstone, and siltstone and contain both Early and Late Cretaceous fossils, which are more widespread in the southern part of the Qiangtang I unit (Plate 1). In several areas the Mesozoic rocks are intruded by numerous small alkalic plutons of probable early Cenozoic age ($\gamma\delta\pi_6$ and $\varepsilon_6$) that are part of a widespread, early Cenozoic alkalic magmatism throughout southeastern Tibet.

Rocks assigned an Eocene age occur in many small and several large basins, but most of the strata do not contain fossils,

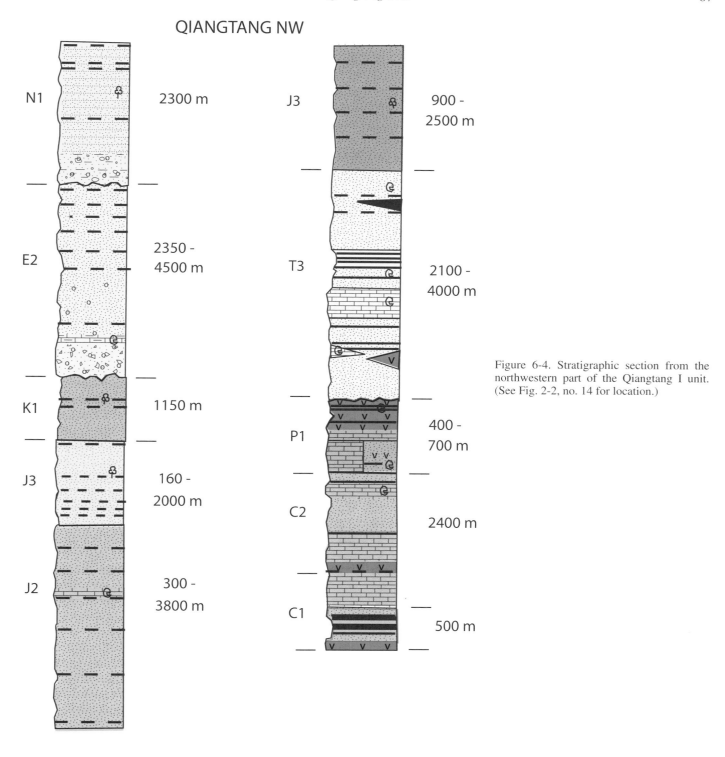

Figure 6-4. Stratigraphic section from the northwestern part of the Qiangtang I unit. (See Fig. 2-2, no. 14 for location.)

and their age is based on the fact that they are red sandstone, mudstone, siltstone, and conglomerate, with beds of lacustrine limestone, and they lie unconformably on all older rocks within the Qiangtang I unit and are correlative lithologically with similar deposits that are fossiliferous. They reach thicknesses of >4000 m in the basins in the northern part of the area (Fig. 6-4).

The Gonjo basin is the best studied of these basins (Plate 1) for which Studnicki-Gizbert at al. (2008) showed that this basin developed during syndepositional shortening. The conglomerate units along the east side show growth strata related to the west-vergent folding and thrusting along that margin. Locally, similar strata are present along the west side of the basin along east-vergent structures. At the north end of the basin, felsic volcanic strata at the top of the section yielded $^{40}Ar/^{39}Ar$ ages of 43 Ma (Studnicki-Gizbert et al., 2008).

The Eocene basins in the northernmost part of the area are particularly important, because they contain fossils of definitive Eocene age. They typically consist of red sandstone, conglomerate, mudstone, and some lacustrine limestone. Thicknesses given for the northwesternmost of these basins is >5000 m. As elsewhere, these Eocene basins in the north lie unconformably on folded Jurassic and Triassic strata. Cretaceous strata conformable with the underlying Jurassic strata lie close by, but are not overlain by, Eocene rocks, but do indicate that the pre-Eocene folding is post- or Late Cretaceous in age.

A small Neogene basin (N1, Plate 1) in the northwestern part of the area consists of nearly 3000 m of sandstone, conglomerate, and mudstone and contains Miocene fossils. Several small areas of fossiliferous Pliocene white tuff and tuffaceous sandstone are present in the southern part of the area, where they lie unconformably on folded Mesozoic rocks. Locally they reach 1000 m in thickness and are moderately folded.

## STRUCTURE OF THE QIANGTANG I UNIT

The structure of the Mesozoic and Cenozoic rocks within the Qiangtang I unit is reasonably clear; however, structural relations along the western boundary of the unit are complicated and remain poorly understood. The structure is best displayed in the multicolored Mesozoic and Eocene strata, but it also involves their Paleozoic underpinnings. Folds and faults trend generally N-S in the south and gently curve to NW to the north (Plate 1). The folds are mostly upright and show little consistent vergence, as do the faults that have general thrust separation (Plate 4, sections 15, 17, 18, 19, 41 [see footnote 1]). Toward the south the folds appear to be tighter and are locally isoclinal and even overturned. In places pre-Triassic rocks are exposed, some in the cores of folds or along thrust faults, but in some areas, as in the northeast, they crop out in erosional windows through broad simple upwarps. It is tempting to try and make the deformation within the Qiangtang I unit a fold-thrust belt detached at depth, but at what level it is detached, and how the pre-Triassic rocks became involved, is not easy to resolve (see below).

At least four periods of deformation can be documented within the Qiangtang I unit (Plate 3). The oldest deformation (or multiple deformations) occurred in the Paleozoic rocks below Lower and Middle Triassic rocks (Plate 3). The second deformation occurred within the Lower and Middle Triassic units (Plate 3), but the nature of that deformation is not well understood, with some indication from the distribution of sedimentary rock types that they may have been at least partly deposited in an extensional setting. The third and perhaps most important deformation occurred in latest Cretaceous to Eocene time, when the entire area of Mesozoic rocks in the Qiangtang I unit was folded into a wide fold-thrust belt. In places in the south the Qiangtang I unit's Mesozoic section is continuous without major unconformities from the Upper Triassic into the Upper Cretaceous and Paleogene rocks, where they unconformably overlie structures that involve Upper Cretaceous strata. In the north, where Cretaceous rocks are rare, a Late Cretaceous age of folding is probable, but it is not easy to directly confirm. The fourth period of deformation occurred during and/or after Paleogene time, when folding affected the Paleogene basins. In some basins, such as in the Gonjo basin (Studnicki-Gizbert et al., 2008), the folding can be shown to be syndepositional. How broadly this fourth period of deformation affected areas beyond the Paleogene basins is unknown, as the folds of the third and fourth periods are generally parallel: a question that remains to be resolved. Only a few Neogene outcrops are present in the area, and where they are the rocks are only moderately folded.

Farther south in the large rectangular exposure of lower and mostly upper Paleozoic rocks, the Paleozoic rocks, like a second large exposure to the south (VI and VII, Fig. 6-2), are strongly folded and faulted on N-S trends, and in the southern area they are metamorphosed locally to amphibolite grade (Plate 1). In both areas the Paleozoic rocks are locally unconformably overlain by both the Lower Triassic volcanic-rich unit, but more commonly they are unconformably overlain by the Upper Triassic non-volcanic section, clearly showing their pre-Triassic and pre–Upper Triassic periods of deformation that were contemporaneous with similar periods within the Jinsha complex to the east (Plate 3).

Understanding how much of this deformation is pre- or post-Paleogene is important for establishing its possible relationship with the India-Eurasia collision, but available constraints are sparse. Where Paleogene rocks are present, such a determination can be made, but over broad expanses of the Qiangtang I unit this is not possible. For example, folds west of the Gonjo basin trend NW and pass to the SE below the unconformity at the base of the Paleogene strata in the basin, and the Paleogene strata are folded along NNW–trending axes. Here the difference in age of folding can be established, but along the east side of the basin the folds trend N-S parallel to the major Paleogene structures in the basin. Farther to the east the folds continue to trend N-S across a width of 25–50 km, but whether they are pre- or post-Paleogene cannot be determined.

### Boundaries of the Qiangtang I Unit

The boundaries of the Qiangtang I unit, as the boundaries of most of the other tectonic units, are important in the regional tectonic interpretation of SE Tibet. The western boundary is taken to be the easternmost fault in a broad zone that consists of an anastomosing series of faults that enclose lenticular bodies of pre-Mesozoic rocks separating the Qiangtang I unit from the Qiangtang II and III units (Fig. 1-1B: It is important to point out here again that the main boundaries of units are shown on Plate 1, as a broad white line that can be hidden or turned off when necessary because it obscures some geological details. This fault zone has been referred to in the past as the *Langcang suture* (see, e.g., Li et al., 1996). However, nowhere along this boundary with the Qiangtang I unit are ophiolitic rocks or rocks that suggest the presence of an oceanic suture. That does not mean it is not an important fault zone, and we will stress its regional importance below. The faults that make up this fault zone are in many places mapped as steep faults, with mainly NE-vergent thrust separation, but in many places thrust faults along this fault zone are mapped as low-angle thrust faults vergent to the NE or E (e.g., sections 15 and 17, Plate 4). (As mentioned before, within Plate 1, low-angle thrust faults are not a commonly recognized feature for reasons that are unclear. In some places we have observed that mapping has generally overestimated the dips on some faults, but this cannot be the whole answer.) Of particular interest are the faults that bound the broad lens of the Middle and Upper Triassic volcanic rocks (Mz1a and Mz1b) east of the boundary fault. The fault along the east side is locally a low-angle thrust fault (Plate 4, section 15), and windows a few kilometers west of the fault expose the Jurassic strata in the footwall (Plate 1). The east side of the lens of Paleozoic rocks is marked by three thrust faults. The fault dips west, carrying Paleozoic rocks above the volcanic rocks, and near its north end it appears to be a folded thrust forming a half-window in its footwall, suggesting that the fault has a low angle with at least several kilometers of displacement. To the south are two other west-dipping thrust faults that might be part of this thrust system and are shown to be low angle. To the west, several narrow basins of Paleogene rocks are involved in isoclinal folds within the hanging wall of upper Paleozoic rocks that suggest that these thrust faults are early(?) Cenozoic in age, an important interpretation in the regional tectonic analysis presented below. Near the southern end of the Middle Triassic volcanic rocks, one large and one small klippe of Upper Paleozoic rocks are thrust over the volcanic rocks. It is assumed that they came from the west, but whether they are derived from the eastern part of the Qiangtang III unit to the west, or from the Lhasa Unit farther west, is unknown; knowing their derivation would be important in unraveling the southward trace of the Bangong suture (see below). A discussion of the southward continuation of the western boundary zone will be presented below.

West of the lens of Triassic volcanic rocks are large plutons and upper Paleozoic rocks that form another lens-shaped body with thrust faults that dip west on the east side, and east on the west side, where they are thrust onto the Qiangtang III unit.

The eastern boundary of the Qiangtang I unit has already been discussed, where it is taken to be the contact between the Jinsha complex and the Yidun Unit, a boundary that is cut out by young faults to the north and modified by numerous younger faults to the south.

The boundary between the Qiangtang I unit and the Lanping-Simao Unit is extremely important, not only for reconstructing the Mesozoic paleogeography but also for tectonic reconstructions and interpretation of the evolution of the Cenozoic intracontinental deformation. Both units are characterized by thick sequences of Jurassic and Cretaceous red beds, and they are often considered to be the direct continuation of one another. Unfortunately, as clearly shown on Plate 1, the critical area where they merge lies where all the units drastically narrow as they approach one another, and possible connecting rocks, while present, are reduced to narrow and highly deformed fault-bounded belts.

Taking the two fault zones along the west and east sides of the large lens of the Middle Triassic volcanic rocks (Mz1a) as the western boundary of the Qiangtang I unit (note: this will be modified; see below), they merge to the south, and the volcanic rocks end. At this locality, all the rocks both to the north and south are bent and offset left laterally along a prominent shear zone, but the shear zone does not extend very far to the NW or SE. Mesozoic and Paleozoic rock units along the eastern side of the shear zone continue southward but also make the same bend to the east. The dips in the Jurassic red beds that lie on both the north and south of this shear zone are subvertical and follow this bend continuously to the south. There is a break in the continuity of the Triassic volcanic rocks (Mz1a) at this bend, but they reappear ~20 km to the south, where they are much narrower. From where the Middle Triassic volcanic rocks reappear to the south of the oblique shear zone, they can be followed continuously several hundred kilometers south, where they are intruded by several small plutons, but here they lie along the western boundary of the Lanping-Simao Unit (Plate 1; Fig. 1-1B). Along this entire narrow belt the volcanic rocks are fault bounded on both sides. They end farther south, east of the middle part of a major mylonitic shear zone, the Chong Shan shear zone. The NW-striking, left-lateral shear zone mentioned above lies south of the western end of the late Cenozoic and active left-lateral Zhongdian fault (see above, and Fig. 2-10). The Zhongdian fault ends to the west within Paleozoic rocks in the western part of the Qiangtang I unit. We interpret these relations as a stepover in the left shear along this NW trend, but there is no evidence that this shear zone continues farther NW. Importantly, the Mesozoic red beds that lie east of the Middle Triassic volcanic rocks in the north belong to the Qiangtang I unit, but to the south they appear to belong to the Lanping-Simao Unit.

Southeast of the left-lateral shear zone, rocks of the Qiangtang I unit continue south, parallel to somewhat similar strata that we assign to the Lanping-Simao Unit. The area is easily distinguished on Plate 1 by the belt of Lower and Middle Triassic rocks (Mz1) that continue to the south, where they end at the north end of the Dian Cang Shan (Fig. 1-1B). For reasons discussed below, we take the narrow, N-S–trending belt of Jurassic red beds (Mz2b) that lies between two belts of Paleozoic rocks (Pzu), the eastern belt consisting of metamorphic and shared volcanic-rich rocks, to be the boundary fault zone. Because of this fault zone the eastern part of the Qiangtang I unit is separated from rocks typical of the Lanping-Simao Unit in this area, where the two units thin drastically, but appear, erroneously in our interpretation, to merge into each other. The relations in the transition area between the two units is complex, but it is our working hypothesis that the faults along the narrow belt of Jurassic red beds are part of major faults along which the Lanping-Simao

Unit moved far to the SE relative to the Qiangtang I unit during early Cenozoic time. These relations will be discussed in more detail with the Lanping-Simao Unit below.

## PRE-MESOZOIC ROCK LENSES SEPARATING THE QIANGTANG UNITS

Normally all the rocks northeast of the Bangong suture are assigned to a single Qiangtang Unit; however, because of major differences in the stratigraphy and structural relations of the rocks west of the Qiangtang I unit, as defined above, we have divided the Qiangtang Unit sensu lato into two additional units, Qiangtang II and III, but have retained the name *Qiangtang*. The separations between Qiangtang I and Qiangtang II and III are marked by large fault-bounded, lens-shaped bodies of largely pre-Mesozoic rocks that lie along an arcuate belt with a WNW trend in the north to a NNW trend in the south (Fig. 1-1B). The northern two lenses consist of metamorphic and plutonic rocks, whereas the southern compound lenses contain mostly upper Paleozoic rocks intruded by large plutons. The large area of Middle Triassic volcanic rocks discussed as part of Qiangtang I is also considered as an additional lens within this zone. The fault-bounded, lens-shaped areas of pre-Mesozoic rocks, including the lens of the Middle Triassic rocks, we interpret to be included within a major broad shear zone that extends from the western boundary of the Qiangtang I unit southward across the Bangong suture, the Lhasa Unit, and into the western side of the Lanping-Siamo Unit (see below). A discussion of this shear zone must await the discussions of all the tectonic units and will be covered in the regional synthesis.

The two northern lenses of pre-Mesozoic rocks consist of high-grade metasedimentary and igneous rocks (A and B, Fig. 6-2). On Chinese maps the metamorphic rocks are assigned different ages, ranging from Proterozoic to Carboniferous, and the ages of their protoliths remain uncertain. For example, in the northern lens (A), rocks north of the plutons in the middle of the lens are assigned a Proterozoic age on one map, but where they cross into the adjacent map to the southeast they are shown as metamorphosed Carboniferous. To the south of the plutons they are shown as pre-Carboniferous metamorphic rocks that consist mostly of strongly folded metamorphosed clastic sedimentary rocks with rare marble. The metamorphic and plutonic rocks are unconformably overlain by unmetamorphosed Triassic strata (Mz2a, Plate 1); thus their pre-Mesozoic age is supported.

In the southern lens the rocks consist of folded metaclastic sedimentary rocks with some marbles. The metamorphic rocks are intruded by several plutons that are assigned a pre-Devonian age on several maps, but when traced south they are assigned an Early Carboniferous age.

Farther south, all the lenses consist of rocks mapped as upper Paleozoic (Pzu on Plate 1; C, Fig. 6-2) metamorphic clastics and with marbles intruded by plutons. The metasedimentary rocks are strongly folded and foliated to varying degrees. Rarely they contain fossils that are Carboniferous, and on Chinese maps the rocks in these lenses are all assigned a Carboniferous age. They are intruded by large plutons shown as Triassic ($\gamma_5^1$).

The plutonic rocks in these lenses have yielded a wide range of old K-Ar and Rb-Sr ages. U-Pb ages for the plutons in the northern lens have yielded ages of 244 and 269 Ma, and the plutons are labeled ($\gamma_5^1$ and $\gamma_4^3$). From the next lens south, a U-Pb age of 210 Ma is shown on the Chinese geological maps, and thus they are shown as Triassic ($\gamma_5^1$). In the northern lens are plutons shown as Jurassic ($\gamma_5^2$), but on what basis is unknown. Information for judging the reliability of these data is not available.

The rocks in these lenses appear to be distinct from rocks to the east below the Upper Triassic unconformity (T3, Plate 3) in the Qiangtang I unit by their rare occurrences of volcanic rocks and the abundance of large plutons. Because most of the upper Paleozoic rocks are metamorphosed, their stratigraphy and correlation with other units remain unclear. The limited U-Pb data for the plutons indicate Permian and Triassic ages, which we consider to be important in the regional synthesis below, where the rocks in these lenses will be interpreted to be correlated with rocks in units much farther south (see below).

## QIANGTANG II UNIT

The Qiangtang II unit is a small, wedge-shaped area of Paleozoic to Eocene strata that lies west of what we interpret to be the northern continuation of the western boundary fault of the Qiangtang I unit and north of a large, lens-shaped area of pre-Mesozoic rocks to their south (Figs. 1-1B and 6-2). The unit becomes wider northwest of Plate 1, extending into central Tibet.

### Stratigraphy of the Qiangtang II Unit

The Mesozoic strata in the Qiangtang II unit have some major differences from the Qiangtang I unit to the east. In the northwestern part of the area of Plate 1, ~400 m of Upper Triassic felsic volcanic rocks (Mz$_{2aI}$), overlain by >600 m of sandstone, shale, and some limestone with Late Triassic fossils (Mz$_{2aII}$), unconformably overlie Carboniferous rocks (Fig. 6-5). The Carboniferous strata consist of ~3000 m of clastic rocks with some felsic volcanics and a unit of basaltic volcanic rocks in the lower part (Fig. 6-5). Both the Triassic rocks and the Carboniferous rocks are within a fault-bounded lens that we place within the Qiangtang II unit (Plate 1; Figs. 1-1B and 6-2). These Triassic rocks are different and contrast significantly with the non-volcanic Upper Triassic rocks to their south in the Qiangtang III unit, and to their east in the western part of the Qiangtang I unit. The contrast occurs across boundary faults and a structural lens.

Jurassic strata unconformably (Mz$_{2bI}$) overlie the Carboniferous rocks, but are in fault contact with the Triassic strata in the Qiangtang II unit (Plates 1 and 3), indicating a post–Late Triassic deformation that is not present elsewhere. The Jurassic rocks consist of ~700 m of fine-clastic strata with a basal conglomerate overlain by >600 m of interbedded sandstone and shale with some limestone interbeds. Both units are assigned a Middle

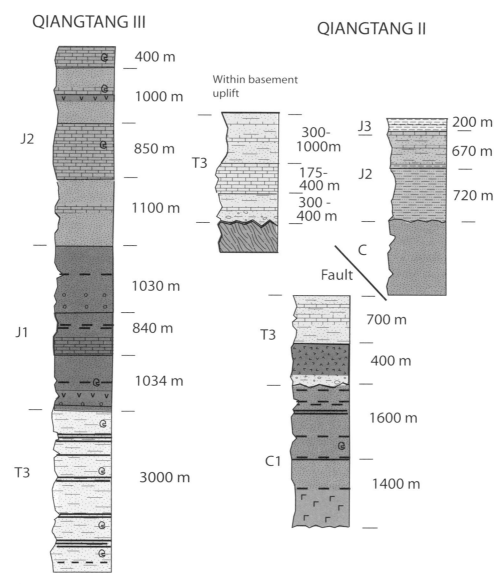

Figure 6-5. Stratigraphic sections for the rocks in the Qiangtang II and Qiangtang III units. (See Fig. 2-2, no. 15 for location.)

Jurassic age, but fossils are present only in the upper beds. The top of the section is 200 m of shale with rare limestone beds and is assigned a Late Jurassic age (Fig. 6-5).

The Upper Triassic strata that unconformably overlie the metamorphic rocks in the northern fault-bounded lens (A, Fig. 6-2: Mz$_{2a}$ in Plate 1) contrast with the volcanic section to their north; they are non-volcanic and begin with ~300–400 m of conglomerate at the base, overlain by fine-clastic strata with rare limestone beds (Fig. 6-5). Above lie ~400 m of limestone capped by >1000 m of clastic strata. All these rocks contain fossils, mostly plant remains, dating the sequence as Late Triassic. These Triassic rocks are difficult to assign to either the Qiangtang II or III unit and may have been transported laterally into their present positions (see below).

### Structure of the Qiangtang II Unit

The Carboniferous rocks (Pzu) within the Qiangtang II unit are strongly folded and are unconformably overlain by the Upper Triassic volcanic-bearing strata. Both rock sequences are folded together in a younger event. Jurassic strata unconformably overlie structures of this younger event (Plate 3). All of these rocks were folded again before deposition of the Eocene beds at the northeast margin of the Qiangtang II unit (Plate 1). The end result is folds and thrust faults that are consistently northwest trending and northeast vergent (Plate 1 and Plate 4, section 37b), but whether the northeast vergence is the result of only the last deformation that involved the Paleogene strata, or partly the result of older deformations, remains unknown. Several of the thrust faults in this folded zone are shown to be low angle, a rarity on most cross sections. The fault on the north side of the metamorphic rocks along the southern margin of the Qiangtang II unit is also north vergent.

## QIANGTANG III UNIT

The Qiangtang III unit lies south and west of the lenses of metamorphic and upper Paleozoic rocks that separate it from the Qiangtang II unit in the north and the Qiangtang I unit in the southeast. Its southern boundary is along a mélange belt (Jm) that we include with the Bangong suture rocks (see below). The unit consists mostly of Mesozoic rocks that are fault bounded and that structurally pinch out to the southeast (Fig. 1-1B); however, the area where they end is unclear, because it lies at the join of three map sheets that do not match across their boundaries. In the pinchout area are Paleozoic and some Mesozoic rocks that may belong to the Qiangtang III unit, but their relations are unclear, and their paleogeographic and structural positions will be discussed with the Bangong suture.

The southeastern boundary of the Qiangtang III unit also remains unclear. To the southeast there are two belts of Mesozoic rocks, a western belt of only Triassic strata ($Mz_{2a}$), and an eastern belt of mainly Jurassic strata ($Mz_{2b}$), overlying a very thin Upper Triassic sequence ($Mz_{2a}$) that in turn unconformably overlies Carboniferous rocks (Plate 1). At first glance it appears easy to place the eastern boundary of the Qiangtang III unit east of the Jurassic belt of rocks, although the relation may suggest that the boundary should be placed at the N-S–striking fault that separates the Jurassic and Triassic belts. The problem will be discussed below, because it depends on the stratigraphic interpretation of these rocks.

### Stratigraphy of the Qiangtang III Unit

Everywhere the boundaries of the Mesozoic rocks in the Qiangtang III unit are faults except for a short stretch along the southeast side just north of where they pinch out and where they are shown to lie unconformably above Carboniferous rocks. Along this stretch the basal unit of the Upper Triassic strata is a conglomerate with plant fossils and is overlain by ~300 m of fossiliferous limestone, followed by several hundred meters of sandstone and shale with some limestone and dolomite with fossils that indicate that the upper part is both marine and non-marine. This part of the section ($Mz_{2a}$) is similar to the basal Upper Triassic units in the Qiangtang I unit. However, the Jurassic strata ($Mz_{2b}$) lie directly on Upper Triassic limestone, and the argillite unit that forms the characteristic top of the Triassic section in the Qiangtang I unit is missing. The argillite unit is present west of the thrust fault along the west side of the Jurassic rocks, where a thick (3000 m), almost complete section of Triassic rocks is present with the overlying Jurassic strata that are present only in the northwestern part of the unit (Plate 1). The Triassic rocks west of the thrust fault are everywhere fault bounded, and no stratigraphic basement for them is exposed. This major change in Triassic stratigraphy suggests that the thin Triassic belt and its overlying Jurassic strata in the footwall of the thrust fault belong to another fault-bounded sequence. The Jurassic strata contain marine limestone >1000 m thick, so these rocks are tentatively assigned to a displaced part of the Qiangtang III unit. The reason for the major change in stratigraphy across the thrust fault may suggest that the fault is an inverted Mesozoic normal fault related to subsidence at a developing passive margin facing the Bangong ocean.

Within a wide belt through almost the entire the Qiangtang III unit is a thick succession of Upper Triassic strata ($Mz_{2a}$), and the section is well exposed and mostly complete. It consists of three units that are somewhat similar in character to the Qiangtang I unit: a basal unit of red, yellow, and tan sandstone, mudstone, siltstone, and conglomerate, overlain by a white to light-gray limestone, followed by a dark-gray to almost black siltstone, sandstone, mudstone, and shale that in places is an argillite. Rare felsic tuffs occur in the basal unit. The thicknesses of these different units vary greatly from a few tens of meters to >1000 m for the basal unit (Fig. 6-5; for the upper unit, the only unit shown in Fig. 6-5, these may be poor estimates because of structural thickening shown by their abundant mesoscopic folds). The lower unit is probably mostly non-marine, the limestone and lower part of the upper unit is marine, and the top part of the upper unit is partly non-marine. This section of rocks is superficially similar to the Upper Triassic of the Qiangtang I unit, but the uppermost dark-gray, fine-clastic unit is thicker and more argillitic. The main difference between the Qiangtang I and III units is in the thick marine Jurassic rocks in the Qiangtang III unit.

In the northwest the Jurassic strata conformably overlie the Upper Triassic rocks and consist of a lower part that contains >3000 m of sandstone, shale, and some limestone that contains Early Jurassic fossils. The Lower Jurassic rocks are overlain by >2000 m of limestone, some with nodular chert, and units of sandstone that contain Middle Jurassic fossils throughout (Fig. 6-5). In the southeast a thick section of clastic rocks is assigned to the Jurassic, which overlies Triassic rocks, but they are dominantly dark gray, and some of these rocks could be part of the Upper Triassic sequence. This sequence also contains abundant limestone that makes their assignment to the Qiangtang III unit possible. What makes this section different from the Qiangtang I unit is that the Jurassic rocks are marine and mainly gray to dark gray and contain abundant limestone, whereas the Qiangtang I unit Jurassic is mainly red beds. The change occurs across the belt of structural lenses of older rocks that separates the two units. This section represents a marine margin that thickens and deepens southwestward toward the Bangong suture.

Cretaceous strata are absent within the Qiangtang III unit, another difference from the Qiangtang I unit. Paleogene strata are rare and are present only as two small belts in the middle part of the unit (Plate 1), where they unconformably overlie the Jurassic rocks (Plate 3).

## Structure of the Qiangtang III Unit

With the lack of Cretaceous and the rarity of Paleogene strata, the ages of structures in this unit are hard to determine. The oldest structures are in the strongly folded metamorphic and upper Paleozoic rocks within the fault-bounded lenses that form the northern boundary of the Qiangtang III unit. They are intruded by plutons whose ages are poorly known. Plutons in the northern lens of older rocks have yielded U-Pb ages of 338, 300, 269, 253, 244, 204, and 196 Ma (ages are taken from the Chinese map sheets). The southern lens contains a pluton with a U-Pb age of 210 Ma. These data might suggest that most of the plutons are of Permian or Triassic age, a suggestion that will be used later in the regional interpretations. The plutons and structures in the lenses lie below the unconformity at the base of the Upper Triassic strata (T3, Plate 3) which gives an upper limit to their age.

In the northwestern part of the unit, Upper Triassic and Jurassic strata are strongly folded and faulted. The folds trend northwest, are often tight, and some are overturned to the SW (Plate 4, sections 17, 37a). Faults dip to the northeast and have dominantly thrust separation. The rocks become progressively younger to the southwest in this southwest-vergent fold-thrust belt. The belt of pre-Mesozoic rocks to the north is thrust to the southwest above the Mesozoic strata. The thrust is mapped as a moderately to steeply north-dipping fault. Along the southern part of the fold-thrust belt is a narrow belt of Upper Cretaceous red beds (see Plate 1, Mz3) that lie unconformably on Jurassic rocks that are part of a mélange belt (Jm) and are related to the closure of the Bangong suture (Plate 1; see below). Thus the southern boundary fault of the Qiangtang III unit is a post–Late Cretaceous, north-dipping thrust fault, but whether this dates the main deformation within this fold-thrust belt is unknown.

To the southeast the relations are almost reversed, with the Triassic strata ($Mz_{2a}$, Plate 1) present along the southwest side of the Qiangtang III unit, and Jurassic strata ($Mz_{2b}$, Plate 1) mainly on the northeast side. The rocks are strongly folded, and fold vergence is both SW and NE, but the thrust faults dip mainly west and are east vergent (Plate 4, section 13). The belt of pre-Mesozoic rocks to their east that belong to the belt of fault-bounded lenses is thrust southwest. Structure within the Lhasa Unit to the west shows that deformations occurred in the Jurassic, middle Cretaceous, Upper Cretaceous, and post-Paleogene time (see below), but whether this also applies to the deformations within the Qiangtang III unit is unknown.

There are only two rare outcrops of strata, with no fossils, that were assigned a Paleogene age within the southern part of the Qiangtang III unit Mesozoic strata. They rest unconformably on Jurassic strata and lie in the footwall of NE-vergent thrusts (Plate 1). Also, within the lens of upper Paleozoic rocks to their east is a long narrow belt of folded Paleogene red beds that rests unconformably on the Paleozoic rocks and is bounded locally by a west-dipping fault along its east side with apparent normal separation. Two slivers of Paleogene strata are present along the northeast side near the southern end of the northern lens of metamorphic rocks, where they are cut by the fault that bounds this metamorphic lens. The relative displacement on this fault is unknown, as the dip on the fault is not indicated. All these relations suggest pre-Paleogene uplift and erosion of the Paleozoic rocks relative to the Jurassic rocks of the Qiangtang III unit, and at least some post-Paleogene shortening, but also some post-Paleogene extension (see below). The ages of these events are poorly constrained.

The belts of older rocks along the east side of the Qiangtang III unit are dominantly thrust to the S or SW along their south and southwest sides and to the N or NE along the north or northeast side. The ages of these faults are unknown except for one place along the NE side of the northern lens of metamorphic rocks, where Paleogene strata are in the footwall of the bounding fault (Plate 1).

Within the Qiangtang III unit are a few small plutons that intruded the Mesozoic rocks and therefore are Mesozoic or Cenozoic in age. This is in contrast to the more extensive plutons that intruded the Lhasa Unit south of the Bangong suture, evidence supporting a southward subduction along the Bangong suture (see below).

The general tectonic position of the Qiangtang III unit represents the westernmost extension of the Qiangtang Unit (sensu lato), but it forms a west-facing margin along the Bangong suture. However, the abrupt change from the thick non-marine Jurassic and Cretaceous red-bed succession in the Qiangtang I unit to the Jurassic marine succession in the Qiangtang III unit perhaps requires careful interpretation. The change could be interpreted to be caused by extension during the Jurassic opening of the Bangong suture (see below), and while that may be true, we will argue below that this change takes place along the belt of faulted lenses of pre-Mesozoic rocks that may be interpreted as a zone of major Cenozoic strike-slip faulting, later modified by shortening. Like many of the units south of the Qiangtang III unit, when the Mesozoic rocks of the Qiangtang III unit are traced southeastward, they end against faults that mark the cryptic Bangong suture (see below). Even though the marine strata of the Qiangtang III unit appear to deepen southwestward into the Bangong oceanic mélange sequence, there are no transitional rocks to the deep-marine environment. We interpret this to mean that the fault at the southwest side of the Qiangtang III strata is a major fault interpreted below as a major strike-slip fault, a relation discussed in more detail below.

## MAJOR PROBLEMS IN THE QIANGTANG UNIT

### Validity of the Separation of the Qiangtang Unit (Sensu Lato) into Three Subunits

The Qiangtang Unit (sensu lato) is a compound unit that we have divided into three parts: Qiangtang I, II, and III. The reasons for this division are because each of these parts contains Mesozoic stratigraphic sections that are in some cases similar to, but mostly distinct from, each other, and the differences take place abruptly across faults or on opposite sides of the fault-bounded lenses of older rocks. Comparisons and paleogeographic study of these different units has not been done in this region and requires a major effort. In our interpretation of these units, although their boundary faults may have older displacement, and have been juxtaposed by major Mesozoic and mainly Cenozoic strike-slip faults, a hypothesis is presented below that requires extensive testing. Additionally, the nature of boundary faults between Qiangtang subunits and the Bangong suture also requires explanation.

The major boundary faults between the Qiangtang units mostly bound large, generally NW-trending lenses of pre-Mesozoic rocks. They consist of both high-grade and low-grade metamorphic rocks, intruded by abundant plutons. The rocks are unconformably overlapped by unmetamorphosed Middle or Upper Triassic strata so that their metamorphism and magmatism are older. The ages of these events is poorly known, and some of the plutons are clearly Paleozoic in age; others are assigned presumably a pre–Middle or pre–Upper Triassic age. The paleogeography of the protoliths of the metamorphic rocks is unknown, but if our interpretation is correct that they lie within a major left-lateral, strike-slip shear zone, the comparison of these rocks with other units southward requires a major regional study (see discussion below).

## Nature of the Basement of the Qiangtang Unit and Subsidence of the Mesozoic Basin

The Qiangtang Unit is underlain by what appear to be highly disparate pre-Mesozoic rocks. The Jinsha complex has been discussed in detail above and may underlie the northeastern part of the Qiangtang I unit (see above). Erosional windows through the Mesozoic rocks in the remainder of the Qiangtang Unit expose a wide variety of Paleozoic rocks, but how all these rocks relate to one another is unclear. Some appear to be parts of the Jinsha complex, but others are distinctly different. It remains for further study to show whether these Paleozoic rocks are part of a larger coherent terrane or belong to an accretionary complex of disparate elements.

The nature of the basement below the Qiangtang I unit would play a major role in understanding the causes of the subsidence and deposition of the thick section of Mesozoic strata, particularly the red beds within the Qiangtang I unit, which overlies the basement complex. Deposition began mainly in the late Late Triassic with a sequence that appears relatively uniform and composed of a similar unit of rocks, but variable in thickness, across the entire basin. The subsidence appears to have begun following the final accretion of the Jinsha complex, but the dynamics of the basin subsidence are unclear. The basin is large and mainly non-marine, and its limits are bounded by faults, but it is probably part of an immense basin that includes the Lanping-Simao to the south and possibly also may have somehow included parts of the Chuxiong basin in the Yangtze Unit. There is evidence within the Lanping-Simao Unit to the south for syndepositional extension (see below), but weaker evidence, mainly in the lateral thickness variations of rock units, is within the Qiangtang I unit. The subsidence of these regions of thick, largely non-marine deposition remains uncertain.

The subsidence within the Qiangtang III unit also began in the late Late Triassic with a similar succession, but it deepened rapidly in the Late Triassic with deposition of a very thick (~4+ km) sequence of argillitic rocks, followed by marine deposition of Jurassic strata. This subsidence is probably related to the formation of a passive margin, although the transition to deep water deposits is missing, along the north side of the Bangong ocean. It is distinctly different from the subsidence that formed the large non-marine depositional region in the Qiangtang I unit.

## Upper Triassic Volcanism within the Qiangtang I Unit

In the northeastern part of the Qiangtang I unit the Upper Triassic strata contain intermediate and felsic volcanic rocks (Plate 1). On Plate 1 they appear to be widespread, but the map pattern expresses thin units exposed over a broad area. Within this area, as well as within the Jinsha complex, some plutons intrude the Upper Triassic volcanic section. The source of the probable subduction-derived volcanism must be sought to the east, as there is no evidence for related subduction activity to the west at this time. Could the magmatic activity be related to spillover from similar age magmatic activity in the Yidun Unit, following accretion of the Qiangtang I unit and the Yidun Unit, or could this be the last gasp of westward subduction in the Jinsha complex prior to accretion of the two units? A third possibility is that the volcanism is related to extension during the beginning of basin subsidence, an interpretation we tentatively favor. The answer is unclear and remains a problem for study. It has important regional implications for the closure of the Jinsha oceanic terrane.

## Ages of Major Shortening within the Qiangtang Unit

The age or ages of shortening within all units in southeastern Tibet is a problem, because often the appropriate rocks to date the deformations are rare or poorly dated. Within the Qiangtang Unit the major shortening of post-Paleozoic age appears to have been Late Cretaceous, Paleogene, and/or post-Paleogene. The problem of how much deformation may have taken place in each event is important but difficult to evaluate; it becomes important when considering what deformation is pre– and post–India-Eurasia collision. Where Paleogene rocks are present such an evaluation can be made, but over broad expanses of the Qiangtang Unit this is not possible. For example, folds west of the Gonjo basin trend NW and pass to the SE below the unconformity at the base of the Paleogene strata in the basin, and the basin Paleogene strata are folded along NNW–trending axes, and syndepositional strata demonstrate that some of the deformation is Paleogene in age. Here the difference in age of the deformations can be established, but along the east side of the basin the folds and faults trend N-S, parallel to the major Paleogene structures in the basin, but farther east the structures continue with the same trend for 25–50 km, but whether they are pre- or post-Paleogene cannot be determined.

In the northern part of the Qiangtang I and II units, similar relations to Paleogene strata occur and present the same problem. There is scattered evidence of syn- and post-Paleogene deformation, but how extensive it may be remains unclear. From the limited data available it appears that Paleogene and post-Paleogene deformation may be very extensive, which has important consequences for the dynamics of the deformation and its relation to India-Eurasia intracontinental convergence. (Again it is useful to point out the interactive nature of Plate 1. The reader could call up just the Paleogene rocks on the map and examine their relations to mapped structures.)

# CHAPTER 7

# *Lanping-Simao Unit*

The Lanping-Simao Unit is a mostly fault-bounded unit between the Lin Cang and Yangtze Units in the south and, farther north, between the Baoshan and Tengchong Units to the west and the Transitional Unit, and the Qiangtang I unit to the east (Fig. 1-1B). It is characterized by a thick section of Triassic to Paleogene sedimentary rocks with some volcanic rocks in the Triassic part of the section. Jurassic to Paleogene red beds make up the greatest thickness of Lanping-Simao deposits and may total >8 km. The unit is ~150 km wide in the south and narrows to ~30 km to the north, where it appears to connect northward through a narrow transition zone with the Qiangtang I unit, but this connection may be more apparent than real, and it requires careful analysis and study because of its tectonic significance. Intrusive rocks are rare except in the southern part of the unit, where Jurassic and Cretaceous plutons are present, where plutons are associated with underlying Paleozoic rocks, and where many small alkalic, shallow-level intrusive rocks of probable Paleogene age are present throughout most of the unit. Pre-Mesozoic rocks are present within several erosional windows below the Mesozoic rocks, but the Mesozoic strata define the unit.

## PRE-MESOZOIC ROCKS AND THE BASAL UNITS OF THE MESOZOIC SEQUENCE OF THE LANPING-SIMAO UNIT

The oldest rocks within the Lanping-Simao Unit are Paleozoic rocks that occur in NW- to N-S–trending erosional windows along anticlines or thrust faults mainly in the south and southwestern part of the unit, and continuous exposures of Paleozoic rocks along the eastern side of the unit (Plate 1 [available on DVD accompanying this volume and in the GSA Data Repository[1]]; Fig. 7-1). In the eastern part of the Lanping-Simao the Paleozoic rocks are unconformably overlain by Upper Triassic strata (T3 on Plate 3), but in the western part of the unit, Lower, Middle, and lower Upper Triassic rocks are grouped together as Mz1 (Plate 1), and they commonly contain volcanic rocks and unconformably (T1, T2, and T3 on Plate 3) overlie Paleozoic strata.

Along the southeastern part of the unit are outcrops of graptolite-bearing shales and fine-grained sandstone, which range from Early to Late Silurian in age (Fig. 7-1, localities I, II, and IV; Pzl, Plate 1). These rocks are found in low-grade greenschist facies, are commonly sheared, and have a phyllitic to scaly foliation. Northward the belt of windows with Silurian rocks ends (IV, Fig. 7-1), and a 150-km-long belt of rocks to the east is shown on Chinese maps as Silurian because of the fossils they contain (V and VI, Fig. 7-1), and they are in fault contact with the Ailao Shan shear zone (Plate 1). These rocks, mapped as Silurian, contain abundant serpentinized ultramafic bodies, dikes, and sills of basalt and are associated with the Shaunggou mafic-ultramafic body, regarded by Xua and Castillob (1994), Yumul et al. (2008), and Jian et al. (2009a, 2000b; Fig. 7-2) as parts of oceanic crust of Carboniferous age (see below). The enclosing strata are shown on Chinese maps as Silurian because of fossils they contain; however, these rocks appear to continue northward by a belt of upper Paleozoic rocks that contain blocks with Late Permian fossils. The upper Permian blocks are within a scaly, sheared matrix that we regard as a mélange belt of late Paleozoic or early Mesozoic age, because the mélange is overlain unconformably by Upper Triassic rocks (see below). Thus we interpret the rocks shown on Chinese maps as Silurian (V and VI, Fig. 7-1) as part of the same mélange belt to the north of Late Permian or early Mesozoic in age. Both belts are shown with the mélange overprint on Plate 1, and the color and labels retain the original age designations of the Chinese units. The fossil content of these rocks suggests that the mélange may contain a wide range of Paleozoic rocks that have been incorporated into it, but the mélange is not Silurian, as shown on some maps. The exposures of Silurian rocks (Pzl, Plate 1) and adjacent areas of upper Paleozoic rocks (Pzu, Plate 1) and Mesozoic rocks are intruded by large plutons. The intruded plutons continue to the southeast into northern Vietnam, where they constitute a region of Indosinian deformation. The distribution of these rocks is confined to the southeastern part of the Lanping-Simao and may represent part of a separate fragment of largely pre-Mesozoic rocks that are part of a collage that underlies much of the Lanping-Simao Unit (an interpretation discussed in more detail below).

The windows in the SE part of the Lanping-Simao, but west of the windows of Silurian rocks (Pzl), contain Devonian to Permian strata (VII and VIII, Fig. 7-1). The Devonian rocks (500 m thick, Pzu) occur only in fault-bounded fragments adjacent to Silurian rocks and consist of fossiliferous graptolite-bearing shale, fine-grained sandstone, and rare limestone (Fig. 7-3). Carboniferous strata consist of ~500 m of fossiliferous limestone, with some coal-bearing sandstone locally. Lower Permian strata unconformably overlie the Carboniferous rocks and contain ~350 m of limestone at the bottom, overlain by >~1100 m of sandstone, shale, and limestone, with local limestone blocks and basalt, and

---
[1]GSA Data Repository Item 2012297, Plates 1–4, is available at www.geosociety.org/pubs/ft2012.htm, or on request from editing@geosociety.org or Documents Secretary, GSA, P.O. Box 9140, Boulder, CO 80301-9140, USA.

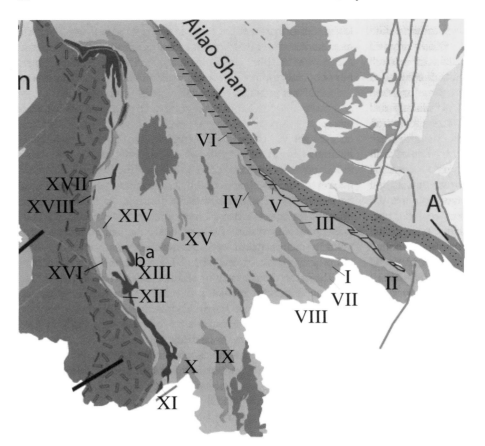

Figure 7-1. Locations of sections of Paleozoic rocks within the southern Lanping-Simao Unit, described in the text.

intermediate to felsic volcanic rocks. Upper Permian strata are 1100 m thick and are composed of non-volcanic sandstone, siltstone, and shale with some limestone at the top. They are unconformably overlain by Upper Triassic strata (T3 on Plate 3).

In the southwestern part of the Lanping-Simao Unit are two exposures of upper Paleozoic rocks (IX, X, Fig. 7-1), both framed or partly framed by a Lower Triassic volcanic assemblage that lies above an unconformity (T1 on Plate 3) and is at the base of Middle Triassic rocks that in turn are unconformably overlain by non-volcanic Upper Triassic strata (Plate 1, Mz1, and Mz2b, respectively) and by Middle Jurassic strata above the third unconformity (Plate 3). In exposure X (Fig. 7-1) Devonian clastics with abundant intermediate volcanic rocks are 2500 m thick and contain plant fossils, indicating a non-marine depositional environment. A thin sliver of Carboniferous limestone is in contact with the Devonian rocks, and they and the Devonian rocks are unconformably overlain by Lower Permian strata composed of limestone, sandstone, and shale with some coal at the top (350–800 m thick). Upper Permian rocks are sandstone, shale, and siltstone with rare intermediate tuff that forms its basal part, which is overlain by limestone and sandstone at the top. Upper Permian rocks are up to 2400 m thick and contain marine fossils in the upper part but non-marine fossils in the lower part. In exposure IX (Fig. 7-1) Carboniferous rocks are the oldest rocks and consist of ~1300 m of dominantly limestone with some sandstone at the base and top, and rare intermediate tuffs. In this window, Upper Permian strata, similar to those in exposure X, unconformably overlie Carboniferous rocks. The Paleozoic rocks in both exposures were folded in pre–Late Permian time and also in pre–Middle Triassic time (T2 on Plate 3). The Middle Triassic rocks that partially frame these two exposures contain volcanic rocks on the east side of the western exposure and along the western side of the eastern exposure, but not on its eastern side. This defines the eastern limit reached by the Early Triassic volcanism in the SW part of the Lanping-Simao Unit.

Within the convex west bend of the Lanping-Simao are numerous exposures of Paleozoic rocks, dominantly Permian in age, that follow the convex structural arc (XII, XIIIa and XIIIb, XIV, XV, XVI, Fig. 7-1). Older rocks are exposed only at localities XIV and XIIIa and XIIIb. Exposure XIV contains rare fault-bounded outcrops of Devonian strata that consist of ~1800 m of fine clastic rocks with rare Late Devonian fossils at the top. At their southern ends, exposures XIIIa and XIIIb contain Carboniferous rocks that consist of 450 m of limestone at the base, unconformably overlain by 800–1200 m of clastic rocks with intermediate volcanic rocks and agglomerate overlain only in exposure XIIIa by Lower and Upper Permian strata. In the smaller of the two windows (XIIIb, Fig. 7-1) Middle Triassic non-volcanic rocks unconformably overlie the Carboniferous strata and define an eastern limit to most of the Middle Triassic rocks and the eastern limit for

Upper Triassic volcanic rocks that lie above Permian strata in exposure XIV. In exposure XIIIb, Lower Permian sandstone and shale (2200 m) are overlain by 1800 m of limestone and Upper Permian strata that consist of 250–630 m of sandstone, siltstone, and shale, overlain by up to 700 m of limestone. The narrow exposure farther east (XV, Fig. 7-1) contains only Permian strata.

To the west of exposures XII and XVI, only Permian strata similar to those in the windows to the east are present, except along the border with the Lin Cang Unit. Here the Permian rocks are strongly sheared in a zone of regional importance (Lancang shear zone, see below), and within the shear zone are faulted slivers of metamorphic pendant rocks from the Lin Cang Unit to the west. Rocks along this fault zone are commonly mylonitized. The Paleozoic rocks in the western exposures are unconformably overlain by mainly Upper Triassic felsic and intermediate volcanic rocks and rare Middle Triassic volcanic strata below (Plates 1 and 3).

A single elongate mafic-ultramafic pluton, the Banpo complex, near the western border of the Lanping-Simao Unit, appears to have intruded the Permian and Triassic volcanic rocks and is unconformably overlain by Jurassic red beds and thus appears to be Triassic in age (Fig. 7-2). This body is mainly gabbro ($\beta_5^1$, Plate 1) and was studied by Jian et al. (2009a), who reported it to intrude Permian rocks and to be unconformably overlain by Upper Triassic strata. This is different from Chinese maps, which show it intrusive into Lower Triassic rocks to the north. The ages obtained from tonalite (285.8 ± 2 Ma) and gabbro (285.6 ± 1.7 Ma) samples within this complex are consistent with a Permian age for these rocks, and the intrusive contact shown on Chinese maps with Triassic rocks may well be related to post-plutonic reactivation of these serpentinized rocks. Not only is this a rare gabbro pluton; it also contains numerous small bodies of serpentinized rocks. These serpentinized bodies have been used in the past to support the interpretation that the fault zone along the west side of the Lanping-Simao Unit is an oceanic suture zone (the Lancang suture) or is related to volcanic arc magmatism from a westward-dipping subduction zone that lay to the east (see Jian et al., 2009a). However, their close association with the large plutonic complex to the west in the Lin Cang Unit (see below) might suggest that they are more directly related to the magmatism in the Lin Cang Unit to the west rather than to remnants of oceanic crust. One of the peculiar features of the Lanping-Simao Unit is its general lack of Mesozoic plutonism, except in the southeast, as pointed out above, particularly when it is adjacent to the large Lin Cang batholith to its west (see below).

Permian strata within the same general N-S–trending belt of Paleozoic rocks continue northward in narrow exposures within the Mesozoic rocks, and particularly along a regionally important shear zone, the Lancang shear zone, at the boundary between the Lin Cang and Lanping-Simao Units. This shear zone appears to step to the east and follow a narrow belt of Permian rocks to the north where the Permian rocks associated with a few small plutons end within the Lanping-Simao Unit, but the fault zone and the rocks on both sides continue and enabled the prominent convex north arc to continue to the west. The volcanic rocks west of the shear zone are as young as Late Triassic and are intruded by Lin Cang plutonic rocks, indicating that some of the plutons of the Lin Cang Unit are as young as Late Triassic and may be the coeval, deeper seated parts of the Triassic volcanic sequence. New work (2004, unpublished, by the Chengdu Institute of Geology and Mineral Resources) indicates that some of these volcanic rocks are as young as Early Jurassic, but these younger parts of the volcanic sequence lie to the south and are not shown to have been intruded by the plutons (Plate 1). The volcanic rocks to the east, as well as the volcanic and plutonic rocks to the west, of the shear zone lie unconformably below Middle Jurassic rocks (J2 on Plate 3). To the east the Middle Jurassic rocks above the unconformity are at the base of the thick succession of Jurassic to Cenozoic red beds of the Lanping-Simao Unit, but to the west the Middle Jurassic sedimentary rocks rest unconformably on the Upper Triassic (also the newly dated Lower Jurassic; Plate 3) volcanic rocks, and also the Lin Cang plutons within synclinal and fault-bounded, N-S–trending belts that may have been synsedimentary grabens or half grabens. The basins south and west of the shear zone contain rocks that are as young as Early Cretaceous (Plate 1), but the Mesozoic succession is not nearly as thick as correlative rocks in the Lanping-Simao Unit. Importantly, these volcanic rocks are exposed on both sides of the shear zone that separates the Lanping-Simao and Lin Cang Units, a relationship that has regional significance and that will be discussed below.

The Lancang shear zone forms the boundary between the Lin Cang and Lanping-Simao Units and makes a second horseshoe bend convex to the west, where it forms the northern boundary of Triassic volcanic rocks; it continues to the west, where it bounds the Lin Cang and Changning-Menglian Units on the north (Plate 1). Northeast of the bends in the Lancang shear zone is a large arcuate anticline, convex to the north, cored by a thick sequence of low-grade, foliated metamorphosed sandstone, shale, and rare limestone that forms part of the Wuliang Shan (Plate 1, Pzm). The rocks are dark gray and so are unlike most of the strata in the Paleozoic sections within the Lanping-Simao. The age of these unfossiliferous rocks is unknown, but they are generally interpreted to be Paleozoic rocks that underlie metasedimentary Mesozoic rocks correlated with the Triassic strata of the Lanping-Simao. The only rocks within the Lanping-Simao that even remotely resemble these metamorphic clastic strata are the Silurian rocks in the southeastern part of the unit. At their southern end the pre-Mesozoic rocks core a southward-plunging anticline and are overlain unconformably by Middle Jurassic strata of the Lanping-Simao Unit. To the north and west the Paleozoic(?) rocks separate into two parts; the northern part forms an anticline that plunges NW beneath Upper Triassic(?) strata, and the southern part ends to the west against a fault. The age of the Triassic rocks here is queried, because they are weakly metamorphosed and without fossils, and their age is based on their lithological similarity to nearby dated Triassic strata. From here northward, only one small area exposes any Paleozoic rocks,

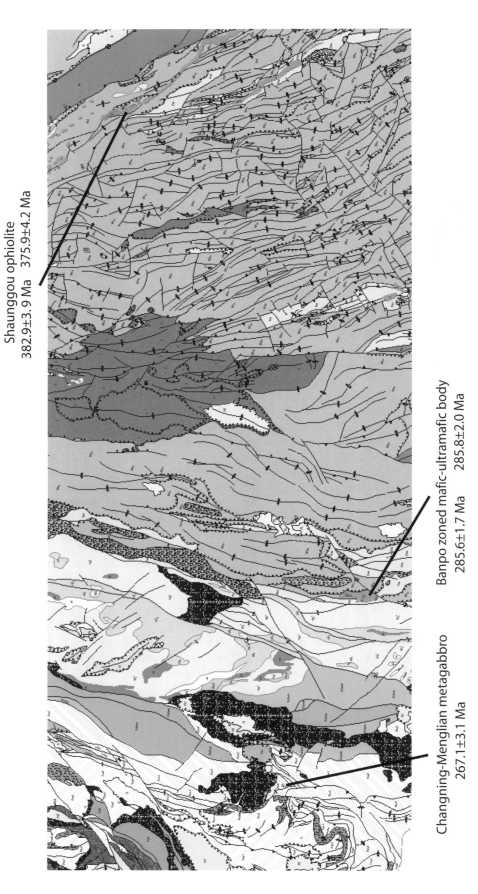

Figure 7-2. Locations and ages of mafic and ultramafic rocks, shown in the southern part of Plate 1 and mentioned in the text.

and here is where Upper Permian sandstone and slate crop out in the northern part of the Lanping-Simao Unit (Plate 1).

In summary the Paleozoic rocks that underlie the Mesozoic rocks in the Lanping-Simao Unit consist of incomplete sections of (1) Silurian deep-water shale and sandstone that may be in part mélange and which are present only in the southeastern part of the unit, (2) fragmentary sections of clastic Devonian strata, (3) Carboniferous limestone and clastic rocks that in the west contain thick units of intermediate and felsic volcanic rocks and agglomerate in their upper parts, and (4) Permian strata with thick units of clastic and carbonate rocks with some volcanic rocks (Fig. 7-3). In some places there is evidence for Late Permian deformation, as constrained by an unconformity within the Permian rocks. The Carboniferous limestone is similar to Yangtze Unit rocks, although the volcanic rocks at the top are dissimilar. The Permian rocks lack any sign of Emei Shan basalt,

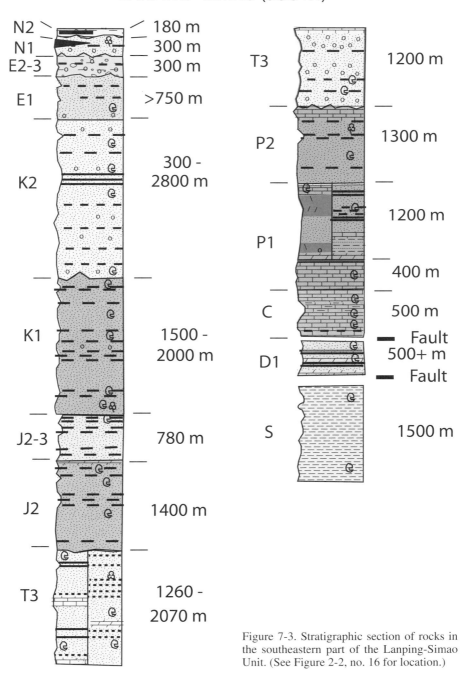

Figure 7-3. Stratigraphic section of rocks in the southeastern part of the Lanping-Simao Unit. (See Figure 2-2, no. 16 for location.)

one of the most characteristic units of the Yangtze platform east of the Lanping-Simao Unit. A belt of mélange that extends continuously for several hundred kilometers along the northeastern side of the unit has yielded fossils as old as Silurian and as young as Permian. These rocks we regard as having formed a late Paleozoic mélange and will be discussed in detail below. All the pre-Mesozoic rocks in the Lanping-Simao contrast with rocks in the Jinsha complex to the north, where Carboniferous volcanic rocks do occur, but the Permian rocks are dominated by marine mafic volcanic rocks many of which appear to be related to arc volcanism. Around the exposures of Paleozoic rocks in the eastern part of the Lanping-Simao Unit are unconformably overlying Upper Triassic strata (T3 on Plate 3), which only rarely contain volcanic rocks in the south. In the west, however, Paleozoic exposures are commonly framed by Lower to upper Upper Triassic rocks that commonly contain abundant volcanic rocks. In the western exposures of Triassic strata, the eastern limits of the abundant Triassic volcanic rocks can be determined, and they do not extend below the central and eastern parts of the Lanping-Simao Unit.

The importance of the Paleozoic rocks in their limited exposures is that they expose some similar and some different sequences of Paleozoic rocks, a pattern that is similar to the Paleozoic rocks exposed in the Qiangtang I unit to the north. Furthermore, the geometry of the unconformably overlying lower Mesozoic rock units suggests the cutting out of different parts of the lower Mesozoic units in a way that suggests a terrane of fragmented crust underlying the main part of the Upper Triassic to Cretaceous sequence of red beds, disrupted partly by extensional tectonism, which may explain the Mesozoic subsidence. The variability of the Paleozoic rocks may also be interpreted to be the result of disparate accreted fragments similar to those discussed for the Qiangtang I unit (see above).

## MESOZOIC ROCKS OF THE LANPING-SIMAO UNIT

Like the Qiangtang Unit, the Lanping-Simao Unit is largely defined by Mesozoic strata. The most characteristic rocks are the Jurassic to Paleogene strata that are largely non-marine red beds and are mostly devoid of volcanic rocks (Figs. 7-3 and 7-4). Similar to the Qiangtang I sequence, there are three important parts of the Mesozoic succession. The lower part consists of Lower, Middle, and lower Upper Triassic strata that in the southwest contain volcanic rocks and are separated by unconformities. New work suggests that some volcanics also are present in the Lower Jurassic sequence in the SW part of the area adjacent to or within the NE part of the Lin Cang Unit (Chengdu Institute of Geology and Mineral Resources, 2002). These volcanic rocks crop out mainly in the southwestern part of the unit. Only rare volcanic rocks are present in Upper Triassic strata in the southeastern part of the unit (Plate 1). Within the Lin Cang Unit to the west the volcanic rocks are very thick, but they thin greatly across the Lancang fault zone and pinch out to the east (Fig. 7-5). Rocks of this age within the central part of the Lanping-Simao Unit are non-volcanic. Above an unconformity at the base of the lower Upper Triassic strata (T3 on Plate 3), volcanic rocks are rare, and this part of the section has some similarity to the Qiangtang I unit, with which it is often directly correlated, but it also has some differences. At the base of Middle and some Lower Jurassic strata is a major unconformity (J2 on Plate 3), above which occurs the thick sequence of red beds. There are other, but more local, unconformities that will be discussed when appropriate, but the three mentioned unconformities (T1-T2, T3, and J2 on Plate 3) divide the section into three important sequences throughout the Lanping-Simao. The presence of the unconformity at the base of the (Lower) Middle Jurassic rocks is different from that present in the Qiangtang I unit to the north, where an unconformity at this level is known but is less regional (Plate 3).

The Paleozoic rocks are overlain in the west by strata that are dominated by mafic and felsic Lower, Middle, and Upper Triassic volcanic and volcanoclastic rocks (perhaps extending into the Lower Jurassic (Fig. 7-5). The volcanic rocks pinch out to the east, where Middle and mainly Upper Triassic sequences are non-volcanic, except locally in the south, along the northeastern margin of the Lanping-Simao (Plate 1). In general, except for the volcanic rocks in the southwestern part of the Lanping-Simao, the Mesozoic rocks are largely non-volcanic with few exceptions (see below). The distribution of volcanic rocks in the southwest suggests lateral thickening and thinning, and this is particularly well expressed below the overlying Lower and Middle Jurassic unconformities; below these unconformities different rock units are present, suggesting possibly tilted and rotated blocks in an extensional structural environment during deposition of the Triassic rocks. Such an interpretation needs to be demonstrated as one possible mechanism for the great subsidence during the Mesozoic in the Lanping-Simao; a similar suggestion was made for the Qiangtang I unit. However, there is little evidence for continued extension during the principal long period of Lanping-Simao deposition.

Triassic rocks all along the northeastern side of the southern part of the Lanping-Simao Unit consist mostly of Upper Triassic sandstone, conglomerate, and siltstone that are mostly non-volcanic (Fig. 7-3). The only exception is in the far SE of the unit, where some intermediate volcanic rocks are interbedded with sedimentary rocks (Plate 1). Lower Triassic rocks are missing, and rare Middle Triassic rocks form a narrow, discontinuous belt in the eastern part of the Lanping-Simao adjacent to the Ailao Shan metamorphic belt (Plate 1). These rocks consist of dark slates and fine-grained clastics that are sheared and may be mélange (Mz1 with mélange pattern, Plate 1, and see below). Also in this narrow belt are large areas of fault-bounded fragments of deformed red beds. Both of these Middle Triassic rock types are unconformably overlain generally by less deformed coarse clastic and some conglomeratic Upper Triassic strata that may reach a thickness of >1000 m (Plates 1 and 3; Plate 4, section 31 [see footnote 1], where they are shown only as folded Middle Triassic, T2). Upward, the unconformably overlying Upper Triassic rocks become generally finer grained, dominated

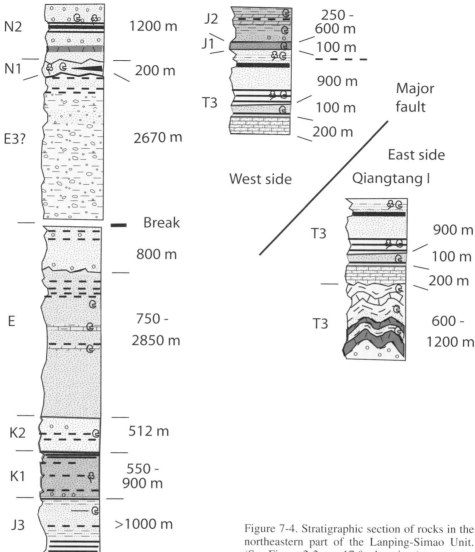

Figure 7-4. Stratigraphic section of rocks in the northeastern part of the Lanping-Simao Unit. (See Figure 2-2, no. 17 for location.)

by sandstone, siltstone, and shale, with some limestone and coal deposits. They contain fossils that indicate both marine and non-marine deposition. The Upper Triassic fine-grained clastic rocks may reach a thickness of >2000 m (Fig. 7-3).

In the northern part of the Lanping-Simao the Triassic rocks are not as thick, and the base of the section is not exposed (Fig. 7-4). These rocks are present mainly along the east side of the unit (west of the major fault zone that bounds the unit to the NE against the Qiangtang I unit, Plate 1). They consist of a lower limestone unit that is up to 900 m thick, overlain by fine-grained sandstone, siltstone, and shale that is up to 1000 m thick, both of Late Triassic age. Whereas these units have some resemblance to strata in the Qiangtang I unit to the north and to the east (see below), more study is needed for these important rocks

to see how they correlate between the two tectonic units. Along the west side of the northern Lanping-Simao Unit a very thick section (5 km?) of mafic volcanics and spillites (T3b, Plate 1) form a N-S–trending belt below the Jurassic unconformity. They contain abundant Upper Triassic fossils and are intruded at their north end by a large complex of mafic and ultramafic rocks—the Gicha body—that contains lenses of serpentinite (Fig. 7-6). Some of the rocks in this mafic and ultramafic complex were studied recently by Jian et al. (2009a), and radiometric ages from zircons from a microgabbro, a plagioclase hornblendite, and a diabase yielded U-Pb ages of $301.2 \pm 2.9$ Ma, $297.1 \pm 2$ Ma, and $281.3 \pm 1.8$ Ma, respectively. At its north end the complex is shown to have intruded Permian volcanic and sedimentary rocks, but at its south end Chinese maps show that it intruded the

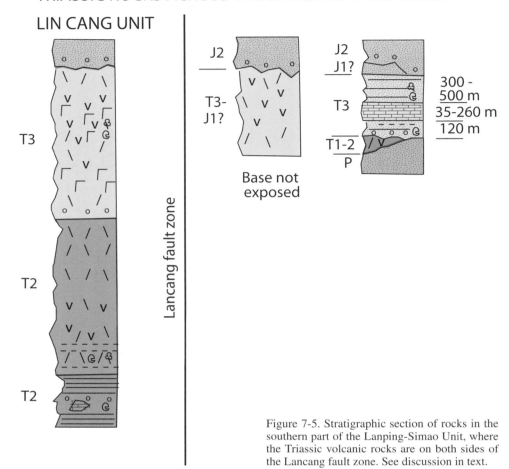

Figure 7-5. Stratigraphic section of rocks in the southern part of the Lanping-Simao Unit, where the Triassic volcanic rocks are on both sides of the Lancang fault zone. See discussion in text.

Upper Triassic spillitic volcanic rocks mentioned above (T3b), a relationship that is inconsistent with the ages published by Jian et al. (2009a). Obviously the field relationships at its south end need to reexamined. The Permian and Upper Triassic spillites are unconformably overlain by Jurassic red beds of the Lanping-Simao, and all these rocks are in fault contact along the western boundary zone of the Lanping-Simao. The spillitic rocks are interpreted to be an intra-arc rift of Late Triassic age. The Gicha mafic-ultramafic complex was interpreted to have been generated in a backarc rift distant from an arc (Jian et al., 2009a), probably the arc represented by the Permian volcanic rocks.

At this latitude the long, narrow belt of fault-bounded Triassic arc volcanic rocks (Mz1a, Plate 1) that generally marks the western boundary of the Lanping-Simao Unit widens and contains evidence for their Middle and Late Triassic age in interbedded fossiliferous clastic strata (Mz1s). Whereas to the north, these volcanic rocks rest unconformably on Permian strata, where they are displaced by the short left-lateral shear zone mentioned with the Qiangtang I unit above, here they rest on a thick sequence of sandstone and shale with some limestone of Middle Triassic age (Mz1s). At this latitude the volcanic rocks are intruded by many small plutons, shown on Chinese maps as Cretaceous ($\gamma_5^3$), but they are without age constraints and may be plutonic bodies related to the Triassic arc. Even though there are numerous plutons in these rocks, there are none to the east in the Lanping-Simao. To the north this belt of Triassic arc volcanic rocks at the west side of the Qiangtang I unit are overlain by Upper Triassic (Mz1b) volcanogenic strata, providing an upper limit on their age. It is a long way north to the west side of the Qiangtang I unit, but these volcanic rocks are mapped as continuous, even though they narrow considerably between the two areas. This section of probable Triassic arc rocks is considered different from the volcanic rocks in the Lower, Middle, and lower Upper Triassic volcanic rocks along the east side of the Qiangtang I unit, because outcrops connecting the volcanic rocks between the two areas are not present. Also the volcanic rocks in the west contain a great thickness of abundant felsic ignimbrites and agglomerates unlike the rocks to the east, which are generally more mafic in composition. These rocks also may contain intrusive bodies of similar hypabyssal composition. The western volcanic arc belt in this northern part of the Lanping-Simao narrows to the south, and farther south it is missing for 150 km at the western side of

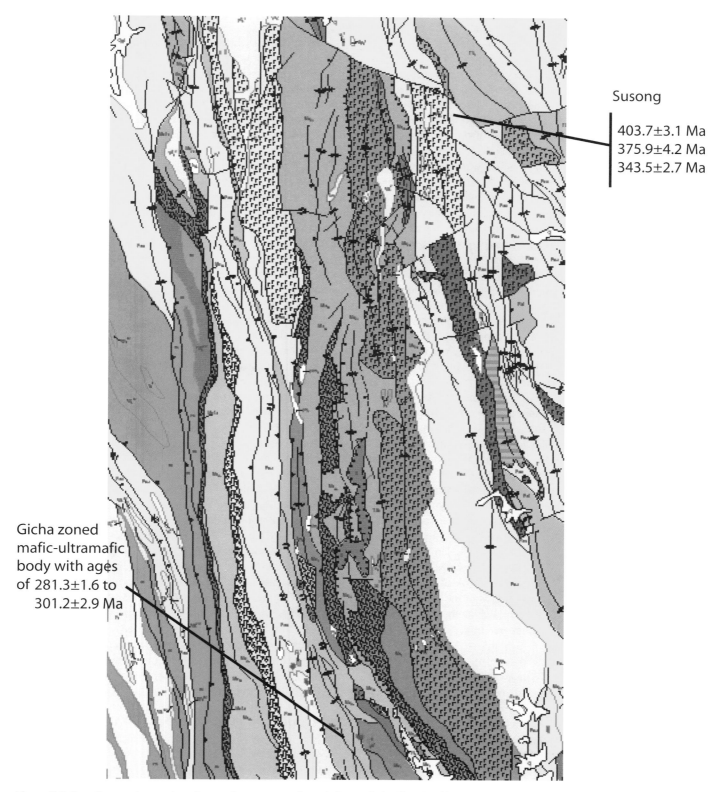

Figure 7-6. Locations and ages given for two important mafic and ultramafic bodies: the Gicha body in the Lanping-Simao Unit, and the Susong body in the Jinsha accretionary complex.

the Lanping-Simao Unit. In our interpretation this volcanic belt reappears to the south in the big, convex east, horseshoe bend in the Lanping-Simao Unit south of the Wuliang Shan.

The main body of rocks in the Lanping-Simao Unit constitutes up to 8 km of Jurassic and Cretaceous non-marine strata that consist of mainly red, pale red, and maroon sandstone, conglomerate, and mudstone (Figs. 7-3 and 7-4). These rocks appear to constitute a conformable succession in the northern part of the Lanping-Simao, but in the southern part of the unit, unconformities appear in the section (for distribution of unconformities, see Plate 3). There is a widespread unconformity at the base of the Lower Cretaceous strata, which is generally a disconformity underlain by the upper part of the Jurassic succession. However, in some places rocks as old as Triassic are below the unconformity. Upper Cretaceous strata also lie disconformably on Lower Cretaceous rocks and locally on pre-Cretaceous strata (Plate 3).

Paleogene strata are exposed along the middle part of the Lanping-Simao Unit, suggesting a generally broad, synclinorial structure for the unit (Plate 1). Paleogene strata in the southwesternmost part of the Lanping-Simao are present in two adjacent synclinal basins. The sections begin with 225–600 m of sandstone and conglomerate, with fossils at the top and bottom that date them as Paleocene, one of the few places where rocks of this age have been determined. They are overlain by >1800 m of sandstone, siltstone, and some mudstone that are fossiliferous at the top and yield an Eocene age. Unconformably overlying the Eocene strata are conglomerate and sandstone 500–1000 m thick, but whose exposures are only in the syncline to the west, where they unconformably overlie also Jurassic and Cretaceous rocks (Plates 1 and 3). Beneath both Paleogene units the Mesozoic rocks are folded, and in and around these basins stratigraphic relations indicate pre-Paleogene and post–Late Cretaceous, intra-Paleogene, and post-Paleogene deformation in syndepositional deposits. The unconformity below the Paleocene rocks in the eastern syncline becomes paraconformable along its east side, indicating that the older folding event was not uniform across the area and may have been syndepositional. Northward the Paleocene rocks continue as thin belts of folded and thrusted rocks and as more equidimensional basins to the east. Both the Upper and Lower Cretaceous strata are overlain by the Paleocene strata, indicating a wide but irregular area of pre-Paleocene folding and faulting (Plate 3).

Several smaller, generally NW-trending outcrops of Paleogene strata are present to the NE near the Laotian border, where they consist locally of >1500 m of red sandstone, mudstone, and some conglomerate. Locally, fossils of Paleocene age are present in the lower part of the section. The Paleogene strata unconformably overlie Lower Cretaceous strata, some with a low-angle discordance, and others with no discordance, and some overlie what appear to be folds in the Cretaceous rocks (Plate 1). Less commonly these strata lie on Upper Cretaceous strata. It still remains to be explored as to whether these relations suggest extensional or compressional deformation, or both, in pre-Paleogene time. Folds are present below the unconformity, but whether this explains all the unconformable relations remains unknown.

The large Paleogene basin in the northern part of the southern segment of the Lanping-Simao, the Jinping basin, contains up to 4000 m of red, maroon, tan, and yellow sandstone, siltstone, mudstone, and conglomerate, with the upper part of the section having generally lighter colors. The lower ~300–500 m contains Paleocene fossils overlain by 1100–1500 m of unfossiliferous rocks regarded as Eocene in age. They are unconformably overlain by >2000 m of unfossiliferous, finer grained strata assigned an Eocene–Oligocene age (Plate 3). The basal part of the section lies on Lower Cretaceous strata, and the contact is shown to be a parallel disconformity. Importantly, the Paleogene strata are unconformably overlain by a few hundred meters of tan, yellow to white, fine-grained sandstone, siltstone, and shale with early Miocene fossils (Plate 3), which date the upper limit to the folding.

In the northern part of the Lanping-Simao Unit is a third large exposure of Paleogene rocks that contains five units: The lower unit consists of 300–2000 m of red conglomerate, sandstone, mudstone, and siltstone, overlain by 800 m of red and maroon mudstone and siltstone. These beds are overlain unconformably by the third unit of 800–1000 m of conglomerate and sandstone (Plate 3). The sandstone in the third unit commonly is massive and is light yellow in very thick to massive beds, forming high cliffs. The fourth and fifth units consist of 700–800 m of paler red sandstone, siltstone, and mudstone with some conglomerate, and they lie conformably above the massive sandstone of the third unit. There are some undiagnostic fossils in the lower units, and Eocene fossils in the fourth unit, but none in the topmost part of the section, which is considered to be Eocene–Oligocene in age. West of the fault along the east side of the Lanping-Simao (see below) the Paleogene rocks lie with low-angle disconformity on Upper Cretaceous strata. This section of Paleogene rocks is stratigraphically similar to the Paleogene rocks in the Jinchuan basin, which overlie rocks of the Zhongza belt of the Yangtze and Transitional Units to the east. This apparent similarity makes for a puzzling situation, as some tectonic scenarios place a major left-lateral Cenozoic strike-slip zone between them (see below).

Small basins of Neogene strata contain yellow-weathering sandstone, siltstone, and mudstone with beds of lacustrine limestone. They commonly crop out along present river drainage systems, suggesting that at least some of the present-day topography formed in Neogene time. These rocks consistently are only gently warped and lie unconformably on older, more deformed strata.

## STRUCTURE OF THE LANPING-SIMAO UNIT

Structurally the Lanping-Simao Unit trends north-northwest and can be divided into broad, arcuate northern and southern segments separated by a narrower convex-east segment in the middle (Plate 1). The northern segment consists of folds and parallel faults that form a gentle concave-east arc, although the northeast side of this arc is bounded by a NW-trending linear fault zone.

The folds and faults within the northern segment are truncated by this linear fault zone, shown clearly by the truncation of the large area of Paleogene deposits (E, Plate 1). The southern segment forms a southward-widening belt of Mesozoic rocks in which the structures have a broad, arcuate shape concave to the northeast, and are again truncated on the northeast by a broad, northwest-trending linear fault zone and mélange (Mz1, Pzu, and Pzl with a mélange pattern, Plate 1). The narrow, central part of the unit is formed by a tighter convex-north structure that also affects the adjacent tectonic units to the south, where it has the NE-striking late Cenozoic Nantinghe left-lateral fault in its core. Structures related to this tighter arcuate structure overprint structures in the Lanping-Simao to the north, forming a complex interference pattern (Plate 1).

Within the southern segment of the Lanping-Simao, all the exposures of Paleozoic rocks below the Mesozoic rocks show intra-Paleozoic deformation to varying degrees. It is only in the southern segment of the unit that such deformations can be observed, because Paleozoic strata are exposed only in one place in the northern segment (Plate 1). In the southeastern part of the southern segment the Silurian strata are well-foliated, metamorphosed to low-grade greenschist facies, and commonly contain abundant quartz veins. The rocks are lithologically monotonous, and the structure within them is largely unknown. They are everywhere in fault contact with upper Paleozoic rocks, so the age of this deformation is unknown, except that the upper Paleozoic strata are not as internally strained, which suggests that an early Paleozoic deformation is recorded in the Silurian rocks. Deformation in upper Paleozoic rocks in some exposures show folds along trends that are not too different from folds in the overlying Mesozoic rocks. These structures in some places, however, are unconformably overlain by Lower and/or Upper Permian rocks, demonstrating late Paleozoic deformation, particularly in the exposures along the west side of the Lanping-Simao (Plates 1 and 3). It is only in the area where the Silurian rocks are exposed that plutons have intruded the Paleozoic and some adjacent Mesozoic rocks. This part of the Lanping-Simao continues eastward into northern Vietnam, where similar relations demonstrate major early Mesozoic deformation, magmatism, and metamorphism (see Carter et al., 2001, and Lepvrier et al., 2004). This area may be underlain by a fragment of crust, different from the rest of the Lanping-Simao to the west, which influenced younger Cenozoic deformation (see below).

Because the Lower, Middle, and lower Upper Triassic rocks, with volcanic rocks, are exposed only in the western part of the southern segment, with rare exceptions, deformation within Early Triassic time can only be documented in this part of the Lanping-Simao Unit. The unconformity at the base of the Middle Triassic, volcanic-bearing units indicates that some folding and faulting took place during late Paleozoic or Early Triassic time. The evidence is limited to the areas where Middle Triassic rocks lie unconformably above older strata (see T1 and T2 on Plate 3). The Middle Triassic rocks were also deformed during pre–late Late Triassic time, as shown by the unconformity at the base of the upper Upper Triassic strata (T3 on Plate 3), where the overlying rocks rest above different Mesozoic and Paleozoic strata. Small plutons near the Lin Cang batholith, in the southwestern part of the Lanping-Simao Unit, intrude both Middle Triassic and Upper Triassic volcanic rocks. Two such plutons are parts of mafic-ultramafic complexes (Banpo and Gicha bodies: Figs. 7-2 and 7-6) that contain numerous small serpentinite lenses. The way the unconformities truncate units below them in some, perhaps in many, cases is suggestive of tilting around a horizontal axis rather than folding, perhaps indicating that the deformation is extensional, a hypothesis that needs considerable testing. The Middle and lower Upper Triassic volcanic units pinch out to the east, and eastward deformation of pre–late Late Triassic age cannot be exactly determined.

Only in a narrow belt in the southeastern part of the Lanping-Simao Unit are there discontinuous lenses of slates and some red beds adjacent to the Ailao Shan mylonitic unit that are Middle Triassic in age (Mz1). The fine-grained strata in this unit contain strongly sheared, slaty rocks that have a mélange character (as they are shown on Plate 1). Within these lenses are steeply dipping, less deformed red beds, but how they relate stratigraphically to the slaty rocks is unknown. These lenses are unconformably overlain by relatively undeformed Upper Triassic rocks, some with interbedded intermediate volcanic rocks. These lenses we interpret to mark a belt of mélange, and strongly faulted rocks of pre–Late Triassic age that mark a fault zone of regional importance (see below).

Northwest of these Middle Triassic rocks is a 200-km-long belt of rocks mapped as Permian (Pzu) and Silurian (Pzl). They contain a highly sheared matrix in which numerous lenses of serpentinite, mafic and ultramafic rocks, and basalt and sheared dikes of basalt are present (Plate 1). This is clearly a belt of mélange, and the age assignment is based on fossils found in the incorporated blocks; thus the date gives only a minimum age for the mélange formation. The ages of the rocks shown on Plate 1 are given by the ages of the fossils found within them, but the mélange overprint pattern indicates that we consider this a belt of upper Paleozoic mélange (Plate 1). Mostly along the southeastern part of the belt the mélange is unconformably overlapped by Upper Triassic strata (Plates 1 and 3), the same Upper Triassic strata that overlie the sheared Middle Triassic slaty mélange to the south, thus putting an upper age limit on mélange formation. Two areas in the southern part of the mélange belt labeled Pzl contain Upper Triassic strata that unconformably overlie the mélange along its northern side, but everywhere else the mélange is in the footwall of a thrust that carries the Ailao Shan metamorphic rocks southwest over the mélange and over the unconformably overlying Upper Triassic strata (Plate 1). In many places the southern boundary of the mélange is marked by a thrust that placed the mélange over Permian or Upper Triassic rocks, and the thrust faults are part of a thrust belt that involves rocks as young as Late Cretaceous. We interpret that there is only one belt of mélange that is probably Permian(?) and Early and Middle Triassic in age that was reactivated in latest Cretaceous or early

Cenozoic time. This structural belt figures prominently in the regional analysis below.

Recent work on samples from one of the mafic-ultramafic bodies along this mélange belt, the Shaunggou body, by Jian et al. (2009a) suggested that it is part of an ophiolite (Fig. 7-2). A plagiogranite and diabase were sampled and yielded zircon U-Pb ages of $381.9 \pm 3.9$ Ma and $375.9 \pm 4.2$ Ma, respectively, indicating a Devonian age. Analyses of the rocks in this complex by Jian et al. indicated that they are part of a normal mid-oceanic-ridge basalt (NMORB) type of an ophiolite, suggesting that there was oceanic crust in this area by Devonian time. Within both the mélange belt and some of the Upper Triassic strata adjacent to it are numerous bodies of basalt, serpentinite, and mafic and ultramafic rocks, one of which is the Shaunggou body studied by Jian et al. (2009a). Most of these rocks occur as fault-bounded bodies within the mélange, but some of them are within Triassic rocks adjacent to the mélange and that unconformably overlie the mélange. These mafic and ultramafic rocks are commonly strongly serpentinized and are probably cold intrusions from the mélange into the unconformably overlying Triassic strata. Many of the blocks within the mélange are likewise serpentinized bodies, showing cold incorporation into the mélange.

Most of the outcrops within the Lanping-Simao Unit are a characteristic succession of Upper Triassic to Paleogene strata, with the bulk of the section above the Triassic formed by non-marine red beds with only rare marine incursions along its western side. Within the southern segment of the Lanping-Simao belt, these rocks are folded and faulted along parallel trends. Many of the faults are mapped as thrust faults, but most of the faults have no proven kinematics, and many are inferred to be thrust faults because they have thrust separations. As in much of this part of southern China, within the numerous fold and thrust belts of the region, mapped low-angle thrust faults are rare, or perhaps unrecognized, for reasons that are not clear. Many of the structures involve the assemblage of pre-Mesozoic Paleozoic rocks that are exposed in the cores of anticlines or along thrust faults. The overall structure appears to be a fold-thrust belt that locally, or more likely regionally, involves pre-Mesozoic rocks (see, e.g., Plate 4, sections 9, 21, 22, 31). Where a basal detachment beneath such a fold-thrust belt would lie remains an unresolved question. The fold-thrust belt has the form of large synclinorium, with the youngest rocks, typically Paleogene (E), in its central part. This feature is emphasized by the fact that the unconformity at the base of the upper Upper Triassic rocks marking the beginning of the thick Lanping Simao Mesozoic section is exposed only in the eastern and western parts of the unit (Plate 3). The regional unconformity at the base of the Jurassic red-bed succession (J2 in Plate 3) is also exposed mainly on its east and west sides.

The main folding and thrusting event within the Lanping-Simao is Paleogene, but structures of pre-Jurassic, intra-Cretaceous, and intra-Paleogene times are known (Plate 2 [see footnote 1]). Locally(?) folds in Upper Triassic rocks are unconformably overlain by Jurassic strata, demonstrating post–late Late Triassic, pre–early Jurassic shortening. In other places below this unconformity, however, units are truncated is such a way as to suggest they were tilted and beveled, as would be expected in an extensional structural setting. Unconformities that involve Cretaceous rocks are mostly present in the southern part of the southern segment of the Lanping-Simao Unit (see Plate 3). Many of the stratigraphic relations across the unconformities that are mapped at the base of the Lower Cretaceous, between the Lower and Upper Cretaceous, and at the base of the Paleogene, are disconformities, but not all are. There are locally truncated folds or faults at these unconformities and beveling of dipping units. In a few places, some folds with overturned limbs are beveled at the unconformity (see relations shown on both Plates 1 and 3). The importance of these structures remains unknown, but most of them are in the southern part of the unit.

Most of the folds and faults in the Lanping Simao appear to be related to Paleogene deformation. The unconformities at the base and within the Paleogene rocks show evidence for pre- and early Paleogene deformation, and across most of the unit the Paleogene rocks are folded mostly along trends parallel to bedding attitudes and faults in the underlying strata. Additionally the Paleogene strata contain thick, coarse conglomerate with cobbles of the older rocks consistent with syndepositional deformation. Mostly the distribution of Mesozoic and Paleogene strata is consistent with their representing a single major stratigraphic succession before deformation, in contrast to the distribution of lower Miocene and younger Cenozoic strata with irregular local distribution and that lie unconformably above the Paleogene and older rocks.

Within the southern segment of the Lanping-Simao Unit, folds are mostly upright and have no general vergence (Plate 4, sections 21, 22, 31). In the western part of the segment the folds are generally open, whereas in the eastern part of the segment they are tighter, with steep, commonly near vertical, fold limbs (Plate 4, sections 21, 22). The magnitude of shortening across this part of the Lanping-Simao is unknown and is difficult to estimate from published cross sections, as our observations indicate a bias toward shallower dips for bedding and steeper dips for the faults. The magnitude of shortening would be an important contribution to the regional tectonic analysis, but it remains unknown.

In the northern segment of the Lanping-Simao only rocks of Late Triassic age or younger are exposed, and the Mesozoic section appears to have undergone less internal deformation during Cretaceous time than the segment to the south. The relations between Paleogene and Cretaceous strata, and within the Paleogene strata, are similar to those for the southern segment (see Plate 3), and the main deformation appears to have been early Cenozoic. The exception is in the western part of the northern segment, where the deformation related to the formation of the Upper Triassic basalt-spillite unit (T3b) and the intrusion of the mafic Gicha pluton appear to have disrupted its characteristic stratigraphic succession. Whereas extensional deformation may have played an unknown role in the southern segment (see above), the formation of these marine mafic rocks that incorporated large masses of Carboniferous limestone may be a more obvious expression of local(?) extensional tectonism. These rocks are unconformably overlain by Jurassic red beds that here form much of the basal part of the northern segment of the Lanping-Simao Unit (Plate 1).

The area between the two segments of the Lanping-Simao show an additional and major departure from the deformational pattern within the unit. The narrowing of the Lanping-Simao Unit is caused by overprinting of the normal northwest structural trends by a major fold in the Wuliang Shan. A thick sequence of greenschist grade, mainly clastic metasedimentary rocks (Pzm, Plate 1), is present in the core of a large anticline whose axial trace is convex to the north. Two small granitic undated plutons, assigned a Jurassic or Cretaceous age, are mapped along the anticlinal axis. The metasedimentary rocks are strongly foliated and contain abundant quartz veining, but they lack fossils. They are in fault contact with the surrounding rocks except at the SE-plunging anticlinal nose, where they are unconformably overlain by Jurassic strata (Plate 3). There are no volcanic rocks present in either the greenschists or adjacent rocks, although the structure lies directly north of a large area of Triassic volcanic rocks (Plate 1). Rocks in the antiform are not easy to assign an age to and are shown with a special pattern on the map. Along their northern side they are thrust north above a well-defined typical Lanping-Simao section of Upper Triassic rocks, with limestone at the base, overlain by clastics, followed by the beginning of the Jurassic red-bed sequence. Some of the rocks in the footwall of the thrust are weakly metamorphosed, many with a well-developed phyllitic foliation, and to which are hard to assign a specific age (T3m). The thrust and associated structures make the same arcuate bend convex north, as does the major anticline and the volcanic and plutonic rocks to the south at the north end of the Lin Cang Unit. The rocks in these younger structures commonly contain a well-developed phyllitic foliation in contrast to the older structures they overprint. We interpret this interference pattern of structures as part of a large fold with a vertical axis superposed on the Lanping-Simao Unit. It is the formation of this fold that narrows the Lanping-Simao Unit and divides it into its two segments. This large vertical-axis fold we interpret to be a Cenozoic structure, probably late Cenozoic in age, because it overprints the early Cenozoic structures and is probably related to a regional NE-SW left-lateral shear (see below) associated with the formation of the Nantinghe fault, which lies to the southeast in the axial surface of this vertical-axis fold but dies out into the fold hinge at its northeastern end. The effects of this late fold do not extend north across the full width of the Lanping-Simao Unit (Plate 1), and the formation of this structure appears to have been decoupled from the more linear structures along the northeastern boundary of the Lanping-Simao Unit.

Determining the magnitude of shortening across the northern segment of the Lanping-Simao has the same problems as in the southern segment. Folds are generally upright except in the western part of the unit, where overturned folds and thrust faults are common (Plate 4, sections 20, 22). The vergence of these structures is both east and west, but east-vergent structures dominate. As in the southern segment, faults parallel the arcuate fold axes, and separations on these faults are both thrust and normal. No kinematic or timing studies have been undertaken for these structures.

Neogene strata are developed along the east side of the northern segment and were deposited in NW-trending, fault-controlled basins of mostly extensional origin with a strike-slip component, and beds are tilted but not folded. In the narrow central and southern segments of the unit the Neogene basins were developed sporadically across much of the unit. The Neogene strata lie unconformably on older rocks and put an upper limit to the pre-Neogene deformation. Unfortunately most of the basins are Pliocene in age, and Miocene strata are rare, so the upper limit of deformation is not well constrained, except in the large Paleogene basin in the southern segment, where early and middle Miocene strata unconformably overlie more strongly folded Paleogene strata (Chen et al., 1995).

## Boundaries of the Lanping-Simao Unit

The eastern and western boundaries of the Lanping-Simao Unit are formed by complex, poly-deformed structures with regional tectonic significance. The western boundary is formed by a fault zone that makes a sharp contact with adjacent units, but separates rocks of the Lanping-Simao on the east from the different tectonic units to the southwest (Fig. 1-1B). At its southern end the fault zone, which we interpret to be the southern extension of the Lancang fault zone, separates a broad terrane of plutonic rocks of the Lin Cang Unit to the west from upper Paleozoic rocks unconformably overlain by mainly Lower Triassic volcanic-bearing strata with rare plutons that belong to the Lanping-Simao Unit to the east. The volcanic-bearing rocks appear to crop out on both sides of the fault zone, and along this fault zone are mafic intrusions that may be the deeper plutonic equivalents of older Paleozoic volcanic rocks. Along the fault zone are strongly folded Paleogene strata that are unconformably overlain by weakly folded Neogene strata. Northward the fault zone is well defined as the boundary between the Lin Cang plutonic complex and Mesozoic rocks of the Lanping-Simao and is marked here by sheared granitic rocks, some mylonitic, on the west side, and by a narrow zone of faulted upper Paleozoic, mostly Permian, and Mesozoic strata on the east side. There is a narrow belt of sheared rocks that structurally imbricate Permian sedimentary and granitic rocks about a vertical axis in a pattern that suggests left-lateral shear (Plate 1). Farther north the Lancang fault zone continues through a broad exposure of Triassic volcanic rocks and is marked by a narrow belt of fault-bounded, sheared, and foliated Permian strata with numerous lenses of sheared limestone. All along this southern stretch the Lancang fault zone is subvertical.

The western boundary marked by the Lancang fault zone continues north in first, a convex north arc, and then a concave-north arc south of the anticlinal structure in the Wuliang Shan. The arcs we interpret to be part of a large Cenozoic vertical axis structure. Recently Heppe et al. (2007) reported a cooling age of 17 Ma (K/Ar) for the pre-Triassic metasedimentary rocks in the core of the antiform. However, they also reported several older ages as old as Middle Jurassic, whose meaning is unclear. The Permian rocks along the Lancang fault to the south end before entering the convex north arc, and the fault zone separates Triassic volcanic rocks on the south from metamorphic rocks of unknown age in the core of the Wuliang structure to the north. Westward, the Lancang fault zone makes a broad arc concave to the east and sharply defines the boundary between the Lanping-Simao red beds from different tectonic units to the west. Triassic volcanics south of the fault in the arcuate structure are missing for ~150 km, and they reappear to the north, where they can be followed northward along the west side of the northern segment of the Lanping-Simao Unit, and then along the west side of the southern part of the Qiangtang I unit (see above; Plate 1). In this northern stretch, rocks to the west are metamorphic and commonly mylonitic, particularly where they are part of the Chong Shan shear zone (see below) at the eastern side of the Baoshan Unit (Fig. 1-1B: Akciz et al., 2008). The Lanping-Simao strata near the western boundary are weakly metamorphosed and deformed to varying degrees, but not as strongly deformed as the rocks to the west. Within the Lanping-Simao the strata are strongly folded for up to 10 km east of the contact, and beds are near vertical and folds commonly overturned, many to the east, and are cut by steeply dipping, east-vergent thrust faults. The Lancang boundary fault zone is generally subvertical to steeply west dipping and appears to cut the less steeply dipping strata and structures in the Lanping-Simao rocks to the east (Akciz et al., 2008). Where the Triassic volcanic rocks are present they are bounded by faults on both sides, but locally they contain Middle Triassic sedimentary rocks (Mz1s) below that place them in the Lanping-Simao Unit.

The most important area for determining the nature of the Lancang fault zone is in the region south of the Wuliang anticlinal structure, where Triassic volcanic rocks within the sharply curved fault zone are present on both of its sides (Plate 1 and Fig. 7-5). The Lancang fault zone is in many places, but not everywhere, marked by a narrow belt of highly sheared Permian strata and taken as the western boundary of the Lanping-Simao Unit, but the Triassic rocks on both sides need to be discussed together. East of the shear zone the Triassic rocks are assigned a Late Triassic age and underlie the unconformity at the base of the Middle Jurassic (Plate 3; Fig. 7-5). Near the shear zone these rocks contain a very thick succession (>3000 m) of felsic, intermediate, and mafic volcanic rocks that contain fossils in the sedimentary interbeds. These rocks are everywhere in fault contact with the sheared Permian rocks, except for two narrow belts to the south where they are shown on older maps as Middle Triassic, but with new mapping they are considered Upper Triassic. In the eastern of the two belts they are shown to lie locally unconformably on the Permian rocks, but in most places no older rocks are exposed below them, and they are unconformably overlain by Middle Jurassic rocks. Farther east of the fault zone the Upper Triassic rocks are fossiliferous but are without volcanic rocks, and they are underlain by a thin unit of Lower-Middle Triassic volcanic rocks that pinch out abruptly eastward, resting unconformably on Permian strata. These Upper Triassic strata are at the base of the very thick, typical red-bed succession of the Lanping-Simao. West of the shear zone the volcanic rocks lie within the Lin Cang Unit, and similar Upper Triassic volcanic rocks are present, but here beneath the Upper Triassic volcanic rocks lies another very thick succession (up to 2700 m) of mainly felsic and intermediate volcanic rocks with up to 800 m of mainly non-volcanic sedimentary rocks below. Both units carry Middle Triassic fossils. Both the Middle and Upper Triassic rocks are intruded by large intermediate plutons of the Lin Cang Unit. These plutons are not present east of the Lancang fault zone. The significance of these relations across the shear zone boundary are important regionally and will be discussed below.

The western boundary of the Lanping-Simao Unit is a continuous fault zone that separates the Lanping-Simao Unit from all the tectonic units to the west, with only rare interleaving of the tectonic units. It is also the separation of plutonic, metamorphic, and mylonitic rocks on the west from mainly weakly to unmetamorphosed sedimentary rocks on the east. This boundary has often been referred to as the *Lancang suture,* but there is little evidence of oceanic rocks along this zone, and we refer to it as the *Lancang fault zone.* It is a fault zone of great importance in the regional tectonic framework of the region (see below). A recent study by Heppe et al. (2007) presented an interpretation based on their field and laboratory study that the Lancang fault zone was the locus of a Carboniferous–Early Permian, west-dipping subduction zone, and during the Late Permian–early Mesozoic the western part of the Lanping-Simao terrane became an extensional region. They also discounted east-dipping subduction along the Changning-Menglian Unit and placed the Lin Cang and Changning-Menglian Units, together with the Baoshan Unit, as one large continental fragment (an interpretation with which we disagree; see below). Although there may have been short-lived Paleozoic subduction within the accreted fragments within the pre-Mesozoic rocks underlying the Lanping Simao, and based on the discussions above, we regard the Lancang fault zone as a long-lived boundary between the Lin Cang and Lanping-Simao Units that are both parts of the accreted Paleozoic elements below the Lanping-Simao Mesozoic strata. We also recognize that east-dipping subduction occurred within the Changning-Menglian belt beneath the Lin Cang Unit during at least late Paleozoic time and into Mesozoic time.

Like the western boundary fault zone, the eastern boundary of the Lanping-Simao is marked by an equally, if not more complex, fault zone. East of the narrow central and southern segments of the Lanping-Simao the mélange belts discussed above that contain Silurian, Permian, and Middle Triassic fossils are in fault contact with metamorphic rocks of the Ailao Shan shear zone, or in the south in contact with Paleozoic rocks that we interpret to belong to the Ailao Shan belt and which have ties to the Yangtze platform Unit (see below). This fault or fault zone we take as the eastern boundary fault of the southern segment of the Lanping-Simao Unit, and we assign the mélange belt to the Lanping-Simao Unit. The mélange is overlain unconformably by Upper Triassic rocks that place an upper limit on mélange formation, and some of these Triassic rocks contain serpentinite bodies that we interpret as cold intrusions of serpentinite from rocks in the mélange below the unconformity. In its northern part, the mélange was thrust southwestward above the Upper Triassic rocks along a thrust belt with SW vergence that involves rocks as young as Late Cretaceous. The mylonitization of the Ailao Shan shear-zone rocks is dated as Cenozoic (see below), and the thrusting of these rocks southwestward above the mélange is also Cenozoic. The reactivation of the mélange belt and the thrusting of the Ailao Shan rocks are regarded as part of a major Paleogene to late Early Cenozoic tectonic event related to the SE extrusion of the Lanping-Simao Unit (Tapponnier et al., 1982; Leloup et al., 1995; Burchfiel et al., 2008b) and thus is interpreted to have a major left-lateral strike-slip component. To the south we interpret the mélange belt to be continued by Middle Triassic sheared strata with Middle Triassic fossils (see above). Neogene strata are mapped as overlying both the mélange and rocks to its north. However, our examination of these rocks indicates that both Neogene and Paleogene strata are present here. The Paleogene strata appear to be much more strongly deformed, but the rocks need detailed mapping to resolve the critical relationships that bear on the timing and magnitude of reactivation of the mélange unit in this area. All the evidence indicates that the rocks along the northeastern margin of the southern segment of the Lanping-Siamo Unit have had a long and complex history, first with mélange formation in pre–Late Triassic time, and then reactivation in both early and late Cenozoic time.

Northwestward the complexity along the eastern boundary continues, as the mélange belt and the Ailao Shan mylonitic belt narrow and end where they are crossed by the Red River. At this river the Red River fault zone, a late Cenozoic and active right-lateral fault zone, which lies east of the Ailao Shan mylonitic belt to the south, cuts across the mélange and mylonitic belt at a low angle, terminating them. It is here that Allen et al. (1984) correctly determined that the Red River fault is active and offsets the Red River by ~5 km right laterally (Wang et al., 1998; Schoenbohm et al., 2006b). To the north the Red River fault zone becomes the boundary between the Lanping-Simao Unit and the Yangtze and Transitional Units to the east. The fault zone can be traced northward as a linear fault zone to the southern end of Erhai Lake (Plate 1). This is the area in which the active extensional Dali fault system in the Transitional Unit on the north abuts the Lanping-Simao Unit (Leloup et al., 1995; Wang et al., 1998; Schoenbohm et al., 2006b). The faults that control the boundary between the two tectonic units are not easy to interpret at the south end of Erhai Lake. The geology shows that the Lanping-Simao Mesozoic rocks end against an E-W–trending fault, where the Red River fault can no longer be followed north with confidence. This E-W–trending fault dips gently ~30° south, and footwall Proterozoic rocks that core the Dian Cang Shan are strongly mylonitized with a lineation that plunges south in a foliation that shows top-to-the-south kinematics (Leloup et al., 1993). The rocks have yielded $^{40}Ar/^{39}Ar$ data that suggest that the mylonitization took place ca. 4.7 Ma (Leloup et al., 1993). A fault separating rocks of the Dian Cang Shan from the Lanping-Simao continues northward along the west side of the Dian Cang Shan, where it dips west at a moderate angle (~45°) and displaces strata, indicating west-side-down movement. Regional relations suggest that this fault should have a component of right-lateral strike slip, but the fault is not continuous and is offset by short, E-W–striking faults. Work has not been done here to demonstrate how these faults interact. The south end of the Dian Cang Shan has the characteristics of a metamorphic core complex with N-S extension along faults that are young and probably active.

The fault zone along the west side of the Dian Cang Shan continues north-northwestward along a straight trace that can be followed along the west side of the belt of Lower Triassic volcanic rocks unconformably overlain by upper Upper Triassic strata that belong to the southward continuation of the southern Qiangtang I unit (Fig. 1-1B; see above). It is here that east of the fault zone the Qiangtang I unit ends against the Dian Cang Shan. The fault zone passes along the east side of several linear basins that parallel the fault and contain Miocene and Pliocene strata that are tilted NE and are now the locus of Quaternary deposition, indicating that the faults were active during late Cenozoic time and are still active today. Near the north end of this fault zone a long, fault-bounded lens of metamorphic rocks (m) formed the Xuelong Shan. These metamorphic rocks contain lenses of serpentinite and have a mylonitic fabric that dips west and shows left-lateral shear sense indicators (Leloup et al., 1995). This lens has yielded $^{40}Ar/^{39}Ar$ ages that range from 35 to 21 Ma (Gilley et al., 2003), but also has yielded some older ages, and the fabric development has been interpreted to be correlative with that in the Ailao Shan to the south; but the faults that bound this lens cut both Neogene and Paleogene strata. The trace of the fault zone continues northward, separating Jurassic and Cretaceous red beds of the Lanping-Simao Unit from Triassic strata that are part of the Qiangtang I sequence on the west and east, respectively. The fault that is the northward continuation of the eastern boundary of the Lanping-Simao Unit lies along the east side of a narrow belt of Jurassic rocks that separates two larger units of upper Paleozoic strata (Plate 1). The Paleozoic unit to the east is metamorphosed to locally amphibolite grade and contains abundant mafic rocks. The Paleozoic rocks to the west are largely devoid of volcanic rocks, and the thin belt of Jurassic red beds that lie west of the boundary fault trace lie depositionally on the Paleozoic rocks. This fault at its northern end lies within the transitional area between the Qiangtang I unit and the Lanping-Simao Unit, and lies along the west side of the Qiangtang I unit, and its northward continuation is subject to interpretation. We argue that the fault zone continues north along the west side of the Qiangtang I unit to merge with faults that form a broad belt of lens-shaped crustal fragments that separate the Qiangtang 1 unit and the Lhasa Unit on the east and west, respectively (see above). This fault marks the modified boundary where major crustal units were extruded to the SE in early Cenozoic time (see below).

In summary, the eastern boundary of the northern segment of the Lanping-Simao Unit is a young fault zone of late Cenozoic age, which in many places shows evidence that it is active. Plate 1 shows that this fault zone, even though young, is the fault that truncates the folds and faults in the Lanping-Simao Unit, juxtaposing them with several other structural units to the east. This young fault zone has reactivated older early Mesozoic and early Cenozoic fault zones of much larger magnitude along which extrusion of crustal and lithospheric fragments took place (see below).

**Paleomagnetic Data from the Lanping-Simao Unit**

New paleomagnetic data by Geissman and Burchfiel (unpublished) provides important evidence for the tectonics of not only the Lanping-Simao Unit but the regional tectonics of the SE Tibet region as well. Inclination data from a robust and reasonably well-distributed paleomagnetic data set from early Cenozoic and late Mesozoic strata from much of the western part of the Lanping-Simao Unit show that they are too steep in their present position, suggesting that the unit has moved southward by 500–1000 km relative to the Yangtze Unit, an amount that is larger than the error associated with the inclination data (Fig. 7-7). In addition the inclinations recovered are too steep and thus cannot be the result of flattening, often associated with sediment compaction. Similar conclusions were reached by Tanaka et al. (2008). From only four localities in Lower and Middle Cretaceous red beds within the Lanping-Simao Unit, their data showed a southward translation of 7° ± 1° between ca. 32 and 17 Ma. The southward latitudinal difference between the Lanping-Simao and Yangtze Units is only the southward component of extrusion, and because the extrusion is to the southeast its magnitude must be closer to 700–1400 km. These data are important in supporting the interpretation that the Lanping-Simao and other tectonic units to the southwest have been extruded by a significant magnitude. The data also suggest that the rocks were rotated clockwise by ~45°–90°, and a more limited data set from the southeastern part of the belt does not indicate appreciable rotation (Fig. 7-8). The boundary between rotated and unrotated rocks lies along a very narrow zone in the central part of the southern segment of the Lanping-Simao Unit, and we infer that this boundary is a previously unrecognized structural-boundary fault zone that we interpret to have been a left-lateral shear zone separating the Lanping-Simao element into two domains with different rotational histories (Fig. 7-8). This shear zone will figure prominently in the interpretation of how the early extrusion of the Lanping-Simao Unit took place (see below). This shear zone was later deformed in late Cenozoic time into a more sinuous trace, similar to deformation that also affected the Lancang fault zone and the folds and faults within the central segment and other parts of the Lanping-Simao Unit. The rotation

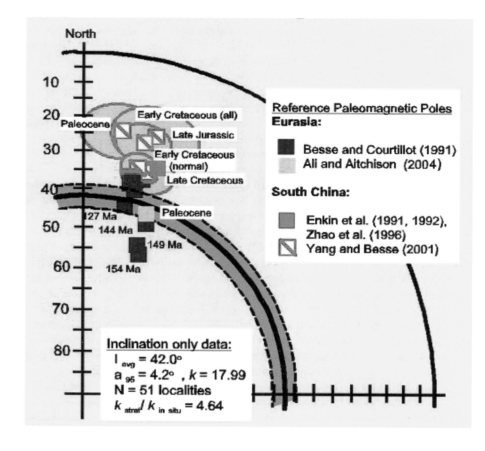

Figure 7-7. Expected paleomagnetic directions and normal polarity for Late Jurassic through Early Triassic time for a locality in the center of the Lanping-Simao fold-thrust belt, based on paleo-magnetic poles derived from results from Eurasia and the South China craton (Yangtze Unit). The inclinations determined for Paleogene and Late Cretaceous rocks of the Lanping Simao plot along a circular arc because they have been rotated, but their inclinations are fairly uniform and indicate that during Paleogene and Late Cretaceous time the Lanping-Simao Unit lay farther north than the Yangtze Unit, whereas today it lies to the south. This indicates a southward motion of the Lanping-Simao Unit of ~500–1000 km relative to the Yangtze Unit. This difference is greater than the error in secular variation normally present in paleomagnetic data. It also cannot be due to flattening related to sedimentary compaction, because the compaction correction is in the wrong direction. Data from Geissman and Burchfiel (unpublished).

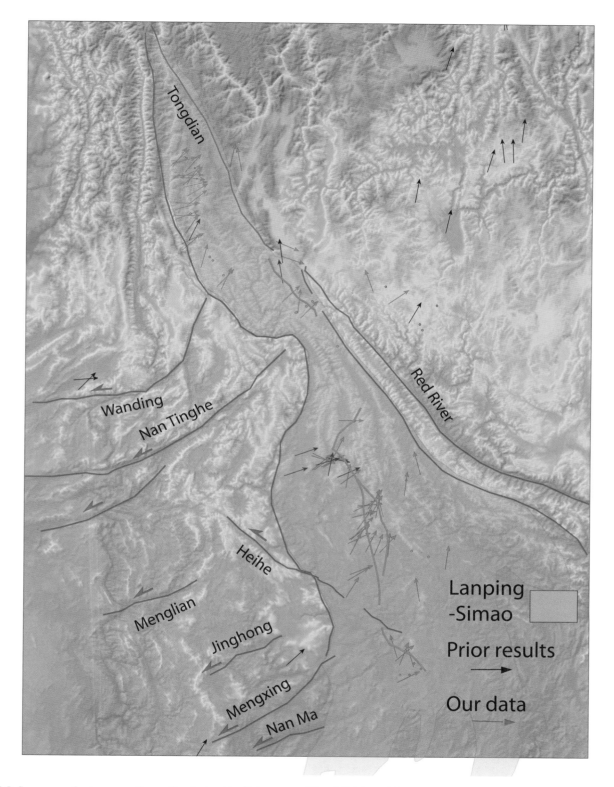

Figure 7-8. Summary of paleomagnetic results obtained by Geissman and Burchfiel (unpublished) for the Lanping-Simao belt. Individual locality mean results are depicted as red arrows the angular deviation of which from north is the magnitude of inferred rotation estimated on the basis of a comparison with expected paleomagnetic directions derived from paleomagnetic poles from Eurasia and South China (black arrows). Important left-lateral faults west of the Lanping-Simao element are also shown. A boundary between unrotated and rotated parts of the southern Lanping-Simao is shown in blue. The rapid change in rotation is narrow and interpreted to be a new unrecognized fault, which is highlighted in blue.

of rocks west of this boundary have led to the formation of crustal fragments that rotated clockwise around the eastern Himalayan syntaxis in early Cenozoic time (see Wang and Burchfiel, 1997) similar to the better documented clockwise rotation that occurred in late Cenozoic time (King et al., 1997; Zhang et al., 2004) but around a different axis of rotation (see below).

**Faults Active during the Early Cenozoic Extrusion Tectonics of the Lanping-Simao Unit**

Paleomagnetic data suggest that the southeastward extrusion in early Cenozoic time of the Lanping-Simao Unit that the unit should be bounded on the northeast by faults along its northeastern side. Traditionally this fault or shear zone is taken as the Ailao Shan shear zone, which is represented by its mylonitic rocks that show left-lateral horizontal shear. However, we will make the case in the regional synthesis below that several faults take up this shear as the rocks of the Lanping-Simao were contemporaneously extruded, internally shortened, and rotated clockwise. Such faults lie not only along the Ailao Shan but also along the mélange belt to its southwest and the newly discovered shear zone within the southeastern part of the Lanping-Simao Unit. This newly discovered fault lies west of the area in the southeastern part of the Lanping-Simao Unit that was intruded by Mesozoic plutons and is underlain by lower Paleozoic strata, and may have been influenced by the pre-Mesozoic geology. The Lancang fault zone was also active in the extrusion process, but like the other early Cenozoic faults and shear zones it was modified by younger deformation in many places.

Other fault zones west of the Lanping-Simao Unit that may have been active during the early Cenozoic extrusion were suggested by Wang et al. (2006). The right-lateral Gaoligong shear zone has been taken as the western boundary for extruding fragments since the early interpretations by Tapponnier et al. (1982, 1986). This shear zone is certainly one of the major contributors to early Cenozoic extrusion, but there are several shear zones that lie west of it within the Tengchong Unit that also show strike-slip displacement, but these have not been studied to know how significant they might have been in the extrusion process.

The northern continuation of these faults and shear zones forms an important part of our synthesis for early Cenozoic extrusion, but very little has been done to investigate these faults, their history, and possible magnitudes of displacement; thus they remain a fertile area for research.

**Late Cenozoic Faults in the Lanping-Simao Unit**

Late Cenozoic faults are present in the tectonic units on both sides of the Lanping-Simao and are part of a system of faults that mostly trend N-S to the north and NE-SW to the south of the unit (Fig. 2-10). The active left-lateral faults of this trend are well known, and they appear to accommodate the clockwise rotation of large fragments of crust (lithosphere?) that rotates around the Eastern Himalayan syntaxis. These faults do not pass through the Lanping-Simao Unit (Wang et al., 1998; Burchfiel et al., 2008b). The major Xianshuihe-Xiaojiang fault system that traverses the Yangtze platform Unit widens to the south, and individual faults lose displacement and appear to end as discrete faults at the northeastern boundary of the Ailao Shan Unit, but the left-lateral shear bends the Ailao Shan and rocks to the south by an amount approximately equal to the magnitude of the displacement on the fault system (Fig. 2-10; Wang et al., 1998). In the discussion above we interpreted the bend in the western boundary Lancang fault zone and rocks on both sides to be the result of late Cenozoic left-lateral shear related to the rotation around the Eastern Himalayan syntaxis and partly manifested by displacement on the Nantinghe fault. Two other active faults south of the Nantinghe fault contributed to the bend in the Lancang boundary fault in the southern segment of the Lanping-Simao Unit. The NW-striking, active left-lateral Heihe fault zone cuts through the eastern part of the Lin Cang Unit and into the Lanping-Simao Unit and ends in an area where folds and faults are bent left laterally on a small scale; these bends probably absorb the displacement. The NE-striking active left-lateral Jinghong fault strikes into the southernmost convex-east bend in the western boundary of the Lanping-Simao Unit, and ends before offsetting the boundary fault, but it may contribute to its bend, just as the Nantinghe fault does more dramatically farther north.

Within the Lanping-Simao Unit are many NE- and NW-striking short fault segments, with the NE-striking faults being more prominent. These faults may be absorbing, in a distributed manner, much of the left-lateral shear that must have passed through the Lanping-Simao Unit in late Cenozoic and Recent time (Zhang et al., 2004; Burchfiel et al., 2008a).

What is puzzling is that other Cenozoic and active faults appear to have had no effect on the boundaries of the Lanping-Simao Unit. For example, the Wanding fault cuts through the Baoshan Unit and displaces several boundaries right laterally ~25 km; however, it ends at the Lancang boundary fault zone (see below). This measured offset is early Cenozoic in age and suggests that the boundary fault zone was active in early Cenozoic time (see also Lacassin et al., 1998). The Wanding fault is now active as a left-lateral fault, but it still does not offset the boundary fault zone. Some NE-striking faults of the active Dali fault system north of the Lanping-Simao Unit also end against the northeast boundary fault zone of the Lanping-Simao. Only the Chenghai fault, the eastern fault in the Dali system, appears to offset the late Cenozoic Red River fault (Schoenbohm et al., 2005), but only by a small amount. These relations are difficult to explain and certainly need more study.

## PLUTONIC ROCKS OF THE LANPING-SIMAO UNIT

The discussion above has focused on the mafic and ultramafic rocks within the Lanping-Simao Unit, and even though the large Lin Cang batholith lies to the west, felsic and intermediate plutons are rare within the Lanping-Simao Unit. This relation requires explanation and will be discussed with the Lin Cang Unit below.

Only in the southeastern part of the Lanping-Simao Unit are there larger plutons. These plutons form the northwestern continuation of the abundant plutons present in adjacent Vietnam. They form a restricted area in a NW-trending belt of plutons that intrude both lower and upper Paleozoic and Mesozoic rocks. Some plutons are unconformably overlain by Middle Jurassic strata and are regarded as Jurassic in age, whereas others intrude across this unconformity and are considered either Late Jurassic or Cretaceous, but none are radiometrically dated. The origin of these plutons remains uncertain, but an understanding of them must be sought in their continuation into the broader area of northern Vietnam, where such plutons and older Paleozoic rocks are abundant.

The entire Lanping-Simao Unit contains numerous small and some large hypabyssal plutons of alkalic character that are probably of Cenozoic age, although few are dated. These rocks appear common also within tectonic units to the west and north (see above) and form a wide belt of similar Cenozoic plutons that can be traced from this part of China to the northwest and into central Tibet (Fig. 16-16). They have been variously interpreted, but we will suggest in the synthesis section below that they are related to the extrusion process of early Cenozoic age as well as to the late Cenozoic elevation of the southeastern part of the Tibetan Plateau (see below).

## MAJOR PROBLEMS IN THE LANPING-SIMAO UNIT

### Apparent Correlation of Key Rock Units across Possible Extrusion Faults

Several critical relations between rocks of the Lanping-Simao Unit and rocks to the east of its eastern boundary fault bear not only directly on the timing and amount of possible extrusion of the Lanping-Simao Unit during early Cenozoic time but also contribute to the general problem of when and on what faults or shear zones the extrusion took place. The belt of Middle to lower Upper Triassic sedimentary and volcanic rocks that form the lower part of the Qiangtang I unit and their relations to adjacent rock units have an important contribution to make to the correlation of the Qiangtang and Lanping-Simao Units. These rocks in the Qiangtang I unit trend N-S, lie west of the Paleogene Jinchuan basin, and have two important relationships:

1. The deformed and metamorphosed rock units of Middle to early Late Triassic age (with fossils) of the Qiangtang I unit are unconformably overlain (T3 in Plate 3) west of the Jinchuan basin by Upper Triassic strata (Mz2b) that are only weakly deformed and form the lower part of the Upper Triassic sequence (Figs. 6-3 and 7-4). These lower Upper Triassic rocks are correlated with similar units (Mz2b) in the Lanping-Simao Unit that lie west of the eastern boundary fault of the Lanping-Simao Unit, but rocks that lie below them are not exposed. The boundary fault separates them and is interpreted to be one of the major faults active during the extrusion process.

2. Two Paleogene units of the Jinchuan basin also unconformably overlie these rocks. The uppermost unit in the Jinchuan basin (Oligocene [?] without fossils) overlies the deformed Triassic rocks and overlaps the fault that juxtaposes the Triassic rocks against older Jinchuan basin strata (Plate 1). Farther south, where the deformed Triassic rocks are in fault contact with the north end of the Dian Cong Shan, the oldest unit of Paleogene rocks (with no fossils) also unconformably overlies both the deformed Triassic rocks, the Permian rocks that underlie them, and rocks of the Dian Cong Shan, tying these major rock units together by early Paleogene time (Plate 1). The Paleogene rocks form a narrow N-S–trending strip with a fault along their west side. Thus if there was a large displacement in early Cenozoic time it must have been on this fault, or a fault zone in this position, because all rocks to the west belong to the Lanping-Simao Unit. However, this oldest Paleogene unit, which consists of maroon mudstone and sandstone, is correlated not only with the oldest unit in the Jinchuan basin but also with the oldest unit in the Paleogene succession in the Lanping-Simao Unit to the west. All the rocks in these three areas are lithologically similar, and the Paleogene sequence in the Jinchuan basin is similar to the Paleogene sequence in the Lanping-Simao basin to the west.

These relations raise several important questions. If all these sequences do correlate, then where could a fault or fault zone of large magnitude have been present during the early Cenozoic extrusion of the Lanping-Simao unit? Do these relations suggest that the main displacement between Lanping-Simao rocks occurred in very early Cenozoic time or even in Triassic time? And did it occur on faults that we have not recognized as part of the extrusion process? The implications for regional tectonics of these unstudied problems are of first-order importance, and clearly additional research is needed.

### Cause of Subsidence of the Thick Mesozoic Strata

One of the major problems in interpreting the Mesozoic strata in the Lanping-Simao Unit is the same for the similar Mesozoic succession in the Qiangtang Unit: What caused the great subsidence within this widespread area of largely non-marine strata? The unconformities at the base and within the Triassic rocks have patterns of overlap that suggest that the rocks below them might have been tilted and rotated by extensional faults during deposition. The mapping has not been done to study this problem in enough detail to yield an answer. Unconformities within the Jurassic and the Cretaceous sections might lead to a similar conclusion. However, the poor continuity of exposure, lack of good marker units, heavy vegetation, and lack of detailed mapping have not been conducive to testing such a hypothesis. Many of these unconformities appear to be disconformities rather than angular unconformities, and some are clearly related to shortening rather than extension.

### Collage Nature of the Paleozoic Rocks

We have emphasized the stratigraphic differences in the Paleozoic strata below Mesozoic strata that characterized the

Lanping-Simao and Qiangtang Units. The geology of these rocks is better exposed in the Jinsha accretionary complex than within the Lanping-Simao Unit, where there is less continuity of the Paleozoic rocks. In the Jinsha complex the geology supports a concept that these rocks do not form a coherent terrane, but rather are a collage of accreted fragments of arc, continental, and various oceanic rocks. Although the geological relations are not as clear in the Lanping-Simao, we suggest that the Paleozoic rocks in this unit are likewise a collage of fragments accreted together by possibly Early Triassic or certainly early Late Triassic time—timing similar to that in the Qiangtang Unit to the north. The limited age of the oceanic rocks along their east side suggests that the Paleozoic rocks were separated from the tectonic units to the east during Paleozoic and early Mesozoic time, a conclusion also reached for rocks in adjacent Vietnam (Lepvrier et al., 2004).

One feature that has not been pointed out, and which may be of major geodynamic importance, is the fact that regionally the areas that we interpret to be underlain by the Paleozoic collage are the areas that are overlain by thick Mesozoic strata, with one exception. Fragments in the collage formed by higher grade metamorphic rocks were intruded by abundant plutons, and with probable Precambrian underpinnings appear to be the fragments that were not subject to the same broad subsidence as the rest of the collage terrane. This appears to be true for the entire region of thick Mesozoic strata. In addition, the larger tectonic units, underlain by continental crust, and laterally with more continuous Neoproterozoic and Paleozoic strata, bounded the Lanping-Simao and Qiangtang Units on both sides and did not undergo the same broad, uniform subsidence, except for subsidence related to thrust-related foredeep loading in the Chuxiong basin. This suggests that the collage underpinnings of the Lanping-Simao and Qiangtang Units behaved differently during early Mesozoic time, when broad, and perhaps weak, extensional deformation affected subsidence in the collage terrane in preference to the more coherent crustal areas adjacent to these units. It must be remembered that in a scenario that involves important early Cenozoic extrusion, the distribution of the collage and the overlying Mesozoic basins did not have the same relative positions that they have now.

## CHAPTER 8

# *Lhasa Unit*

The Lhasa Unit occupies much of the northwestern part of the area of Plate 1 (available on DVD accompanying this volume and in the GSA Data Repository[1]) and can be traced regionally to the south through Burma (Myanmar) into the Tengchong Unit, which is its southern continuation (Fig. 1-1B; see below). It is characterized by three belts of rocks: a northeast belt dominated by Mesozoic sedimentary rocks; a central belt dominated by Upper Paleozoic sedimentary and metasedimentary rocks and metamorphic rocks of uncertain protolith, all extensively intruded by Mesozoic plutons; and a southern belt dominated by Mesozoic and Cenozoic plutonic rocks with screens of metamorphic rocks of uncertain protolith (Plate 1). The distribution of rocks suggests that these belts form a crude north-dipping crustal section. The southern belt contains a narrow WNW–trending belt or shear zone, with rocks as young as Mesozoic, that lies along the eastern continuation of the active Jiali fault zone and is here called the *Parlung fault zone*. The plutons within the southern part of the Lhasa Unit are the eastward continuation of the Gangdese batholith belt of southern Tibet, but plutons in the central and northern belts may have contributions from subduction zones that are not the Yarlung-Zangbo subduction zone. To the southeast the northern belt narrows and becomes more extensively intruded by plutons.

The western boundary of the Lhasa Unit as shown on Plate 1 is well defined by the arcuate structure formed by the culmination at the eastern Himalayan syntaxis, of which only a small part is present in the northwestern part of the map area (Fig. 1-1B and Plate 1). This arcuate structure (the Syntaxial Unit) contains the eastward continuation of the Yarlung-Zangbo suture zone of southern Tibet that traditionally marks the southern limit of the Lhasa Unit, and west of the suture zone are rocks that correlate with the High Himalayan crystalline belt. The northeastern boundary of the Lhasa Unit is complex, poorly defined, and subject to interpretation. We draw a tentative northeastern boundary at the Bangong suture, which, as shown on Plate 1, contains the largest ophiolite body in the map area. Its eastern continuation is subject to considerable interpretation, and to its south is a large lenticular body of metamorphic and plutonic rocks whose tectonic position needs discussion. These metamorphic and plutonic rocks are bounded on the south by a narrow, discontinuous belt of mélange that we include within the Lhasa Unit, which, in its eastern part, contains blocks of mafic and ultramafic rocks and serpentinite, but which to the west contains only blocks of sedimentary rocks. The Bangong suture and the southern narrow belt of mélange have been disrupted by younger structures that are hard to interpret and will be discussed in more detail below. The Bangong suture is truncated to the east by a major fault zone and becomes cryptic to the southeast, where it is continued by the Nujiang suture, also largely cryptic (see below).

Several zones of rocks within the Lhasa Unit (sensu lato) may mark other sutures, so the Lhasa Unit may be composed of several smaller fragments. Most of these zones are only suspected suture zones of late Paleozoic or Mesozoic age.

## ROCK UNITS OF THE LHASA UNIT

The oldest rocks within the Lhasa Unit are Precambrian on the basis of geochronological data, such as U-Pb ages for zircons, Rb-Sr whole rock ages, and Sm-Nd model ages of metamorphic and igneous rocks that have yielded ages as old as 2200–1800 Ma (see, e.g., Hu et al., 2005). These rocks consist of gneiss and schist that locally contain kyanite and sillimanite (m on Plate 1). They form generally NW-trending, elongate, highly foliated, and folded bodies, most abundant in the southwestern part of the Lhasa Unit. The metamorphic rocks compose the host rock for large plutons that formed batholiths of Mesozoic and Cenozoic age. In many places they are migmatitic and may represent the deeper parts of a batholith. The protoliths of most of these metamorphic rocks are unknown.

Northwest- to north-south–trending belts of Devonian through Permian sedimentary and metasedimentary rocks, most abundant in the central part of the unit, form nearly all the Paleozoic rocks in the Lhasa Unit. The only exception is in the narrow, WNW–trending Parlung fault zone in the southwestern part of the area. Within this fault zone are several fault-bounded slices that consist of several hundred meters to >1600 m of foliated low-grade sandstone, siltstone, conglomerate, and dolomite with some felsic and intermediate volcanic rocks that are assigned a Neoproterozoic age (Z, Fig. 8-1, and Sinian, $Z_2$ on Plate 1) based on their lithological similarity to known Neoproterozoic rocks that occur elsewhere, but beyond the limits of the Lhasa Unit. Within this same belt are three fault-bounded outcrops of ~400 m of fossiliferous limestone and dolomite of Early and Late Ordovician age (O1, 2, 3, Fig. 8-1, and Pzl on Plate 1), but because of their limited outcrop, very little can be determined about the significance of these strata.

Devonian and Carboniferous strata lie within this narrow fault zone; however, they, along with Permian strata, form the

---

[1]GSA Data Repository Item 2012297, Plates 1–4, is available at www.geosociety.org/pubs/ft2012.htm, or on request from editing@geosociety.org or Documents Secretary, GSA, P.O. Box 9140, Boulder, CO 80301-9140, USA.

## Chapter 8

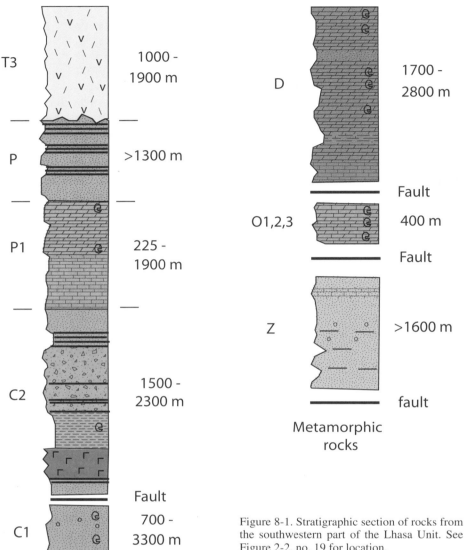

Figure 8-1. Stratigraphic section of rocks from the southwestern part of the Lhasa Unit. See Figure 2-2, no. 19 for location.

broad belt of upper Paleozoic rocks (Pzu) in the central part of the unit (Pzu, Plate 1). The upper Paleozoic rocks are folded and metamorphosed by large plutons that intrude them, so thicknesses given for these rocks are certainly structural thickness (Fig. 8-1). Fossiliferous Middle and Upper Devonian strata, along with lowermost Carboniferous rocks, consist of 2000+ m of limestone. Overlying them are >3300 m of fossiliferous Lower Carboniferous rocks that consist of sandstone and shale with some conglomerate and limestone and a thick unit of basalt. Intermediate and felsic volcanic rocks make up the upper third of the succession. Some conglomerate beds contain blocks up to 200–300 m long of limestone. Middle Carboniferous to Lower Permian strata are >3700 m thick and consist of sandstone, locally associated with basalt, and limestone at the base. Above the limestone is conglomerate (diamictite?), followed by sandstone and shale, with a thick unit of limestone with sandstone at the top with Early Permian fossils. The sequence within the Carboniferous is laterally variable, and some sections contain mainly sandstone with little limestone, but in other places contain mafic, intermediate, and felsic volcanic rocks, which are abundant. Upper Permian strata are mainly sandstone and shale, with some limestone. Much of the variability in the stratigraphic sections of Carboniferous and Permian rocks is probably due to folding, faulting, and metamorphism, which makes constructing continuous sections or lateral correlations difficult at best, so a detailed stratigraphic study of much of the upper Paleozoic rocks is lacking.

Diamictite and diagnostic fossils regarded as belonging to the Gondwana biota, with elements of Permian Gondwana flora and Late Carboniferous *Stepanoviella* fauna, are present, but the rocks also contain some Cathaysian biota in the same stratigraphic section that make a specific "mixed biota." This biota suggests that the Lhasa Unit was originally part of Gondwana or a terrane adjacent to Gondwana during late Paleozoic time (Li et al., 1985; Bureau of Geology and Mineral Resources of Xiang Autonomous Region, 1993). The stratigraphy of the upper Paleozoic rocks becomes very important tectonically in the SE part of the Lhasa Unit. In fact it is critical for determining where the major faults and sutures (e.g., Bangong and Nujiang sutures) lie, and where the eastern boundary of the Lhasa Unit should be placed (see below).

Mesozoic strata are widespread in the northeastern part of the Lhasa Unit, although key areas of Mesozoic rocks are present in small outcrops in other parts of the unit. Discontinuous belts of Triassic rocks occur mainly in three areas: in the north, adjacent to the Bangong suture; south of the lens-shaped body of metamorphic rocks in the north-central part of the Lhasa Unit; and in several NW-SE–trending belts in the eastern part of the Lhasa Unit (Plate 1). The two belts of Triassic rocks ($Mz_{2a}$) in the north and north-central parts of the Lhasa Unit are similar in lithology but different from the Triassic strata to the southeast. These two belts consist of a lower clastic unit with >900 m of sandstone, limestone, shale, and some conglomerate near the base. Fossils from these strata are of Late Triassic age. Above is more than 1200 m of fossiliferous limestone that is also of Late Triassic age (Fig. 8-2). Both units are faulted, making thickness estimates uncertain. The northern of the two belts is entirely fault bounded, whereas the southern belt is unconformably overlain by Lower Jurassic non-marine to marine sandstone and mudstone with coal, and it is >800 m thick. The lower Jurassic strata ($Mz_{2b_2}$) form the base of a thick section of Jurassic and Cretaceous strata ($Mz_{2b_1}$), characteristic of the northern part of the Lhasa Unit (Fig. 8-2). Locally they contain abundant blocks of limestone and sandstone, and in places they may be interpreted to be mélange or olistostromal deposits (see below).

Within the large area of metamorphic rocks (m, Plate 1) between these belts of Triassic rocks are two small fault bounded lenses of Triassic rocks ($Mz_{2a}$?) that lie within narrow zones of upper Paleozoic rocks bounded by metamorphic rocks on both sides. These rocks consist of sandstones with fossils of Late Triassic age, but the rocks are unlike those in the two larger belts of Triassic rocks, and their relations to those two belts remain unknown. Most of the outcrops of Triassic rocks are either bounded by faults or intrusive contacts. There is also one small, isolated outcrop of sandstone with limestone at the top that is mapped as Triassic ($Mz_{2a}$) ~35 km NE of the Syntaxial Unit in the western part of the Lhasa Unit. It rests unconformably on Permian strata (Plates 1 and 3 [see footnote 1]).

To the southeast are several areas of Triassic rocks that are different from the rocks described above. They consist mainly of sandstone, siltstone, shale, and rare limestone, which, when traced farther to the southeast, make up the upper part of an Upper Triassic succession whose base contains interbedded intermediate volcanic rocks ($Mz_{2a}$, Plate 1 and Fig. 8-3). The upper non-volcanic part of this sequence makes up the bulk of the section, and near the top it contains Late Triassic fossils. The basal volcanic-bearing part of the Triassic sequence rests unconformably on folded Carboniferous and Permian rocks, one of the few places where the base of the Triassic rocks is exposed (Plates 1 and 3). The most southeasterly trending belt of Triassic rocks in the southeastern part of the Lhasa Unit is a N-S, fault-bounded belt that consists of >4000 m (structural thickness?) of mainly sandstone and shale with some units of limestone and dolomite ($Mz_{2a}$ on Plate 1, but T3ch in Fig. 8-4 for details). There are no volcanic rocks in this belt, which dips moderately to steeply to the east and is cut by numerous N-S–striking faults. It is unfossiliferous and is assigned a Late Triassic age on the basis of its lithological correlation to fossiliferous clastic units to the NW. This belt, when traced northward, becomes faulted into the southeasternmost part of the Bangong suture (T3wp in Fig. 8-4; Plate 1). By placing these Upper Triassic rocks into the Lhasa Unit, we infer that the southward trace of the Bangong suture lies to the east of these rocks, an inference that needs to be discussed in some detail below.

Jurassic rocks consist of two groups of rocks, a thick succession of marine rocks that makes up a wide belt of Jurassic strata in the northern part of the Lhasa Unit, and narrow, local deposits of red beds to the south. The wide belt in the north can be divided into two parts: a thicker (up to 3000 m) part with abundant flysch-like strata in the north, and a thinner (2000–2500 m), sandier part to the south (both labeled $Mz_{2b_2}$, Plate 1). These two parts are separated by a belt of Lower Cretaceous strata ($Mz_{2b_1}$, Plate 1). The northern part consists of interbedded sandstone, siltstone, and shale that unconformably overlie Upper Triassic rocks (J2 on Plate 3) and that display sedimentological features that in some parts are true flysch. At the northern end of the large lens of metamorphic rocks surrounded by Mesozoic strata the Lower Jurassic rocks lie unconformably above the metamorphic rocks (Plates 1 and 3). They contain Early Jurassic fossils at the base and Late Jurassic fossils at the top. They border the complex fault zone of the Bangong suture in the north. Toward the east, where the large lens of metamorphic rocks is present, the Jurassic rocks become narrower, and depending upon the structural interpretation, may end against a major fault. On the south side of the lens of metamorphic rocks the Jurassic rocks contain abundant large blocks of limestone and some sandstone (w on Plate 1). They form a belt of block-laden Jurassic strata, which, when traced east, appears to continue as mélange with abundant blocks or inclusions of serpentinite (Jm, Plate 1). The Jurassic rocks south of the mélange belt continue eastward, where they are truncated against a major fault along the mélange zone. The southern part of the Jurassic sequence consists of more coarse sandstone with some limestone; it lacks finer grained material and has no flysch characteristics, but locally it contains abundant blocks of sedimentary rocks (w, Plate 1) in several places. It is intruded by numerous plutons, and its southern limits are mostly plutonic contacts; in a few places the Jurassic strata, here

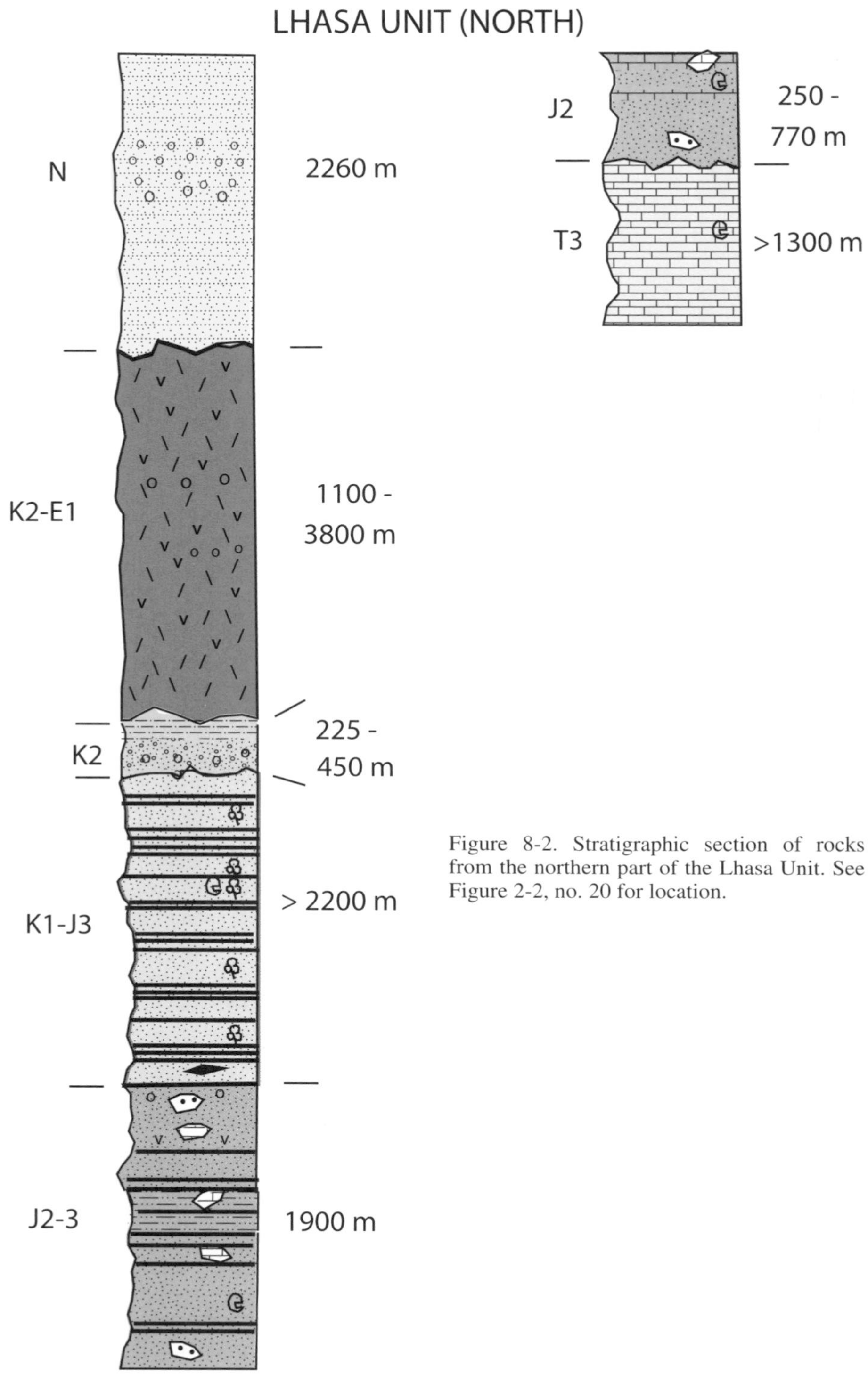

## LHASA UNIT (NORTH)

Figure 8-2. Stratigraphic section of rocks from the northern part of the Lhasa Unit. See Figure 2-2, no. 20 for location.

Middle Jurassic, rest unconformably on Upper Paleozoic rocks (Plate 3). There is no Triassic present along this southern margin of Jurassic strata, but to the east are small remnants of Jurassic rocks unconformably overlying the volcanic-bearing Upper Triassic strata (Plate 1).

A discontinuous belt of Jurassic strata ($M2b_x$) unconformably overlies upper Paleozoic rocks east of the large exposure of metamorphic rocks, referred to as a lens above, that is important in addressing the nature of the Bangong suture configuration. The Paleozoic rocks below them are assigned a Carboniferous age

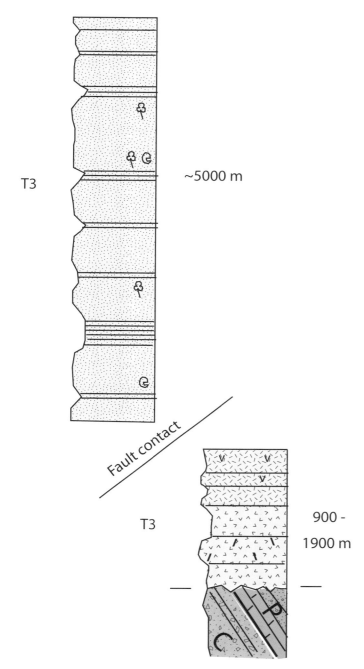

Figure 8-3. Stratigraphic section of the Triassic clastic and volcanic rocks of the southeastern part of the Lhasa Unit. See Figure 2-2, no. 21 for location.

the age of the plutons is poorly established, but they are assigned a Jurassic age on Chinese maps. Thus, the Paleozoic rocks should be considered as part of the large, structurally fault bounded lens that includes the metamorphic rocks. Middle and Upper Jurassic strata ($Mz2b_x$, Plate 1) lie unconformably above upper Paleozoic rocks. They consist of a basal conglomerate overlain by 800–1600 m of sandstone and shale, which in turn is overlain by 380–500 m of limestone. Both of these rock units are Middle Jurassic and mostly marine. The uppermost rocks are >380 m of sandstone and shale of Late Jurassic age. In the east-central part of the belt the Upper Jurassic fine clastics rest directly on the upper Paleozoic rocks. These sections all contain fossils. Importantly, these sections are different lithologically from the Jurassic flysch-like strata that are adjacent to the mostly fault-bounded lens of metamorphic and Paleozoic rocks. The Mesozoic rocks northeast of the Paleozoic rocks in the lens-shaped body consist of a thick section of Triassic and Jurassic, mostly marine rocks belonging to the Qiangtang III unit and are lithologically unlike the $Mz2b_x$ unit. The paleogeographic position of these Jurassic rocks ($Mz2b_x$) becomes important when evaluating the position of the Bangong suture in this part of the Lhasa region (see below).

Cretaceous sedimentary rocks occur in two parts: an older part that gradationally lies above the Jurassic section, and a younger part that lies unconformably on all older rocks (Fig. 8-2). The older Cretaceous strata consist of >2000 m of dark-gray, gray, and tan sandstone, mudstone, and shale. They lie conformably above similar rocks that contain Late Jurassic fossils near their base but Early Cretaceous fossils at the top ($Mz2b_1$, Plate 1). There is no obvious lithological contact that separates these rocks, and they are shown as a color contrast on Plate 1. The fossils are both marine and non-marine. This Upper Jurassic to Lower Cretaceous section contains coal beds at several levels. It shows shallowing of the marine to a terrestrial depositional environment above the flysch-bearing section below.

In the same northern part of the Lhasa Unit, and extending across the Bangong suture, is a section of red sandstone, mudstone, and conglomerate, with thin beds of yellow lacustrine limestone that rests unconformably on all of the older units ($Mz_3$/E, Plate 1; K2 in Fig. 8-2). In some areas these strata contain felsic volcanic rocks. The thickness of this section is uncertain, because these strata are shown on some maps as entirely Paleogene, but new fossil finds have shown that most of the lower part of these strata are Upper Cretaceous. They are at least 1000 m thick in some places.

In the southern part of the Lhasa Unit are small areas of red sandstone and shale that unconformably overlie mostly upper Paleozoic rocks. They occur mainly along the narrow, NW-trending Parlung fault zone, in which some lower Paleozoic rocks are present, and most of their contacts are either fault or intrusive contacts with younger Mesozoic plutons. They are unfossiliferous and are regarded as either Jurassic or Cretaceous (Mz3) or in places, both. Because of their limited exposure their relations to the Mesozoic rocks to the north are unknown.

and are in fault contact with the metamorphic rocks to the west. The Paleozoic rocks are joined together with the metamorphic rocks by plutons that intrude both rock units and the fault that separates them. Thus, at the time of intrusion, the metamorphic and upper Paleozoic rocks were part of one unit. Unfortunately

# Chapter 8

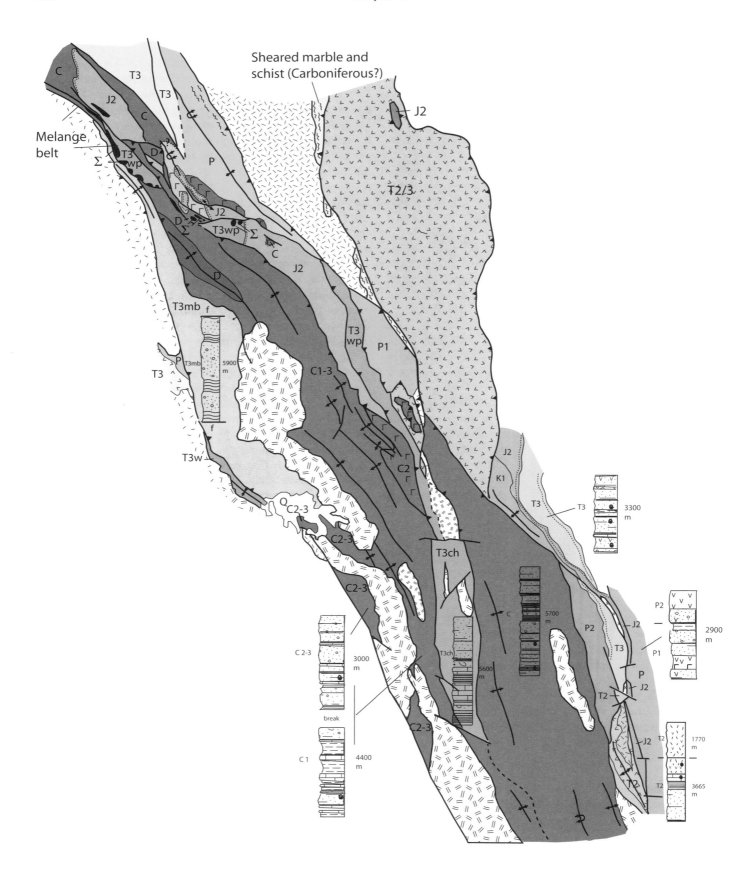

In the central part of the Lhasa Unit, south of the lens of metamorphic rocks, is a section of >3000 m of felsic to intermediate volcanic tuff, agglomerate, and flows interbedded with volcanogenic sedimentary rocks ($Mz_3$). The unit is shown lying unconformably on Lower Cretaceous strata and Mesozoic plutonic rocks. At the only place where we visited these rocks in the field the contact mapped as an unconformity was clearly a low-angle fault, with rocks in its hanging wall dipping vertically into the contact. These rocks are assigned a Late Cretaceous to Paleogene age (K2–E1; Fig. 8-2), but there are no fossils or radiometric ages to support this age assignment.

Paleogene strata (E, Plate 1) are present mainly in the northern part of the area near and above the Bangong suture. They consist of several hundred meters of red sandstone, mudstone, conglomerate, and lacustrine limestone. These rocks were originally mapped as Paleogene, but many of them were later assigned a Late Cretaceous age on the basis of new fossil evidence; thus their thickness and distribution is somewhat in doubt. They are probably Paleogene where they rest unconformably on the Upper Cretaceous red beds as well as on ophiolitic rocks in the Bangong suture zone.

Neogene strata occur as narrow, NW-trending basins in the north-central part of the Lhasa Unit. They consist of sequences 1000–2000 m thick of tan to yellow sandstone, shale, mudstone, and conglomerate that lie unconformably on older units (Plates 1 and 3). They are unfossiliferous, so their ages are uncertain.

Serpentinite and mélange units are present within the Lhasa Unit, depending upon how the unit is defined and how its boundaries are structurally interpreted. The northern boundary of the Lhasa Unit is marked by the Bangong suture (ophiolite and Jm mélange), but it can be followed only partway to the southeast before it is truncated by a major fault. This boundary will be treated separately below. South of the large lens of metamorphic rocks in the north-central part of the Lhasa Unit is a narrow belt of mélange (Jm) that contains numerous blocks of serpentinite in a sheared matrix of phyllitic shale and sandstone with some limestone that has yielded Jurassic fossils (Plate 1). This belt, when followed to the west, appears to grade into rocks mapped as Jurassic fine clastics that contain blocks of only limestone, where they are regarded as a unit within the general Jurassic sequence of the Lhasa Unit, and they end within the Jurassic strata farther west. The mélange with serpentinite blocks can be followed to the southeast, where it becomes faulted against Triassic rocks to the east; the same belt of fault-bounded Triassic rocks can be followed between large areas of Carboniferous rocks in the eastern part of the Lhasa Unit (Fig. 8-4). At this point the mélange belt and the Bangong suture can no longer be confidently traced to the south. Details of an argument that the Bangong suture becomes a cryptic suture and continues to the south will be presented in the discussion on the Bangong suture below.

Within the southern part of the large metamorphic lens in the northern part of the Lhasa Unit is a NW-trending fault-bounded belt consisting of slivers of Carboniferous and Permian sedimentary rocks that divides the lens into two parts. It can be followed to the NW along the northern margin of the lens to where it is shown as unconformably overlain by lower Jurassic strata. Along these Paleozoic rocks are the fault slices of Triassic rocks of unknown affinity ($Mz_{2a}$?) mentioned above. Along several of the faults along the margins and within this belt are slices of serpentinite. How this potential mélange(?) belt can be incorporated into the tectonic framework of the Lhasa Unit will be discussed in more detail below. Recent work by Sha et al. (2009) and Zhang et al. (2010b) reported a high-pressure granulite assemblage of rocks within the metamorphic rocks just to the east of the belt of Paleozoic rocks. The assemblage yielded pressures of 11–13 kb with temperatures of 600–750 °C. These rocks were overprinted by a later amphibolite facies metamorphism that yielded pressures of 6 kb and temperatures of 500–600 °C, but unfortunately no ages have been determined for either metamorphic assemblage. At the northwest end of the large lens of metamorphic rocks is a second belt of Upper Paleozoic strata with slivers of serpentinite and blocks of sedimentary rocks. Like the larger belt to the east it is mainly fault bounded, but it is also overlapped by Lower Jurassic strata.

In the southern part of the Lhasa Unit, northwest of where the narrow Parlung fault belt of Paleozoic and Mesozoic rocks ends, are four outcrops of ultramafic rocks and serpentinite. They occur within fault zones in granite plutons and metamorphic

Figure 8-4. Generalized map in the area of the possible southern continuation of the southern segment of the Bangong suture. Much of this area is within very high, rugged topography with glacial cover; thus the relationships between units are poorly known. The southeastern extent of the Bangong suture, as we interpret it, is marked by abundant serpentine bodies that lie within mélange (purple) and also within units mapped as Triassic (T3wp) and Devonian (D). Farther southeast there are no mapped serpentine bodies, and the unit mapped as J2 may contain some mélange. Triassic units (T3wp, T3ch, T3mb, and T3w), shown in various hues of orange, have different sequences of strata, and how they correlate is unknown. However, they are dominantly clastic sections and are different from the volcanic-rich sections in the Qiangtang I unit to the east, and they are interpreted to belong to the Lhasa Unit. The broad areas of Carboniferous strata (C1-3 and C2-3) are strongly deformed and metamorphosed to varying degrees, but they are dominated by thick clastic sections with few fossils, whose stratigraphy is poorly known, and are intruded by abundant plutons whose ages are poorly known (shown as mostly Cretaceous). These sections are most similar to those of the Lhasa Unit. All the contacts between rock units are faults, many interpreted as thrust faults, and the southeastern continuation of the Bangong suture in this area must be cryptic. The thick, volcanic-rich Triassic sections (T2/3 and T2) and their overlying Jurassic (J2) and Cretaceous (K1) strata are typical for the Qiangtang I unit. Thus the cryptic southeastern extent of the Bangong suture should lie either east of, or along, the faults bounding the Triassic strata of the Lhasa Unit (T3wp and T3ch), or, more likely, east of the Carboniferous strata and along the fault west of the P2 unit, which underlies the Qiangtang I rocks and continues southwest of the Triassic volcanic-rich rocks (T2). P—Permian; f—fault.

rocks. Their occurrence has been used to suggest the presence of a suture zone in this region that might continue to the SE along the narrow Parlung fault (Geng et al., 2005). The evidence is at best equivocal for a suture zone in this part of the Lhasa Unit, and further mapping is clearly needed.

## STRUCTURE OF THE LHASA UNIT

The belts of rocks and structures within the Lhasa Unit trend NW-SE, curving to north-south to the east except in the southwest, where the rocks are dominated by more or less large irregular or equidimensional plutons. Crudely, from southwest to northeast, the belts of rocks become younger, with the exception of the large lens of metamorphic rocks in the north-central part of the unit. In the west the metamorphic rocks have been assigned to the Lhasa basement (mLB), where they have yielded U-Pb ages from zircons, Rb-Sr whole rock ages, and Sm-Nd model ages of metamorphic rocks as old as 2200–1800 Ma (Hu et al., 2005), and many are high-grade kyanite- or sillimanite-bearing rocks. These rocks are continuous with rocks that form wall rocks within the Gangdese batholith to the west, and these rocks and their associated migmatites may locally represent the roots of the Gangdese batholith. These rocks are sharply truncated at the steeply east-dipping arcuate faults that form the eastern boundary of the Syntaxis Unit, where underlying rocks of this adjacent Syntaxial Unit are lower grade metasedimentary schist and gneiss (sa on Plate 1; see below). Near the Syntaxis Unit foliation, the Lhasa Unit rocks dip steeply eastward, away from the syntaxial culmination (see below).

Eastward of the Syntaxial Unit, several NW-striking, SW-dipping faults carry metamorphic rocks of the Lhasa basement northeastward above Devonian and Carboniferous metasedimentary rocks. The dominantly upper Paleozoic rocks to the east are folded and cut by faults, many of them thrust faults that dip both SW and NE, with the SW-dipping faults apparently dominant. It should be noted that all the thrust faults shown within the Lhasa Unit and elsewhere are shown on cross sections as generally steeply or moderately dipping, and mapped low-angle thrust faults are rare (Plate 4, section 12 [see footnote 1]). The way that the thrust faults are mapped through areas of high topographic relief suggests that they are indeed steeply dipping. In the central part of the Lhasa Unit another belt of metamorphic rocks (m, Plate 1) is brought up along faults that thrust them southward over rocks as young as the red beds of Mesozoic age along the narrow, NW-trending Parlung fault zone. The central part of the Lhasa Unit is dominated by folded and faulted upper Paleozoic rocks intruded by numerous plutons of Mesozoic age. Some of the plutons are cut by NW-striking faults, but most of them appear to intrude both the folds and faults within the upper Paleozoic rocks. The age of these plutons is poorly known because there have been no systematic studies of their ages. They are shown on maps as both Jurassic and Cretaceous, and those in the southwest near the syntaxis include abundant plutons assigned a Cenozoic age. A recent study by Chui et al. (2009) indicates that some of them are Cretaceous (133–110 Ma).

The western part of the Lhasa Unit is dominated by metamorphic rocks of unknown protolith, and upper Paleozoic metasedimentary rocks and plutons continue to the southeast, where they form the eastern part of the Lhasa Unit. To the southeast the faults and folds, and the belts of rocks, become narrower, more elongated, and strike more N-S, and the different rock types become more structurally interleaved (Plate 1). Before the rocks enter Burma (Myanmar) to the south, faults, folds, and rock units are all parallel. All faults are steeply dipping, and folds are near isoclinal and upright. Faults bounding rock units dominantly show thrust separation, but normal-separation faults are also present. Some faults are shown as right-lateral strike-slip, but evidence is not given. Some of the patterns of the anastomosing faults do suggest right-lateral shear. The metamorphic rocks are assigned a Proterozoic age, but the basis for this is unknown. Only a 10–15-km-wide belt of rocks continues along the Burma (Myanmar) border to the south, where they join the Tengchong Unit.

Where the Mesozoic rocks are present in the northern part of the Lhasa Unit, more can be interpreted about the timing and style of deformation. The Mesozoic rocks and their structures trend NW-SE, but to the east they narrow and end at the eastern boundary of the Lhasa Unit. This boundary, however, is hard to determine in detail and will be discussed below.

Pre–Upper Triassic deformation can be documented only in the southeastern part of the Lhasa Unit, where the volcanic-bearing Upper Triassic strata unconformably overlie folded Lower Permian and Carboniferous rocks (Plate 3). Later, both the Triassic and Carboniferous rocks were folded together, and map relations show plutons that intrude both upper Paleozoic and Triassic rocks. Small areas of Middle Jurassic strata unconformably overlie the Triassic rocks and the plutons that intrude them, and bracket the later deformation to be Late Triassic or Early Jurassic. This also indicates that some of these plutons are pre–Middle Jurassic and Late or post–Late Triassic in age, which would make them some of the oldest plutons in the batholithic belt, but the plutons are not radiometrically dated. These relations occur in such a small area that it is uncertain how they may figure regionally.

In the northern part of the Lhasa Unit, folded Upper Triassic rocks are unconformably overlain by Lower-Middle Jurassic strata. The same Jurassic strata unconformably overlie the metamorphic rocks of unknown protolith at the northwest end of the large lens of metamorphic rocks. The overlap by the Jurassic strata of the two different rock units suggests that Upper Triassic deformation may have been important. It is only in this area where such relations can be demonstrated (Plate 1). In many places Middle Jurassic strata overlie pre-Triassic rocks, yielding only an upper limit to the age of deformation (see Plate 1).

There is strong evidence for an important and widespread Cretaceous deformation. The continuous Jurassic to Lower Cretaceous sequence in the northern part of the area is folded, thrusted, and overlain unconformably by Upper Cretaceous red beds (Plate 1 and Plate 4, section 38). The structures trend NW-SE, parallel to the northern boundary of the Lhasa Unit, formed by the Bangong suture, and have a fold-thrust-belt style. They are probably detached at depth, but at what depth is unclear. Folds are shown as generally upright, but thrust faults are dominantly south vergent (Plate 4, sections 14, 38). Upper Cretaceous red beds overlie these folded rocks unconformably, demonstrating a deformational event within the middle part of the Cretaceous. The unconformity at the base of the thick volcanic succession of Upper Cretaceous–Paleogene rocks that rest on Lower Cretaceous rocks in the central part of the Lhasa Unit also suggests a Late Cretaceous event. However, care must be taken as to the nature of this contact, because our limited observations indicate that at least some part of this contact is a fault (see above). This contact is locally mapped as intrusive, indicating that some plutons in the batholith are Late Cretaceous or early Paleogene in age. Farther to the SE, another small unit of Upper Cretaceous red beds lies unconformably on Lower Cretaceous rocks, but the limited occurrence of these Upper Cretaceous rocks makes the establishment of the regional distribution of the Middle Cretaceous deformation difficult.

The Upper Cretaceous red beds are also folded, and in a few places they are unconformably overlain by Paleogene red beds, demonstrating a Late Cretaceous or earliest Paleogene deformational event. Most of the evidence for this event lies near or within the Bangong suture zone (Plates 1, 3, and 4, section 37a). Again, care must be taken in trying to establish the distribution of this event, because some of the rocks originally assigned to the Paleogene have yielded Late Cretaceous fossils.

The fault contact between the Lhasa and Syntaxis Units at the northwestern boundary of the Lhasa Unit began to develop in Late Cretaceous time, when Tethyan oceanic crust was subducted beneath the Lhasa Unit. This was followed by a complex history of faulting within the Syntaxial Unit, which extended into late Cenozoic time with reactivation of many of the earlier formed structures (see below). This age is based on regional correlations that place the unit at the top of the Syntaxial Unit as the lateral equivalent of the Yarlung-Zangbo suture (Geng et al., 2005). This was followed by the northeast-directed underthrusting of the High Himalayan crystalline belt (northern part of the Indian continent) represented on Plate 1 by the Namche Barwa Group below the Lhasa Unit.

Latest Paleogene or earliest Neogene deformation is demonstrated at only one place, where Neogene rocks (N, Plate 1; largely undated) rest unconformably on folded Upper Cretaceous–Paleogene volcanic rocks in the central part of the unit. Neogene basins are generally present only locally, and where they are their strata are unconformable on older rocks, but this does not date the exact age of deformation. The Neogene rocks are folded, some strongly with dips near vertical, indicating a late Neogene deformation. More importantly, some of the large, NW-striking faults in the north-central part of the Lhasa Unit truncate the Neogene strata. These and many other NW-striking faults also cut Upper Cretaceous or Paleogene rocks. The age of faults is thus poorly constrained and subject to interpretation, but it is clear that they play an important role in the Cenozoic tectonic history of not only the Lhasa Unit but of the whole region as well (see below).

The ages of the different plutons in the batholith are poorly known, but local relations between some plutons and structural and stratigraphic units can be used to assign some ages. Some examples have been given above. Many of the plutons along the southern margin of the folded Jurassic rocks in the central part of the Lhasa Unit cross-cut the structures interpreted to be of Middle Cretaceous age; thus these plutons are at least Late Cretaceous in age (Plate 4, section 14). Other small to large plutons intrude into the fold-thrust belt of Middle Cretaceous age also indicate that these plutons are at least of Late Cretaceous age (Plate 1). Until careful study of the ages of plutons and their relations to structures have been determined, about all that can be concluded is that deformation has been largely one of intra arc shortening of probable Jurassic, Cretaceous, and early Cenozoic age, but the distribution of these deformations remains largely unclear, particularly in the southern part of the Lhasa Unit, where plutons dominate and Mesozoic rocks are rare.

The narrow, NW-trending Parlung fault in the southwestern part of the Lhasa Unit has generally been interpreted to be an important structure of different types and ages. The Mesozoic red beds in this belt are poorly dated and lie unconformably on Paleozoic rocks. The fact that the red beds are continuous along this belt, and the different older units are not, suggests that at least some of the structures are pre-Mesozoic in age. The structure in this belt is complicated, with most of the units in thrust contact with one another. Thrust faults dip both NE and SW, and all units, including the red beds and some plutons locally, are involved. Rarely, plutons intruded rocks within the belt, including the red beds. The faults within this belt are linear and appear to have considerable lateral continuity, and even where they diverge from this belt some can be followed into or through the plutons. Our cursory examination of the belt near the Parlung River suggests that some of the faults are active (Plate 1). Ratschbacher et al. (1994, 1996), based on old K-Ar dating of biotite by Zhang et al. (1980), suggested that the belt of right-laterally sheared rocks was active at ca. 20 Ma, and a more recent $^{40}Ar/^{39}Ar$ study by Lee et al. (2003) yielded data that can be interpreted as constraining the age of shearing as somewhat younger (15.4 ± 1.1 Ma and 16.7 ± 1.0 Ma), indicating that the Parlung fault had a history of late Cenozoic to present deformation. The right-lateral active Pugu fault splays southward from the Parlung fault east of Namche Barwa (Plate 1) and has yielded $^{40}Ar/^{39}Ar$ ages from mylonitic rocks that range from 11.7 ± 0.1 Ma to 17.1 ± 0.4 Ma, indicating that these two faults were probably active at the same time and remain active today.

Not only has this belt of sheared rocks along the Parlung fault had a history of late Cenozoic deformation, but that also the faults along this belt and their relation to Mesozoic and Paleozoic strata within the belt mentioned above, suggest that deformation along the Parlung fault has had a long history of apparently localized deformation, and that the belt is unique because of the presence of Neoproterozoic and early Paleozoic rocks not present elsewhere in the Lhasa Unit. The pre–late Cenozoic history of the rocks along the Parlung fault remains largely unstudied. Projected to the NW, this belt might pass through the several outcrops of ultramafic rocks and serpentinites along NW-striking faults that cut through metamorphic rocks of unknown protolith and some plutonic rocks. This has suggested to some workers that this belt marks a suture zone that contained oceanic crust (Geng et al., 2005). These authors interpreted this belt to have been an oceanic area that separated the Lhasa Unit into two separate continental fragments until it closed sometime in the Mesozoic. This interpretation must await future mapping and study. If this was an oceanic area of late Paleozoic or Mesozoic age, subduction within this zone could have been responsible for some of the plutonism within the Lhasa Unit, as Geng et al. (2005) suggested.

This narrow belt also lies close to the eastern projection of the Jiali fault in southern Tibet, west of the area of Plate 1, and many workers have considered the Parlung fault to be an eastern continuation. The Jiali fault has been used as a prominent fault in regional tectonic interpretations that suggest that the fault is related to major Cenozoic eastward movement of Tibet (Armijo et al., 1986, 1989) and could have as much as >200 km (Ratschbacher et al., 1996) or 450 km of right-lateral displacement (Armijo et al., 1989). The Jiali-Parlung fault zone may have played an important role in, and the eastward extrusion of, parts of Indochina (Replumaz and Tapponnier, 2003). In some interpretations the faults have been considered to be the westward continuation of the Ailao Shan shear zone of SE China (Armijo et al., 1989; Tapponnier et al., 2001). That interpretation is highly unlikely, based on the mapping shown in Plate 1. For one thing, the Ailao Shan shear zone had a right-lateral shear sense only in the late Cenozoic, when the demonstrable slip was only a few tens of kilometers. Even though the Jiali-Parlung fault zone has been subject to considerable regional interpretation, there has been little geological mapping or evidence presented to support regional interpretations. There are several long, continuous NW-striking faults other than the Jiali-Parlung fault in the SE part of the Lhasa Unit. Selecting which fault is the eastern continuation of the Jiali fault needs more attention. All these faults, where continued to the east, turn south, where they pass into the Tengchong Unit within China (the Parlung fault becomes the Gaoligong shear zone? See below) and have possible correlatives in Burma (Myanmar) (Lee et al., 2003).

Suffice it to say at this point that the NW-trending belt of Mesozoic and Paleozoic rocks along the Parlung fault zone has been the locus of long and apparently focused deformation. Something had to localize the deformation in this narrow belt. At present we suggest that there was a break within the crust of the Lhasa Unit that must have led to the preservation of Neoproterozoic and lower Paleozoic rocks along this belt, as they are absent elsewhere within the Lhasa Unit. The initial break localized succeeding shortening deformation of several ages, and the thrusting within the zone and along its margins brought up higher grade metamorphic rocks on both sides, suggesting transpressional deformation of probable pre– and/or post–red bed (Mesozoic) age. The sense of shear at these times remains unknown.

**Major Unresolved Problems**

*How Many Fragments Are within the Lhasa Unit?*

The Lhasa Unit in most syntheses is shown as one large tectonic unit. However, the geology suggests that it may have consisted of several units that were accreted together before its final incorporation into Eurasia. The narrow, NW-trending Parlung zone, which contains fault-bounded pieces of lower and upper Paleozoic rocks in the southwestern part of the Lhasa Unit, has been suggested to mark an oceanic suture (Geng et al., 2005). The evidence for its possible oceanic character is based on the presence of several isolated fragments of mafic and ultramafic rocks along the northwestern projection of the zone (see above and also Plate 1). The timing of possible closure in this zone would be given by the ages of the plutons that intruded these rocks. Although the plutons are assigned a Jurassic age, there is no radiometric age for these plutons. There are no mafic or ultramafic rocks present within the narrow fault zone to the southeast, and all the fragments of Neoproterozoic and lower and upper Paleozoic rocks are of shallow-water character. The undated Mesozoic red beds (Mz3) that lie unconformably on the older rocks would suggest closure sometime within the Mesozoic if it is a suture zone.

Within the large, lens-shaped body of metamorphic and igneous rocks in the northern part of the Lhasa Unit are two narrow, fault-bounded belts of upper Paleozoic strata with abundant lenses of serpentinite. These two belts are also associated with fault-bounded slices of Triassic strata of unknown affinity ($Mz_{2a}$?). If these belts represent fragments of one or more sutures, they would have been closed by Early Jurassic time, as both belts are unconformably overlain by Lower Jurassic strata. The possible lateral extent of these short belts of rocks is unknown. Whether they might represent older parts of the Bangong suture, or separate sutures, remains unknown, but we suggest below that these belts are rifts related to the opening of the Bangong suture. Zhu et al. (2010) suggested, on the basis of evidence mainly west of the area of Plate 1, that the Lhasa Unit was not a single unit but consisted of at least two or three units separated by oceanic rocks of late Paleozoic age. Whether this is true still requires much future research.

# CHAPTER 9

# Bangong Suture Zone

The Bangong Suture is one of the major suture zones in Tibet proven to have oceanic crust (Fig. 9-1; Cao et al., 2005; see also a recent review by Shi et al., 2007, 2008). The suture zone trends east-west across central Tibet and projects eastward into the area of Plate 1 (available on DVD accompanying this volume and in the GSA Data Repository[1]) in its northwestern part, where it contains the largest ophiolite body in the area. It separates the Lhasa Unit from a unit we assign to the Qiangtang III unit. The two units have very different Mesozoic stratigraphy, as would be expected if the Bangong Suture is a major Tethyan suture. Although the Bangong Suture is a prominent structure and forms the boundary for the units on either side, in the area of Plate 1 its continuation to the southeast becomes unclear, as are some of the rock units associated with the boundaries of the suture. South of the large ophiolite body, at least one and possibly two other belts marked by mélange and mafic and ultramafic rocks could be considered to be parts of the Bangong Suture, but in our interpretation these belts belong to a broad extensional region along the Lhasa northern continental margin during the opening of the Bangong ocean (see below). We suggest that the continuation of the oceanic suture to the south passes into the Nujiang suture, but it is cryptic for much of its trace.

The Bangong Suture is well developed and well defined in the northwestern part of the area of Plate 1. A large body of mafic and ultramafic rocks ~100 km long and 10–15 km wide crops out and contains all the rock types characteristic of an ophiolite and a related subduction complex: basal ultramafic rocks, overlain by mafic and ultramafic cumulates, some sheared with sheeted dikes above, feeding into pillow lavas and basaltic flows overlain by deep-marine sediments. The entire ophiolite body is surrounded by mélange (Jm, Plate 1). The top of the ophiolite body is unconformably overlain by a thin unit of Middle Jurassic marine rocks (J, Plate 1). Both north and south of the ophiolitic rocks are belts of mélange that consist of sheared shale, sandstone with some limestone, and numerous blocks of white limestone, marble, sandstone, and serpentinite. These rocks are widest and best developed southeast of the main body of the ophiolite, where the mélange matrix with sub-vertical foliation also encloses large lenses of basalt and metamorphic rocks in addition to serpentinite. The mélange belt is unconformably overlain in well-exposed outcrops by Upper Cretaceous red beds with conglomerate (Mz3 and Mz3/E). The mélange matrix contains fossils as young as Late Jurassic, and Upper Cretaceous red beds unconformably overlie the ophiolite body at many places; thus closure of the oceanic crust in this area is bracketed between Late Jurassic and Late Cretaceous. If the south-vergent fold-thrust belt south of the suture in the Lhasa Unit is related to the closure of the suture, then activity along the suture must have extended into middle Cretaceous time.

Upper Cretaceous rocks that overlie the ophiolite and mélange are folded and faulted, and the unconformably overlying Paleogene rocks are likewise folded and faulted. This shows continued reactivation of the suture zone at several times, making it difficult to assign structures a definite age. In other parts of the suture, Neogene rocks are also cut by faults that splay from the suture zone, indicating that activity along the suture extended into late Cenozoic time; some of the faults may be active today (Plate 1). The Bangong, as well as other suture zones within southeastern Tibet and SE Asia in general, are loci of continued deformation long after their initial formation and closure. This is one of the problems of unraveling their structural evolution without detailed mapping. The continued reactivation of the Bangong Suture is no exception and has made it difficult to follow along strike, particularly to the SE within the area of Plate 1.

The ophiolite pinches out to the northwest, and the body narrows and pinches out to the east, but the adjacent mélange with lenses of ophiolitic rocks continues to the east for another 25–30 km before it ends. Where the mélange is still recognizable to the east, Triassic and Jurassic rocks of the Qiangtang III unit lie to the north. The mélange can be followed south for another 75 km, and the Triassic rocks on its north side for an additional 200 km (Plate 1 and Fig. 9-1); both rock types are in fault-bounded belts. South of the mélange is a belt of Triassic (Mz2a) and Jurassic (Mz2b$_2$) strata characteristic of the Lhasa Unit, but it ends at about the same place where the mélange ends. From this point, farther to the southeast, Devonian and Carboniferous rocks are in fault contact with the Triassic strata of the Qiangtang III unit to their north. Because these upper Paleozoic rocks are part of the large complex lens that includes metamorphic rocks and upper Paleozoic plutonic rocks that continue to the northwest, where they lie south of the large ophiolite body that marks the Bangong Suture and are unconformably overlain by the Jurassic strata of the Lhasa Unit, they would logically belong to the Lhasa Unit. Thus, southeast of where the mélange belt of the Bangong Suture ends, the Lhasa and Qiangtang Units are separated by a fault zone that marks a modified or cryptic suture. This fault zone dips west and has a thrust separation in the north, but it dips east and has a normal separation in the south. Just before the mélange belt ends, faults along it also displace Paleogene

---

[1]GSA Data Repository Item 2012297, Plates 1–4, is available at www.geosociety.org/pubs/ft2012.htm, or on request from editing@geosociety.org or Documents Secretary, GSA, P.O. Box 9140, Boulder, CO 80301-9140, USA.

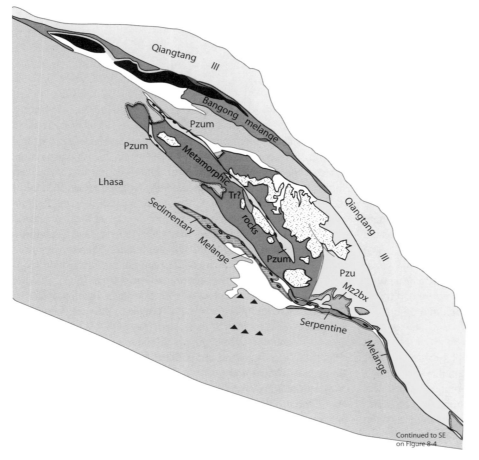

Figure 9-1. Generalized geological map of the Bangong and related mélange between the Lhasa Unit and the Qiangtang III unit. The large ophiolite body (black) crops out in the northern belt of mélange (dark blue) and is the location of the Bangong ocean. The narrow pink unit above the ophiolite is a Jurassic limestone that lies unconformably above the ophiolite. The southern mélange belt contains serpentinite and mafic and ultramafic rocks, but when traced northwest contains only sedimentary clasts (light blue). The Jurassic clastics and flysch (gray) to the south contain several areas with only sedimentary olistostromes (black triangles within the gray). The red and yellow areas are metamorphic and upper Paleozoic rocks, respectively, of the Lhasa basement. They contain narrow belts of Paleozoic rocks with slivers of serpentinite (mélange of unknown age?). The green areas labeled Mz2bx are outcrops of Jurassic shallow-water rocks of unknown affinities (see text for discussion). Pzum indicates belts of upper Paleozoic rocks with bodies of serpentinite and slivers of Triassic(?) rocks, a possible older mélange belt unconformably overlapped by Jurassic strata.

strata, indicating that these faults are a younger modification of the suture.

A second belt of mélange (Jm, Plate 1) with abundant serpentinite lenses crops out south of the belt of Paleozoic rocks regarded as belonging to the Lhasa Unit (Fig. 9-1). The mélange matrix in this belt contains Jurassic fossils; thus it has characteristics not unlike the mélange assigned to the Bangong Suture to the north. This mélange continues northwest, where it lies at the southwestern margin of the large body of metamorphic (m) and plutonic rocks. The northwestern part of this belt of mélange appears to continue as a belt within the Lower and Middle Jurassic flysch of the Lhasa Unit before it can no longer be followed as a discrete unit. Where it continues to the northwest into the flysch the blocks are all sedimentary, largely limestone, and they lack serpentine (Fig. 9-1). To the southeast the southern mélange belt, still with abundant blocks of serpentinite, continues for ~150 km, where it ends along thrust faults that involve both Paleozoic and Triassic rocks. These Triassic rocks are parts of two long, N-S–trending belts of clastic rocks regarded as belonging to the Lhasa Unit (Fig. 8-4). Neither the mélange nor any rocks characteristically part of the mélange are present farther south. Unfortunately along the southernmost part of this mélange belt the rocks on both sides are metamorphosed upper Paleozoic strata (Plate 1), assigned Carboniferous and Permian ages (see Fig. 8-4), but the sections have not been studied in enough detail to know if they can be distinguished as belonging to either the Lhasa or the Qiangtang Unit (Plate 1).

South of the southern mélange belt the Jurassic strata of the Lhasa Unit (Mz2b$_2$) contain blocks of sedimentary rocks that are interpreted to be olistostromes. They are rocks that we interpret to have been deposited during rifting and basin formation, related to the extension of the basement during opening of the Bangong ocean to the north (see below).

Middle and Upper Jurassic rocks (Mz2bx, Fig. 9-1 and Plate 1) are mapped as unconformably overlying the Paleozoic rocks north of the serpentine-bearing mélange and at the end of the mélange belt (Jm), and where fault-bounded Triassic rocks (Fig. 8-4, T3wp) are mapped in association with serpentinite in what appears to be the continuation of the southern mélange belt. At this southeastern end of the mélange belt the Triassic strata associated with serpentinite are mapped as unconformably overlain by Middle and Upper Jurassic rocks (J2, Fig. 8-4, but Mz2b$_2$ on Plate 1). If this can be substantiated it indicates that at least part of the mélange is older than the Late Jurassic–Early Cretaceous age established elsewhere (see above). This is a critical point and will be elaborated in another context below. Regionally two sutures of different ages project into this area from the south, the Nujiang and Changning-Menglian sutures of late Mesozic and later Paleozoic–early Mesozoic ages, respectively (see above); thus where they project into this area is a fundamental problem that remains to be resolved (see below).

Geological relations along the southern mélange belt indicate that this belt has been strongly modified by post-suturing deformation. Faults along much of the mélange belt involve the Upper Cretaceous red beds, indicating that the belt was reactivated (Plate 1). Near the northwestern end of the mélange that contains serpentine blocks are three large masses of metamorphic rocks that overlie the mélange, and they have been mapped as thrust sheets coming from similar rocks to the north. They appear to us to be better interpreted as large landslide masses that have incorporated serpentinites at their bases. In the same area are several NW-striking faults that bound this mélange belt and that also involve Neogene strata, indicating late Cenozoic reactivation along this part of the mélange belt.

The conundrum here is that this southern belt of mélange ends both to the northwest and southeast, apparently with elements of the Lhasa Unit on both sides. Several questions arise. Is this southern belt actually a piece of the Bangong Suture? If it is, how did it attain its present position? Where does the cryptic suture marking the Bangong Suture in the north go to the south? The faults that appear to form the cryptic suture of the Bangong Suture in the north merge into faults that mark the southern mélange belt at their southeast ends and continue southeastward. From the end of these faults and mélange belts, where does the suture or sutures go to the south?

To further complicate the tectonic picture, within the central part of the large body of metamorphic, upper Paleozoic, and plutonic rocks are two zones of faults that bound two belts of metamorphosed upper Paleozoic rocks (Pzum, Fig. 9-1) that contain abundant slivers of serpentinite and one fault-bounded fragment of Triassic strata (Tr[?], Fig. 9-1) of unknown affinity. East of the eastern fault that bounds these belts, recent work has shown the presence of high-pressure rocks (see above); however, they have not been dated, so their relations to a possible suture remain unknown. It is unclear whether these belts of Paleozoic rocks might be a fragment of another Mesozoic suture or an older suture, or even whether this belt marks a suture at all. At their northwestern end, both belts are unconformably overlain by Early or Middle Jurassic flysch; thus if this is a suture, it is older than the Bangong Suture and must have closed sometime before the Middle Jurassic. Again the question arises as to how many of the known sutures from the south may be represented in the mélange belts in the Bangong region. Because we cannot find evidence to support a major suture to the east within the *western* part of the Qiangtang Unit, it indicates to us that both southern sutures may be represented in the mélange belts of the Bangong region. Otherwise the older sutures may have been disrupted and lie in the collage below the Mesozoic strata of the Qiangtang Unit, a question to be addressed in the synthesis below.

The question remains whether the large lens of metamorphic, plutonic, and upper Paleozoic rocks really belongs to the Lhasa Unit. In its northwestern part it is unconformably overlain by Lower and Middle Jurassic flysch at three places that clearly belong to the Lhasa Unit. However, the upper Paleozoic rocks in its southeastern part are unconformably overlain by Middle Jurassic rocks of shallow-marine environments, with a basal conglomerate overlain by up to 1600 m of fine-grained sandstone and shale, followed by 500 m of fossiliferous limestone, and capped by more sandstone and shale (Mz2bx on Plate 1 and in Fig. 9-1). These strata continue in several large, discontinuous exposures to the southeast, where they end in the footwall of a thrust carrying Triassic rocks assigned to the Lhasa Unit (T3wp, Fig. 8-4). It is not clear whether these Jurassic strata belong to the Lhasa or the Qiangtang III unit. If they belong to the marine Qiangtang III unit, the large lens of older rocks is somehow a compound fragment whose parts have been juxtaposed with the mélange belt in as yet some undetermined way that will have a bearing on whether the two belts of mélange on both sides of it belong to the Bangong Suture, and if so, how they became structurally separated. In our interpretation, both the Paleozoic rocks and the unconformably overlying Jurassic strata are parts of an extended continental margin of the Lhasa Unit and are related to the Bangong Suture, an interpretation presented below.

The polarity of the Bangong subduction during closure was probably south or southwest underthrusting of oceanic crust below the Lhasa Unit, but the evidence is somewhat equivocal. Structures within the suture zone have been strongly modified by post-suture closing, so they cannot be used with confidence. Foliation within the mélange belt directly underlying the unconformable Upper Cretaceous strata, where they have not been affected by younger deformation, is sub-vertical and of little use. The ophiolite body faces top-to-the-north, but it is a large block in the mélange. Near Dengqen the top of the body is on its east side and is defined by unconformably overlying Middle Jurassic strata that appear to have been deposited on an erosional surface. Thus this ophiolite body might best be interpreted as an ophiolite formed along a continental margin, i.e., an opening ophiolite rather than one formed at a spreading center in the deep ocean (see below). The Qiangtang III unit to the north may represent a south-facing margin, but evidence for transition to deeper water sedimentary rocks is not present and, in our interpretation, is missing along a major fault zone of probable Paleogene age. The flysch-bearing units in the Lhasa Unit indicate a deeper water environment to the south, suggestive of south vergence for the subduction. The fold-thrust belts on both sides of the suture show south vergence, but the age of these structures may be late or even post-suturing. How this relates to the initial direction of subduction remains unclear. However, the strongest evidence is the abundance of plutons in the Lhasa Unit to the south in comparison with the rarity of plutons to the north, which is suggestive of south or southwest subduction during the Late Jurassic and Early Cretaceous, closing along this part of the Bangong Suture. If this is the correct vergence of subduction the plutonic rocks within the Lhasa Unit formed within two different subduction environments: the north-vergent subduction in the Yarlung-Zangbo zone to the south, and the south-vergent subduction in the Bangong zone to the north.

## ORIGIN OF THE BANGONG SUTURE

Taking the area that contains the Bangong Suture and mélange belts as a whole, the question of whether there is one disrupted belt of mélange that defines the Bangong Suture, or if there is more than one belt of mélange related to the opening of the Bangong ocean. We interpret the geological relations outlined above to support a continental margin origin for the Bangong ophiolite and the other belts of mafic and ultramafic mélange and olistostromal mélange, which contain only sedimentary clasts similar to the conditions described for the Iberian and Newfoundland continental margins by Péron-Pinvidic and Manatschal (2009). They interpret these margins as the result of extensional stretching of the continental crust during opening of the Atlantic Ocean to be similar to low-angle, simple-shear detachment type of extension that is well known in the Basin and Range Province of western North America (Wernicke, 1981). The detachments place fault blocks of continental crust above lithospheric mantle, some of which is exposed at the surface. This forms a wide, subsiding margin with half grabens, some locally underlain by mafic and ultramafic rocks (see also Jagoutz et al., 2007) separating more elevated blocks of continental rocks. The relief along the grabens would be locations for development of olistostromes that were shed into the graben sediments. In our interpretation, only the main belt with the large ophiolite body was the site of the Bangong ocean, and the belts to the south were narrow grabens locally floored by mafic and ultramafic rocks, such as the southern mélange belt, which became sheared with the flysch sediments during closure. The southern mélange belt described above (Fig. 9-1), when traced to the northwest, became floored by thinned continental rocks with only olistostromes of sedimentary rocks incorporated into Jurassic sediments. The southernmost exposures of olistostromes within the Jurassic flysch would be more local grabens or half grabens. The belt of shallow-water Jurassic strata (Mz2bx, Figs. 8-4 and 9-1) that unconformably overlies Paleozoic rocks north of the southern mélange belt we interpret to have been deposited on an uplifted fault block along the extended margin that belonged to the Lhasa Unit. This fault block must have been subaerially exposed before subsiding to only shallow depths. Thus the entire width of the southern mélange belts is within the extended Lhasa continental margin adjacent to the Bangong ocean.

The polarity of subduction along the northern mélange belt of the Bangong Suture is probably southward, but the southern mélange belts were probably not sites of subduction but merely loci of closed, narrow grabens associated with shearing and mélange formation during their closure. Our hypothesis, presented later, will argue for major strike-slip faulting of Mesozoic and Cenozoic ages that has disrupted these belts, but this hypothesis remains speculative.

# CHAPTER 10

# *Syntaxial Unit*

The Syntaxial Unit is present in the northwestern part of the area (Plate 1 [available on DVD accompanying this volume and in the GSA Data Repository[1]]) and contains rocks and structures related to the large eastern Himalayan syntaxis, but only a small eastern part of the syntaxis is shown on Plate 1. It forms a concave-west arcuate pattern of structures that make faulted contact with the Lhasa Unit to its east (Geng et al., 2005). Within the arc are two major tectonic rock assemblages that consist of the Namche Barwa Group to the west and framed on the east by an arcuate belt of rocks belonging to a complex shear zone that contains rocks belonging to the Yarlung Zangbo suture zone of the Himalaya, and rocks that may correlate with other structural elements of the Himalaya and the adjacent Lhasa Unit. Rocks and tectonic units within the shear zone generally dip steeply away from the Namche Barwa rocks and form a large, east-plunging antiform that plunges east beneath the adjacent Lhasa Unit (Fig. 10-1).

## NAMCHE BARWA GROUP

The Namche Barwa Group was divided by Geng et al. (2005) into three units within the core of the syntaxial arc: the Zhibai, Paixiang, and Duoxiongla units. The Zhibai unit forms a narrow, fault-bounded unit in the central part of the core ($NB_z$, Plate 1) and consists of felsic gneiss with areas of garnet clinopyroxenite, garnet amphibolite, and kyanite garnet monzonite gneiss that yield a high-pressure granulite metamorphic assemblage. Burg et al. (1998) determined the conditions of metamorphism to be pressures of 0.8–1.0 GPa and temperatures of 720–760 °C, and Zhong and Ding (1996), Liu and Zhong (1997, 1998), and Ding and Zhong (1999) reported pressures of 1.4–1.8 GPa and temperatures of 750–890 °C. The peak metamorphism for these high-pressure granulites gave ages that ranged from 45 to 69 Ma (Geng et al., 2005). Burg et al. (1998) reported U-Pb ages from zircons at 1312 ± 16 Ma from clinopyroxenite, and 1770 Ma from metapelite, indicating a Precambrian age for the protolith of the Zhibai rocks. The Paixiang unit crops out in the northern part of the Namche Barwa Group ($NB_p$, Plate 1) and consists of felsic gneiss with minor diopside- and forsterite-bearing marble, clinopyroxenite, and scapolite-diopside marble. This unit is strongly small-folded and is interpreted to be a structurally higher part of the group. The Duoxiongla unit is a migmatitic banded felsic gneiss with minor amphibolite, and it forms the southern part of the Namche Barwa Group ($NB_d$, Plate 1), where it locally forms a broad, open antiform.

The Namche Barwa Group can be traced to the west, where the rocks are continuous with the High Himalayan crystalline zone in the Himalaya (Fig. 10-1). Granitic intrusions yielding U-Pb ages of 525–552 Ma are present within the Namche Barwa Group (Pan et al., 2005), which supports a correlation with rocks of the High Himalayan crystalline zone, where rocks of similar ages are common. This correlation is consistent with their structural position below the arcuate enclosing shear zone, which is correlative with the Yarlung-Zangbo suture zone.

## CONCAVE-WEST ARCUATE SHEAR ZONE

A shear zone, from a few kilometers to 15 km in width, with numerous fault slices of different rock types, forms a concave-west arc that borders the Namche Barwa Group on three sides (Plate 1). Beyond the map area, rocks of this zone can be traced to the west into rocks of the Yarlung-Zangbo suture zone (Fig. 10-1), and we regard the shear zone to be the continuation of the suture zone and will refer to it as such. However, some rocks in this shear zone may represent other units of the Himalaya, and its tectonic history contains deformational events other than those commonly seen in the Yarlung-Zangbo suture zone in the Himalaya. Foliations in both metamorphic and mylonitic rocks, and faults within this suture zone, generally dip steeply away from the Namche Barwa Group rocks, but locally along the western margin they are overturned and dip steeply eastward, although the overall structure is a broad, northeast-plunging antiform (Plate 1 and Fig. 10-1).

The main rock types in the shear zone are greenschist-facies metasedimentary rocks that locally reach amphibolite grade and are foliated and commonly mylonitized especially near fault zones. They form a distinct zone of low-grade rocks that contrast and can be easily separated from the adjacent high-grade rocks of the Namche Barwa Group as well as plutonic rocks and their wall rocks of the Gangdese batholith, which are exposed on either side of the arcuate Yarlung-Zangbo suture zone. Many of the greenschists have fine-grained sedimentary protoliths with associated quartzite (Sa), micaceous quartzite, dark phyllite, and marble (η). Plate 1 shows fault-bounded bodies with abundant quartzite and mixtures of greenschist metasediments. The shear zone locally contains large and small bodies of amphibolite, mafic and ultramafic rocks, and basalt locally containing pillows. Some of the larger bodies contain rock associations that suggest they are parts

---

[1]GSA Data Repository Item 2012297, Plates 1–4, is available at www.geosociety.org/pubs/ft2012.htm, or on request from editing@geosociety.org or Documents Secretary, GSA, P.O. Box 9140, Boulder, CO 80301-9140, USA.

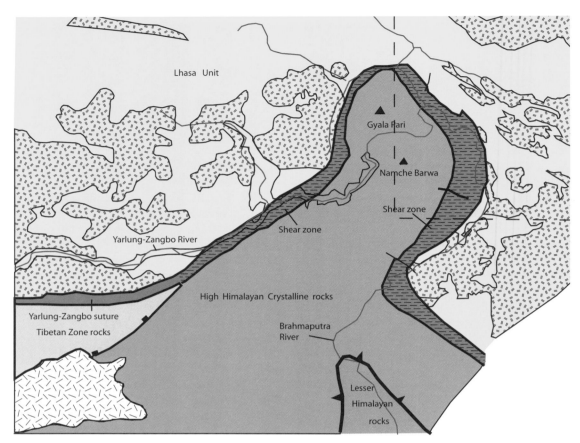

Figure 10-1. Generalized tectonic map of the eastern Himalayan syntaxis. The Yarlung-Zangbo suture and its continuation into the shear zone that frames the syntaxis is shown in green. It surrounds the uplifted antiformal syntaxis area, separating the Lhasa and High Himalayan Crystalline Unit. Plutons of the Gangdese batholith are patterned. The easternmost part of the South Tibetan detachment is shown with the squared pattern on the hanging wall, which contains the Tibetan zone rocks. The area of Plate 1 is shown by the dashed line in the upper right corner. Discussion of this area is given in the text.

of an ophiolite sequence (Op). Geochemical data reviewed by Geng et al. (2005) suggest backarc and supra-subduction settings for some of the ophiolitic rocks, whereas other rocks point to an arc setting. It is not surprising that such a range of rock types is present within this major suture zone, which represents closure of the Tethyan oceanic terrane between India and Eurasia. West of the map area, rocks of the Tibetan Zone, a sedimentary sequence that ranges from lower Paleozoic to Cretaceous, lie above the South Tibetan detachment zone (Burchfiel et al., 1992; Fig. 10-1), a large, north-dipping, north-vergent, largely Miocene-age, low-angle normal fault. West of the area of Plate 1 the detachment projects into the Yarlung-Zangbo shear zone along the northwestern side of the syntaxial antiform (Fig. 10-1), but how it relates to faults within the shear zone remains unknown. If it does enter the syntaxial shear zone, then rocks of the Tibetan zone and the detachment fault may be present within the Yarlung-Zangbo shear zone, an interpretation supported by juxtaposition of low-grade rocks in the shear zone above high-grade rocks within the core of the antiform. Also within the shear zone are tectonic slices of rocks that were derived from both the Namche Barwa Group (WNj) below and Lhasa units (Wlb) above (Plate 1).

Fault zones separate major rock packages, and both the faults and foliations within the Yarlung-Zangbo suture zone mostly dip moderately to steeply below the rocks of the Lhasa Unit, although steep dips are present, suggesting local overturning of the contact. Many of the faults are mylonitic, with lineations that are both parallel and downdip to the strike of the zone, particularly along the faults that bound its NW and SE margins.

## INTERPRETATION OF THE SYNTAXIAL UNIT

We interpret the broad geological relations in the syntaxial area to support the following polygenetic and diachronous sequence of events: (1) collision of India, represented by the Namche Barwa Group, with the Lhasa Unit along the Yarlung-Zangbo suture zone in early Cenozoic time, followed by (2) intracontinental convergence between the Indian plate below and the Lhasa Unit above, causing removal of the lower crust beneath the

Gangdese batholith of the Lhasa Unit along the evolving intracontinental subduction zone marked by the Yarlung-Zangbo shear zone; (3) development of the South Tibetan detachment fault, which juxtaposed the high-grade rocks of the Namche Barwa Group and the low-grade rocks of the Yarlung-Zangbo suture zone and which may also contain rocks of the Tibetan zone of the Himalaya; (4) formation of the antiformal structure, accompanied by high-grade metamorphism of the Namche Barwa Group, by lower crustal flow into an extensional region caused by the indentation of the NE corner of India and the beginning of uplift of deeper crustal levels, and formation of the antiform by uplift, causing steepening of originally gently north-dipping, low-angle structures, and normal shear at their contacts; and (5) final rapid exhumation and uplift of the Namche Barwa Group, enhanced by erosion when the Brahmaputra River captured the Zangbo River and its tributaries within the broader region around the syntaxis; during this time the northern part of the syntaxial antiform was modified and extended farther to the north (Stewart et al., 2008). Because only a small part of the Syntaxial Unit is present in the map area it is necessary to consider the entire area of the syntaxial antiform west of the map area to understand its tectonic setting and evolution (Fig. 10-1).

In the area of the eastern syntaxis the time of the India and Eurasia collision is poorly known. Aikman et al. (2008) presented evidence that shortening related to the early stages of collision took place within the eastern Tibetan zone rocks before 44.1 ± 1.2 Ma, suggesting that the time of collision could have been similar to that suggested from data for the central and western Himalaya and adjacent Tibet, which supports collision at ca. 50 Ma (see Rowley, 1996, 1998, for discussion). Pre-collisional convergence at the eastern syntaxis is manifested by plutons of Cretaceous (ca. 133 and ca. 70 Ma) and early Cenozoic (70–40 Ma) ages in the Gangdese Andean magmatic arc formed along the southern margin of the Lhasa Unit as the Neotethyan oceanic lithosphere north of the Indian continental crust was subducted northward (Booth et al., 2004, 2009). The batholith that cores this arc in the Lhasa Unit surrounds the syntaxial unit on three sides, but no plutons of Cretaceous or early Cenozoic age are present within the arcuate Yarlung-Zangbo shear zone, or the Namche Barwa Group, that form the core of the Syntaxial Unit (Fig. 10-1). Because the Syntaxial Unit contains no plutons and dips below the batholithic rocks, except in a local area west of Namche Barwa where the zone dips steeply to the east (Booth et al., 2004; see Fig. 10-1), the lower part of the batholitic crust must have been sheared off and subducted. These data come from our study of the rocks near Milin on the NW side of the syntaxis and the adjacent Lhasa Unit as well as from geological data from maps by others (Fig. 10-1). Along the NW side of the Yarlung-Zangbo shear zone the Lhasa Unit contains abundant plutonic rocks that are largely intermediate to felsic in composition, with rare gabbroic and other mafic plutons. They are separated by screens of metasedimentary and metaigneous rocks, whose protoliths can sometimes be identified as Paleozoic or Precambrian(?). The plutons and their wall rocks commonly contain abundant migmatitic zones with strongly folded foliations. Adjacent to the Yarlung-Zangbo shear zone the Lhasa Unit rocks contain muscovite schist with garnet, sillimanite, and kyanite. The geologic relations suggest that all these rocks were in the lower part of the Gangdese batholith. In contrast, the adjacent Yarlung-Zangbo shear zone consists of greenschist facies variably mylonitized metasedimentary rocks. They contain a well-developed foliation, commonly small-folded, with a sub-horizontal, northeast-bearing lineation and a superposed younger downdip lineation. The shear sense in the S-C fabrics related to the horizontal lineation indicate a top-to-the-south sense of shear. The S-C fabric, related to the downdip lineation, indicates a west-side-down sense of shear. The contact with the Namche Barwa Group is marked by an abrupt increase in metamorphic grade to upper amphibolite facies, and rocks near the contact in the Namche Barwa Group are kyanite-bearing gneisses.

Thus the Yarlung-Zangbo shear zone rocks are a low-grade unit in tectonic contact between two much higher grade units. The fact that the shear-zone unit lies above the antiformal core, surrounds it, and continues into the Yarlung-Zangbo suture zone indicates that the shear zone was originally a continuous, thin, sheet-like body before it was domed by formation of the syntaxial antiform. Relative upward movement of the antiformal core is recorded by the youngest mylonitic fabric with a downdip, NW-bearing lineation and down-to-the-northwest S-C fabric shear sense adjacent to its contact with the Yarlung-Zangbo shear zone. This shear zone has likewise moved vertically against the plutonic complex in the Lhasa Unit. By unfolding the antiform, the older horizontal lineation within the Yarlung-Zangbo shear zone can be interpreted as a top-to-the-south thrust movement related to the underthrusting of the Yarlung-Zangbo shear zone and the Namche Barwa Group rocks below the Gangdese batholith. This lineation has been used to suggest post-collisional indentation of the core rocks of the antiform by left-lateral shearing along its NW side, and reported right-lateral shearing on its SE side, to form the arcuate structure of the syntaxis. However, the structures within the Lhasa Unit directly NE of the syntaxial core exhibit a linear, NW-SE trend inconsistent with a major indentation (see below). The lack of plutons in the Yarlung-Zangbo shear zone and the Namche Barwa Group indicates that the underthrusting is part of the post-collisional convergence following final subduction of the Indian plate oceanic crust.

The cause of the young antiformal uplift and exhumation of the deep crustal rocks of the Namche Barwa Group has been the subject of numerous hypotheses. The antiformal structure of the Syntaxial Unit has been interpreted as (1) a northward indentation of the High Himalaya rocks into the Lhasa Unit, bounded by strike-slip shear zones of opposite sense on the two flanks of the antiform (e.g., Ding et al., 2001; Geng et al., 2005); (2) a broad antiform that involves the entire crust and part of the lithosphere during compressional folding (Burg and Podladchikov, 2000; Burg et al., 1997, 1998); (3) formation by lateral flow of lower crust and vertical diapirism of low-density lower crust influenced by rapid erosion along the Zangbo-Brahmaputra River (e.g., Zeitler et al., 2001); and (4) formation by lateral flow of weak lower crust into a dilatant area created by rotation of crust around the corner of Indian continental crust (see below).

The Syntaxial Unit is one of two similar structures that lie at the eastern and western corners of the Indian continental crust formed during post-collisional, intracontinental indentation of India into Eurasia (Zeitler et al., 2001). The regional geological setting of these two syntaxes suggests that their formation is related to deformation at the corners of the indenter. Important geological data that must be considered in any origin for the eastern Himalayan syntaxis include (1) the Namche Barwa Group has been exhumed from deep crustal levels during the past ~10 m.y. and that 15–20 km of crust has been removed in the Namche Barwa area in the last 3–4 m.y., yielding exhumation rates of 3–5 mm/yr (Burg et al., 1997); (2) the core rocks contain dikes, sills, and small bodies of melts that have yielded U-Pb ages from zircon that range from ca. 11 to 2.9 Ma and which have been related to decompression melting (Burg et al., 1997; Ding et al., 2001); (3) there has been extensive young exhumation of a large area in the northern part of the syntaxial antiform, extending north of the antiform into the Lhasa Unit (Stewart et al., 2008); this area yielded (U-Th)/He ages younger than 2 Ma, and a smaller area within the antiform yielded $^{40}Ar/^{39}Ar$ ages younger than 2 Ma (Stewart et al., 2008). Stewart et al. (2008) also showed that 50% of the sediment within the Brahmaputra depositional system comes from this area, which represents only 2% of the drainage area of the river system. These data suggest very rapid exhumation rates that average 7–21 mm/yr.

From these data it can be argued that erosional reduction in crustal thickness can lead to focused exhumation of deeper crustal rocks through reducing the overburden and enhancing the flow of weak, lower crustal rocks into the region below the area of reduced overburden pressure. In fact, it can be similar to the autogenerating process of salt dome formation in which the increase in the height of the growing low-density salt dome increases the pressure differential between the dome and the surrounding country rocks. Although this may help to explain the rapid exhumation of deep crustal rocks in the area around the knickpoint in the Tzangbo–Brahmaputra River at Namche Barwa, it does not explain the geometry of the antiformal structure that extends at least another 40 km to the south (Fig. 10-1). Whereas rapid-focus erosion may help to localize the ductile low crustal flow, it may not be the only mechanism for creating the entire antiform.

The fact that both antiformal structures at Himalayan syntaxes occur at the corners of the India indenter suggests that there may be a dynamic cause related to the indentation process. Shen et al. (2001), in their numerical model of the dynamics of the eastern syntaxis, showed that at the indenter corner a region of extension develops. This region is where the eastern syntaxial antiform occurs. It may be that this region of extension localizes the flow of ductile lower crust to cause uplift and exhumation of deeper structural elements, a process that may have started ca. 18–25 Ma with the intrusion of some small plutons. Uplift may have been initially enhanced by development of the South Tibetan detachment that may be traced into the Yarlung-Zangbo suture in early to middle Miocene time, beginning the exhumation of the deeper crustal rocks before formation of the antiform. The detachment may have formed, at one time, at the base of the Yarlung-Zangbo suture zone, juxtaposing the high-grade rocks of the Namche Barwa Group with the overlying greenschist-facies rocks of the Yarlung-Zangbo shear zone. These contacts were later modified by the rise of the core of the eastern syntaxial antiform by flow in ductile lower crust into the area of extension. Final enhancement of exhumation may have been related to the capture of the Yarlung-Zangbo River by the Brahmaputra River. This final event may explain the extension of the antiform to the north in its northwestern part, where the most recent rapid exhumation and uplift occurred around the Namche Barwa and Gyala Pari mountain peaks (Fig. 10-1). Thus the formation of the syntaxial antiform, in our interpretation, is polygenetic and diachronous and includes (1) the dynamics at the corner of the India indenter, (2) formation of the South Tibetan detachment, and (3) capture of the Yarlung-Zangbo River by the Brahmaputra River. Rocks within the Lhasa Unit and the Yarlung-Zangbo suture zone were further deformed with uplift of the Namche Barwa core rocks by steepening and additional shear along the flanks of the uplift. Some faults adjacent to the shear zone in the Lhasa Unit follow the arc of the Syntaxial Unit and were probably also formed during uplift of the core rocks. Farther to the northeast the structures in the Lhasa Unit strike uniformly NW-SE and do not appear affected by the formation of the syntaxial uplift, and it is for this reason that we do not favor an indentation model.

Northward subduction of the oceanic crust beneath the Lhasa Unit during Mesozoic time was the most likely cause of the magmatism within the Gangdese batholith, which continues through the Lhasa Unit (Plate 1) into the Tengchong Unit to the south. U-Pb radiometric ages of the plutons within Plate 1 are rare, but Chui et al. (2009) dated a number of plutons that yield a range of Early Cretaceous (133–109 Ma) and early Cenozoic (57–66 Ma) ages. Many of the plutons in the area of Plate 1 are shown on the Chinese map as early Cenozoic ($\gamma_6^1$), but without confirming radiometric ages. Confirmation of the ages of these younger plutons is essential for understanding the age of early Cenozoic subduction at the eastern Himalayan syntaxis. How much of the magmatism within the northern part of the Lhasa Unit is related to northward subduction is unknown, as some of it may have been related to southward subduction at the Bangong suture to the north (see above) or even closure along the narrow, northwest-trending belt of Paleozoic rocks in the southeastern part of the Lhasa Unit, as suggested by Geng et al. (2005).

# CHAPTER 11

# *Tengchong Unit*

The Tengchong Unit lies west of the Baoshan Unit and is separated from it by the Gaoligong shear zone. To the west the Tengchong Unit passes into Burma (Myanmar), where little is known of it. It also can be traced northward along international borders into the Lhasa Unit, which is its northern continuation (Fig. 1-1B).

## ROCKS OF THE TENGCHONG UNIT

Unlike the other tectonic units as shown in Plate 1 (available on DVD accompanying this volume and in the GSA Data Repository[1]), the Tengchong Unit consists mostly of metaigneous and metasedimentary rocks, commonly strongly sheared and overlain by a late Cenozoic to active volcanic field (Kan et al., 1996; Qin et al., 1996). The oldest rocks within the Tengchong Unit are within the metasedimentary and metaigneous rocks that form large areas of schist and gneiss, with sandstone, shale, and local carbonate protoliths and plutons of different compositions in its western part (Plate 1, Plzm and γm). These rocks are strongly foliated and metamorphosed from greenschist to amphibolite grade with no fossils. They are interpreted to be metamorphosed lower Paleozoic strata (Plzm, Plate 1), although some workers suggest that abundant Precambrian rocks are present, but little is known of their protoliths (Bureau of Geology and Mineral Resources of Yunnan Province, 1990). The ages of most of the foliated plutons is unknown, but are suggested to be Mesozoic and Cenozoic. Lying within this broad metamorphic terrane are isolated areas of metasedimentary rocks that are fossiliferous and form fragmentary parts of Paleozoic and Mesozoic sections (Fig. 11-1). The oldest dated rocks have yielded Upper Silurian (Pzl) and Devonian (Pzu) fossils in the western part of the unit. Silurian rocks are mainly limestone, and Devonian rocks are sandstone with some limestone, overlain by a limestone unit with fossils of Early Devonian age. In the eastern part of the unit the sedimentary rocks are mainly Carboniferous and consist of >2 km of sandstone, shale, and siltstone with some conglomerate and limestone at the base and 200–300 m of limestone near the top; fossils occur mainly in the middle part of the sequence. Thick, isolated unfossiliferous sections of sandstone, shale, and limestone are correlated lithologically with Carboniferous rocks in the Lhasa Unit to the north. Approximately 200–300 m of Lower Permian limestone lies above the Carboniferous in the northernmost part of the unit. Fossiliferous Middle Triassic limestone ~800 m thick unconformably lies above both Permian and Carboniferous strata, with Upper Triassic shale and siltstone at the top. At the northernmost outcrops of the Tengchong Unit (shown within Plate 1) along the Burma (Myanmar) border are red sandstone and conglomerate that lie unconformably on Permian and Carboniferous rocks. They are assigned a Jurassic(?) age but are without fossils.

Within the Upper Carboniferous and Lower Permian rocks is a pebbly mudstone that is interpreted to be of glacial origin. Fossils of the *Stepanovielle* fauna, regarded as diagnostic fossils for Gondwana, have been reported (Bureau of Geology and Mineral Resources of Yunnan Province, 1990), and a more recent study by Jin et al. (2011) has shown the Gondwana origin for the Tengchong Unit from a section in the northeastern part of the unit (Metcalfe, 2006; Jin et al., 2011).

Plutonic rocks make up nearly one third of all the rocks in the Tengchong Unit, but few of them are dated. Some are strongly foliated (γm, Plate 1), particularly those in the west and along the eastern and southeastern sides of the area, where they are affected by the mylonitization associated with the Cenozoic Gaoligong shear zone. Many of the plutons in the central part of the area are not mylonitic. The plutons are assigned ages on Chinese maps that range from Jurassic to early Cenozoic, without geochronological control, and are regionally the southward continuation of the Gangdese batholithic belt within the Lhasa Unit (Ji, 1998).

Neogene strata are present in numerous elongated fault-controlled basins in the eastern and southern parts of the Tengchong Unit. In the north these strata trend generally N-S, but in the south they trend NE-SW. Most of these basins are grabens or half grabens, and each basin has its own stratigraphic section. Many of the basins are pull-apart basins and owe their origin to stopovers along strike-slip faults. The sedimentary rocks are fossiliferous in many places, with the oldest rocks dated as Miocene (Mu et al., 1987), but most of the basins contain Miocene, Pliocene, and Quaternary strata. These rocks consist of ~1 km of conglomerate, sandstone, siltstone, and mudstone in some basins, and coal beds are particularly important in both Miocene and Pliocene strata. Some basins, particularly in the western part of the unit, contain thick units of basalt, and Neogene basalt directly overlies metamorphic and igneous rocks in the western part of the unit. The basalt is mainly Pliocene and Quaternary; at least 20 basaltic cones dot the area, and volcanism remains active today (Qin et al., 1996).

---

[1]GSA Data Repository Item 2012297, Plates 1–4, is available at www.geosociety.org/pubs/ft2012.htm, or on request from editing@geosociety.org or Documents Secretary, GSA, P.O. Box 9140, Boulder, CO 80301-9140, USA.

## STRUCTURE OF THE TENGCHONG UNIT

The arcuate, convex-east structural grain of all the rocks in the Tengchong Unit is obvious from the map (Plate 1). It is part of the main arcuate structure in all of the tectonic units in this part of SW China around the eastern Himalayan syntaxis (Plate 1). Plutons are elongated in comparison with those in the main Gangdese batholith to the north and form a prominent part of the arc. The foliation and folds in the metamorphic rocks also conform to the arcuate shape, and the mylonitic rocks of the bounding Gaoligong shear zone also constitute the arc, as do the structures in the Baoshan Unit to the east (Plate 1). The ages of the folds and faults in the pre-Neogene rocks that follow the arcuate regional pattern probably are mostly late Mesozoic to early Cenozoic on the basis of cross-cutting relations with the plutons and the correlation of deformation within the Lhasa Unit in south-

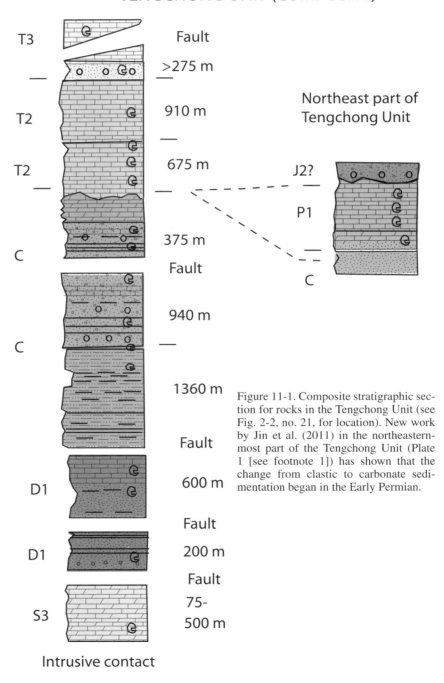

Figure 11-1. Composite stratigraphic section for rocks in the Tengchong Unit (see Fig. 2-2, no. 21, for location). New work by Jin et al. (2011) in the northeasternmost part of the Tengchong Unit (Plate 1 [see footnote 1]) has shown that the change from clastic to carbonate sedimentation began in the Early Permian.

ern Tibet. The relationship between the plutons and structures is important for dating the deformation, although few of the plutons are dated; those that are dated were dated by older methods; thus the constraints on their ages of deformation are poor. Only the Gaoligong shear zone of early Cenozoic age was recently studied, by Akciz (2004), who showed that mylonitization had ceased by 18.5 Ma, but when the mylonitization began has been elusive to determine. A synkinematic leucogranite in the shear zone has yielded U-Pb ages of ca. 65–79 Ma (Akciz, 2004), but these ages have been difficult to interpret. The mylonitic rocks make the arcuate bend, and it is tempting to consider this arc as a vertically plunging fold related to the indentation of India into Eurasia, but at present there is no definitive evidence that it is related, and such a project is a clear target for future study.

Within the metamorphic rocks that make up much of the western part of the unit are many generally N-S–striking faults and foliated shear zones, and within the metamorphic and foliated plutonic rocks are numerous shear zones, some mylonitic. These faults and shear zones may have played an important role in the deformation around the syntaxis. There is probably a large magnitude of distributed shear within these rocks that accommodated the Cenozoic shearing and rotation abound the syntaxis, but such an analysis has not been made. One shear zone, the Nabang shear zone, is in the western part of the area, with a NNE strike. This shear zone contains a lineation that plunges 3°–25° NNE within a steeply dipping L-S mylonitic tectonite with a right-lateral shear sense, which was dated by Wang et al. (2006).

Hornblende and biotite from shear zone rocks gave a $^{40}Ar/^{39}Ar$ age of 32.8 ± 0.2 Ma and 28.1 ± 0.2 Ma (early Oligocene), respectively, consistent with ages for many of the other shear zones within the region (Fig. 11-2). The Nabang shear zone has a similar age as the Gaoligong shear zone to the east that is considered by many workers (e.g., Replumaz and Tapponnier, 2003) to be the western boundary of crustal fragments extruded during early Cenozoic time. However, the Nabang shear zone, and many of the other parallel shear zones within the Tengchong Unit, although not mapped separately, suggest that rocks west of the Gaoligong shear zone were also involved in the early Cenozoic extrusion process, but to what degree is unknown.

An interpretation of the late Cenozoic structure of the Tengchong Unit was presented by Wang et al. (2008), and this structure was interpreted by Wang and Burchfiel (1997) to be related to late Cenozoic clockwise rotation around the eastern Himalayan syntaxis. Structures are dominated by N-S– to NE-SW–striking faults in the eastern and southern parts of the unit, respectively (Plate 1), within a transtensional setting where the N-S faults have a dominant normal, left-lateral, strike-slip component, with an increasing extensional component to the south, and the NE-SW–striking faults have a dominant left-lateral, strike-slip displacement (see also Wang et al., 2008). Many of the numerous Neogene and Quaternary basins in the area (Plate 1) are pull-apart basins along releasing bends. These basins are most abundant in the eastern and southeastern parts of the Tengchong Unit. This transtensional environment

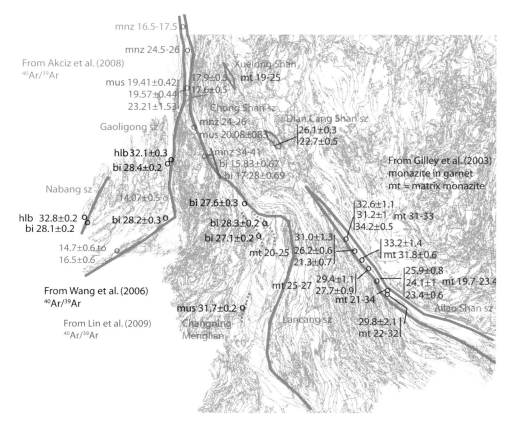

Figure 11-2. Ages (Ma) of shearing within major shear zones (sz), determined from different radiometric dating techniques: mus—muscovite; mnz—monazite; mt—matrix monazite; hlb—hornblende; bi—biotite. Data from Akciz et al. (2008), Gilley et al. (2003), Wang et al. (2006).

extends eastward of the Tengchong Unit into the Baoshan Unit, where it is less pervasive (see below). Some of the basins are present along the Ruili fault, which is an active left-lateral fault that reactivated the older southern boundary faults of the Tengchong Unit. In fact the entire eastern boundary of the Gaoligong shear zone has been reactivated by younger brittle faults. At its southern end, north of the east end of the Ruili fault, the shear zone has been reactivated by normal faults, but to the north the young brittle faults along which the Nujiang River flows do not appear to be active. The late Neogene and Quaternary volcanism is related to this extensional system, but the volcanism does not extend east of the Tengchong Unit. This area of volcanism lies above a major thermal anomaly in the upper mantle (Li et al., 2008).

Within the Tengchong Unit the northern part of the Gaoligong shear zone is dominated by large plutons, metamorphosed Paleozoic strata, and large tracks between plutons of foliated metasedimentary and metaigneous rocks of uncertain age. Northward, this unit passes through Burma (Myanmar) and continues into southern Tibet, where it merges with the Lhasa Unit (see below). The southern part of the unit is characterized by foliated strata of probable Neoproterozoic age, which strikes northeast and curves to the north, making a large arcuate, convex-east arc to merge with the Gaoligong shear zone to the north.

## MAJOR PROBLEMS FOR THE TENGCHONG UNIT

Commonly the Gaoligong shear zone is taken as the western boundary of the crustal fragments extruded from eastern Tibet during early Cenozoic time (Replumaz and Tapponnier, 2003), and it may have a large, but unknown, magnitude of displacement. However, there may be additional shear that is accommodated to its west, such as along the unmapped Nabang shear zone, just as shear is accommodated within the Yangtze Unit east of the Ailao shear zone to the north (see above). Thus within the sheared and foliated rocks of the western Tengchong Unit there may be a large amount of right-lateral shear yet to be discovered that is related to early Cenozoic extrusion tectonics.

The late Cenozoic and active volcanism is probably related to lithospheric extension in the Tengchong Unit. Tomographic studies by Li et al. (2008) show the lithosphere below this area to have very slow upper-mantle velocities, indicating high temperatures at shallow depths. This area lies at the northeast corner of the Indian plate where it indents Eurasia. The Burma (Myanmar) oceanic slab is subducting eastward, and rapid clockwise rotation is also taking place, causing significant lithospheric thinning beneath the Tengchong Unit. How the lithosphere evolved during the past ca. 15 Ma and how it relates to the volcanism constitute a problem ripe for study.

## CHAPTER 12

# *Baoshan Unit*

The Baoshan Unit lies to the west of the Lanping-Simao and Changning-Menglian Units and east of the Tengchong Unit (Fig. 1-1B). It forms the northern continuation of the regional Sibumasu continental fragment, defined by Metcalfe (1986, 1988, 2006), and which lies mostly in Indochina to the south. We include here a discussion of the Chong Shan and Gaoligong shear zones, as they define major shear zones that bound the Baoshan Unit on the northeast and west, respectively. To the north the Baoshan Unit pinches out between these two shear zones. To the south its eastern contact with the Changning-Menglian Unit is complicated and will be discussed in detail.

## STRATIGRAPHY OF THE BAOSHAN UNIT

The stratigraphy of the Baoshan Unit is distinctive and is easily separated from the adjacent tectonic units (Figs. 12-1 and 12-2). The oldest rocks consist of ~10 km (a structural thickness) of generally weakly metamorphosed uppermost Neoproterozoic to Middle Cambrian, mainly siliciclastic and carbonate rocks (Bureau of Geology and Mineral Resources of Yunnan Province, 1990). Ordovician, Silurian, and Lower Devonian strata consist mainly of fossiliferous shallow-water siliciclastics and some argillaceous limestone and shale. From Middle Devonian to the end of the Early Carboniferous, various fossiliferous carbonate units were deposited. The lower part of the Upper Carboniferous strata contains diamictite, along with turbidite, conglomeratic sandstone, and siltstone (Jin et al., 2011), and the upper part contains black mudstone and siltstone. These Upper Carboniferous deposits are regarded as glacial in origin. The Early Permian Woniusi basaltic lava flows, with pillow lavas and tuffaceous intercalations, form a distinctive unit in the upper part of the Baoshan section (Bureau of Geology and Mineral Resources of Yunnan Province, 1990). Gondwana fossils, *Stepanovielle* fauna, and *Glossopteris* flora are present in this unit (Bureau of Geology and Mineral Resources of Yunnan Province, 1990), and a more detailed study by Jin et al. (2011) documented the Gondwana faunal affinity of the Permian strata. These basalts (shown as a separate unit, Pβ, Plate 1 [available on DVD accompanying this volume and in the GSA Data Repository[1]]) have been related to rifting and the beginning of separation of the Baoshan continental fragment from the Gondwana margin, although data supporting this interpretation is poorly documented. Lower Permian(?) red beds lie disconformably above the Woniusi basalts. In the southwestern part of the area near the boundary with the Tengchong Unit, Permian rocks are mapped as resting unconformably on strata of Cambrian to Devonian age, the only place where this occurs in the Baoshan Unit, and a place where there are many differences in the interpretation of this part of the section; for example, Wang et al. (2008) interpret this contact to be a west-dipping normal fault (see below).

Mesozoic strata rest disconformably on Lower Permian and locally Carboniferous rocks, and Upper Permian and Lower Triassic strata are missing below the overlying Middle Triassic limestone in some places (Plate 3 [see footnote 1]). Upper Triassic strata are present mainly in the eastern and southern parts of the Baoshan Unit, and in their eastern part they contain mafic volcanic rocks in their basal part (Mz2b, Plate 1), overlain by up to ~1500 m of sandstone, shale, and conglomerate, with limestone in their lower part. Toward the southeast and south, volcanic rocks become abundant in the Upper Triassic strata (Mz2, Plate 1) and are of mafic, intermediate, and felsic compositions; locally they are >1000 m thick (Fig. 12-2). These Upper Triassic strata have abundant fossils, and the section is both non-marine and marine. Jurassic rocks rest unconformably on both Middle Triassic strata in the west and Upper Triassic strata in the east, but at a few localities the Upper Triassic rocks are mapped as continuous with Lower Jurassic strata (Plate 3). The Jurassic rocks are present mainly in the westernmost part of the Baoshan Unit, and southeast of a major, NE-striking Cenozoic fault that separates the Baoshan Unit into northern and southern parts (Plate 1). The Jurassic rocks in the two parts of the Baoshan Unit are different. In the west a basal conglomerate is overlain by sandstone and mudstone, with a thin (200 m) basalt unit near the base that is in turn overlain by thick (1000 m) marine limestone at the top (Fig. 12-1). The basal conglomerate rests on different units at different places. It rests with paraconformity(?) on both Middle and Upper Triassic strata over much of the area, but in the southwest, just south of the boundary between the Baoshan and Tengchong Units, it rests on Permian strata. In this area are rocks that are suggested, but not proven, to be Cretaceous (Mz3, Plate 1), and within these rocks are small bodies of serpentinite and serpentinized mafic and ultramafic rocks. The youngest Mesozoic rocks in the Baoshan Unit are assigned a Late Jurassic age, but fossils are rare. The eastern section contains thick conglomerate at the base, overlain by silty and shaly limestone at the top. Fossils indicate these rocks to be Middle Jurassic and partly marine (J2 on Plate 1).

The eastern limit of the Baoshan Unit requires some discussion, because in the east Jurassic rocks, and locally Upper

---
[1]GSA Data Repository Item 2012297, Plates 1–4, is available at www.geosociety.org/pubs/ft2012.htm, or on request from editing@geosociety.org or Documents Secretary, GSA, P.O. Box 9140, Boulder, CO 80301-9140, USA.

## BAOSHAN TECTONIC UNIT

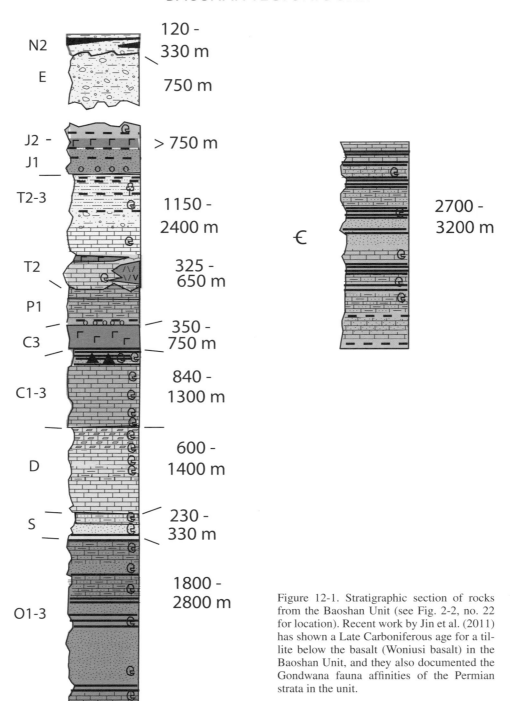

Figure 12-1. Stratigraphic section of rocks from the Baoshan Unit (see Fig. 2-2, no. 22 for location). Recent work by Jin et al. (2011) has shown a Late Carboniferous age for a tillite below the basalt (Woniusi basalt) in the Baoshan Unit, and they also documented the Gondwana fauna affinities of the Permian strata in the unit.

Triassic rocks, lie on somewhat metamorphosed rocks as old as late Neoproterozoic and Cambrian (Pzl2, Plate 1). These older rocks we interpret to be part of the Baoshan Unit, but they have a very different structural position from the rest of the Baoshan Unit (see below).

Paleogene (Eocene?) strata occur only in the eastern part of both southern and northern parts of the Baoshan Unit. They consist of mostly brown, tan, and red coarse conglomerate and sandstone, forming generally narrow N- to NW-trending basins. These rocks rest unconformably on folded and faulted Mesozoic

## SOUTH BAOSHAN

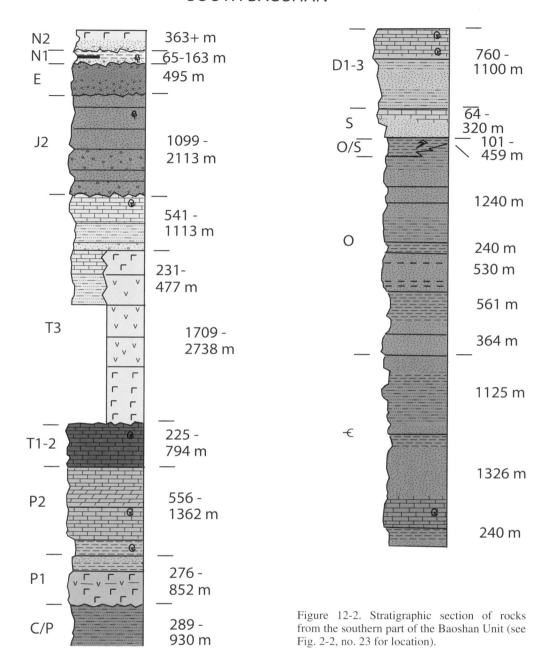

Figure 12-2. Stratigraphic section of rocks from the southern part of the Baoshan Unit (see Fig. 2-2, no. 23 for location).

and upper Paleozoic rocks but are themselves folded, although less so than the underlying rocks. Unfortunately none of these deposits assigned to the Eocene contain fossils, and their age is based on correlation with lithologically similar fossiliferous rocks in adjacent areas.

Neogene deposits are present in some irregularly shaped basins and consist of tan, yellow, and light-brown conglomerate, sandstone, and mudstone, and some basins contain abundant coal. Plant fossils date these rocks as Pliocene. The Neogene rocks are

gently folded in some places, and in other places they are cut by active high-angle faults. Quaternary deposits are present in some large basins many of which are young or active grabens that commonly occur as pull-apart basins along strike-slip faults.

The original paleogeographic position of the Baoshan Unit is controversial. Both paleomagnetic and paleontological data have been interpreted to indicate that the Baoshan Unit was derived from either Gondwana or from a more northerly origin, but the fossils and the sedimentology suggest that it is a Gondwana-derived element. For a summary, see Ueno (2003) and Metcalfe (2006). The Baoshan Unit is bounded by oceanic sutures, largely cryptic, both to the west and east, and was accreted into the SE Asian collage in Mesozoic time (see below).

## STRUCTURE OF THE BAOSHAN UNIT

The Baoshan Unit is bounded by major mylonitic shear zones on the west and northeast, and by complex and poorly understood fault zones on the southeast. The northeastern (Chong Shan shear zone) and western (Gaoligong shear zone) mylonitic shear zones merge to the north, and the unit ends (Plate 1), but how they merge is just now being unraveled (Akciz et al., 2008; see below). Internally the unit is strongly folded and faulted and is divided into two parts by the NE-trending Cenozoic Wanding fault (Fig. 12-3). In the northern part of the Baoshan Unit the structures trend generally N-S, but near its southwestern boundary they bend sharply to the southwest and are intruded by three large and two smaller granitic plutons shown as Triassic ($\eta\gamma_5^1$), Jurassic ($\gamma_5^2$), and Cretaceous ($\gamma_5^3$) in age, but whose ages are only poorly known. Looking broadly at the distribution of major rock units in the northern area, they form several large, N-S, first-order anticlinoria underlain by latest Neoproterozoic (Pzl) and lower Paleozoic rocks (Pzl) as well as synclinoria underlain by

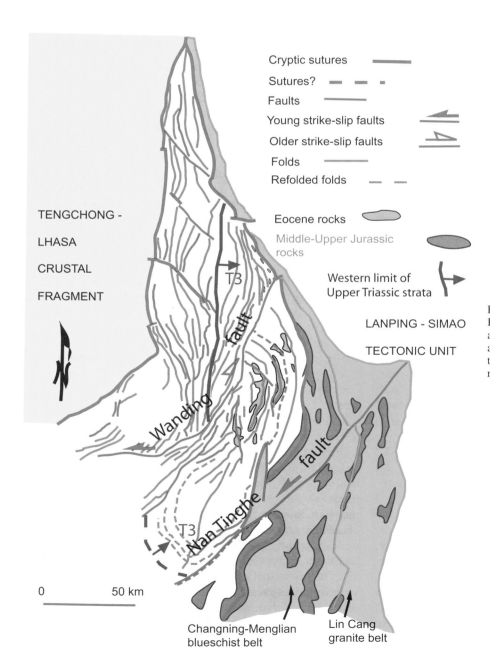

Figure 12-3. Generalized structure map of the Baoshan Unit, showing the major structures and geological elements that were deformed and offset by post-Eocene displacements on the Wanding and Nan Tinghe faults, and other relations discussed in the text.

Mesozoic rocks (Mz1 and Mz2), except in the southwest, where there is a broad area underlain by latest Neoproterozic and lower Paleozoic strata intruded by several large plutons. In this part of the Baoshan Unit, Mesozoic rocks lie only to the northwest adjacent to the NE-trending boundary between the Baoshan and Tengchong Units. In the southern part of the Baoshan Unit the structures are more complicated. In the east, structures and rock units make a convex-east arc, and in the southwest they form what appears to be a simple, large, doubly plunging, NE-trending antiform cored by lower Paleozoic and possibly uppermost Neoproterozoic strata. However, Triassic and upper Paleozoic strata contain folds around the western plunging nose of this antiform, forming a refolded set of folds. Along the southern flank of this large antiform, thrust faults dip both SE and NW, and some of them are low-angle thrust faults that tend to be more common near the boundary, with a belt of low-grade metamorphic strata to the east. Importantly the older rocks in the core of the antiform are not unconformably overlain by Middle Jurassic or Paleogene rocks, which lie unconformably on upper Paleozoic and Mesozoic rocks along the flanks of the antiform (Plate 3), a relationship that is also true for the large first-order folds in the northern part of the Baoshan Unit, suggesting that these large structures could be Cenozoic.

To the southeast a belt of low-grade metamorphic strata 15–25 km wide (Pzl2, Plate 1) lies along the eastern side of typical Baoshan rocks and mimics its broad, convex-east curvature. These strongly cleaved and folded low-grade rocks have been assigned different ages on different maps, but all maps assign them to either a latest Neoproterozoic and/or early Paleozoic (Cambrian?) age. Most of the rocks are greenschist grade, but in the south, south of the NW-trending left-lateral Heihe fault that offsets all the major units to the east, they reach biotite and staurolite grade (m, Plate 1). The phyllite and schist, whose protoliths consist of >2300 m of sandstone, some arkosic, plus siltstone and shale, with rare limestone, are separated by faults from the characteristic Baoshan sequence to the northwest and the Changning-Menglian Unit to the east (Plate 1). In one locality in the southernmost extent of these rocks near the Burma (Myanmar) border, Ordovician fossils have been recovered from the upper part of the section, and the lower grade section of clastic rocks is assigned an Ordovician age at the top and a Cambrian age for the lower two-thirds. In the same belt of low-grade rocks, and lying east of the faults separating them from the unmetamorphosed Baoshan Unit, is a zone of highly faulted and sheared uppermost Neoproterozoic-Cambrian(?) schist (Pzl2, Plate 1), with a section of fine-grained clastic rocks with chert ~2100 m thick (again, all thicknesses are probably structural thicknesses) that contains one conodont fossil locality that yielded an Ordovician age. These rocks overlie, in rare outcrops, strata that are possibly correlative with the Ordovician-Cambrian section mentioned above. All these metasedimentary rocks resemble the thick lower Paleozoic–uppermost Neoproterozoic rocks in the Baoshan Unit, but they are higher grade and are more strongly deformed than the rocks to the west. Our brief investigation of these rocks suggests that they have a protolith consistent with the lower part of the Baoshan succession; thus we regard them as part of the Baoshan Unit (Pzl2), but their fault-bounded structural position requires further discussion (see below).

Near their southeastern and northern ends the low-grade metamorphic rocks are intruded by plutons at of Triassic ($\eta\gamma_5^1$, Late Triassic, based on K-Ar and Rb-Sr ages of 220 and 210 Ma, respectively; from the Chinese map sheet) and Mesozoic ($\gamma_5^2$, $\gamma_5$) age, respectively, and one highly foliated pluton along their eastern boundary that is considered a deformed Triassic granite. The contact of these rocks with the unmetamorphosed Baoshan rocks to their west is invariably faulted and hard to interpret, but the contact is certainly polygenetic, for in places the faults are covered by Jurassic sedimentary rocks, and in other places the faults cut the Jurassic rocks. This belt of low-grade rocks strikes into Burma (Myanmar), but it reappears to the south in China across the left-lateral Heihe fault, where it forms a gentle, arcuate eastward belt, but here the rocks reach amphibolite grade, locally containing staurolite. The fault along the eastern margin of this metamorphic belt we recognize as the tectonic boundary between the Baoshan and Changning-Menglian Units (see below).

The entire Baoshan Unit is strongly deformed by folds that are mainly upright, and faults mapped as thrust faults that are probably genetically related to the folds, although normal faults are also present. Two scales of folding stand out from the map: the first-order anticlinoria and synclinoria mentioned above, and abundant second-order, smaller folds and faults that parallel them (Fig. 12-3). Where the Permian rocks were originally mapped as resting unconformably on older Paleozoic strata in the southwestern part of the area near the contact with the Tengchong Unit, a relationship suggesting a pre-Permian deformation, this relationship has been reinterpreted as a low-angle normal fault (Wang et al., 2008).

In the northern and western parts of the Baoshan Unit the unconformity at the base of Triassic rocks is largely a paraconformity and does not indicate major deformation prior to deposition of Triassic strata. A similar relation is present between the Middle Jurassic and mainly the Middle or Late Triassic rocks that also suggests only minor pre–Middle Jurassic movements, as they are also usually in paraconformable contact (Plate 3). The Jurassic strata are as young as Late Jurassic in the western part of the Baoshan Unit, and the major deformational event that affected this part of the unit can be bracketed only as post-Jurassic and pre-Paleogene (Eocene?), but Paleogene strata are present only in the southeastern part of the Baoshan Unit, where these strata do unconformably overlie folds that contain Upper Jurassic strata. The cross-cutting granitic plutons in the southwestern part of the area could provide an upper limit to the main deformation. These plutons are shown to intrude the structures within the belt of lower Paleozoic strata, and they have been assigned both Paleozoic and Cretaceous ages on Chinese geological maps, but their ages remain poorly known, and at present they yield no constraints on the age of the main deformation within the Baoshan Unit.

In the southeastern part of the Baoshan Unit are more and different constraints on the age of deformation. In the unmetamorphosed part of the Baoshan Unit, Middle Jurassic strata are rare, but where they are present in the southernmost part of this unit they are unconformable on Upper Triassic rocks, and they even rest on rocks as old as Permian, indicating pre–Middle Jurassic deformation. However, in general Middle Jurassic and Paleogene strata do not lie above the folded lower Paleozoic rocks in the large anticlinorial structures; thus it is unclear if this means these large structures are post–Middle Jurassic or even Paleogene in age. In the eastern part of the southern Baoshan Unit, Paleogene strata rest unconformably across most of the structures; however, they are also folded, and the folds have similar arcuate trends to the older structures, but they do not have exactly similar trends. Similar relations are present in the northern part of the Baoshan Unit, but only north of the eastern end of the Wanding fault. Thus shortening events in the southeastern and northern parts of this unit are usually considered to be pre–Middle Jurassic, followed by folding of late Paleogene–pre-Neogene age, but the constraints on the ages of the deformations within the main body of the Baoshan Unit are poor.

If the low- to high-grade rocks that we interpret to be originally an eastern part of the Baoshan Unit are considered, the deformation history becomes more complex. The low-grade rocks are unconformably overlain by unmetamorphosed Upper Triassic strata, but only in the southernmost part of the low-grade belt (Plate 3). The plutons that intrude this belt cut across some of the structures, and the pluton in the southernmost part of the belt also intrudes Upper Triassic rocks (Mz2) and is unconformably overlain by Middle Jurassic strata (J2, Plate 1). This belt may have been deformed at an older time from the rest of the unmetamorphosed part of the Baoshan Unit. The faults that juxtapose the unmetamorphosed Baoshan and Changning-Menglian Units with the low-grade belt are covered in many places by Middle Jurassic strata. So even if the two different parts of the Baoshan Unit have a somewhat different deformational history, they were joined together by Middle Jurassic time, and post–Middle Jurassic deformation affected both units together. Whereas the Middle Jurassic strata do not unconformably cover lower Paleozoic rocks in the main part of the Baoshan Unit, they do unconformably overlie the belt of metamorphic rocks of early Paleozoic age, considered part of the Baoshan Unit (Pzl2); thus these two sequences appear to have a somewhat different deformational history. Middle Jurassic strata in both units (but Middle Jurassic rocks are present only in a limited area in the southern part of the unmetamorphosed Baoshan Unit) are folded, but less so than the underlying rocks. The low-grade unit is also unconformably overlain locally by Cretaceous and Paleogene red beds in their southern part near Burma (Myanmar) and north of the Heihe fault (Plate 1).

The extent of a syn- or post-Paleogene and pre-Pliocene period of folding is poorly constrained, because the rocks assigned an Eocene age are largely unfossiliferous, and their age is assigned on the basis of correlation with lithologically similar rocks in adjacent units, and on their relation to early Cenozoic faults (see below). These rocks occur mainly in the eastern part of the Baoshan Unit and are folded and faulted along trends that are subparallel to the folds in the underlying, more strongly folded Mesozoic rocks (Fig. 12-3). These folds also follow the convex-east arc, and they indicate that the arcuate structure within the Baoshan Unit is a Cenozoic feature probably related to early Cenozoic strike-slip faulting, vertical-axis rotation, and associated deformation.

The difference in the folding in different parts of the Baoshan Unit is probably related to proximity to either its western or eastern boundaries, which are interpreted to mark suture zones. The Nujiang suture in the west and the Changning-Menglian suture in the east underwent different times of suturing: Late Jurassic–Early Cretaceous, and Triassic, respectively. These relations are discussed in more detail below.

**Strike-Slip Faults within the Baoshan Unit**

The Baoshan and adjacent units were affected by important strike-slip faults of both early and late Cenozoic age. The major, curved, NE-striking Wanding fault, which separates the Baoshan Unit into two parts, is one of several NE-striking faults within this unit, and is part of a system of NE-striking faults that are present across the entire southwestern part of the map area (Plate 1; Fig. 2-10). The regional distribution and interpretation of these faults will be considered in more detail below. Within the Baoshan Unit the Wanding fault is the most important. It separates the unit into two parts, and offset rock units across the fault show that its major displacement was right-lateral (Plate 1 and Fig. 12-3). The arcuate, convex-east structural trends near its northeast end show the right-lateral offset within the Baoshan Unit, as does the western extent of Upper Triassic rocks north and south of the Wanding fault (Fig. 12-3). The magnitude of the offset is ~25–30 km, but importantly the eastern end of the fault does not offset the western-boundary fault zone of the Lanping-Simao Unit (Plate 1), a relationship that will be discussed below. The age of right-lateral faulting is post-Paleogene, but it is interpreted to be early to middle Cenozoic on regional evidence (Lacassin et al., 1998). This Cenozoic system of faults in the region, of which the Wanding fault is one, has a complex origin. Many NE-striking faults affect the Baoshan Unit, but how many of these may be part of this early Cenozoic system remains unknown. Such faults could be very important tectonically; for example, one such fault, the Nan Tinghe fault, forms the present westernmost boundary between the unmetamorphosed and low-grade-metamorphosed parts of the Baoshan Unit and will be discussed below.

Many of these NE-striking faults, the Wanding fault included, are now active left-lateral faults; the sense of displacement reversed sometime during the late Cenozoic, but the timing is not well constrained (Lacassin et al., 1998). The evidence is clear, as numerous left-lateral stream offsets and related pull-apart basins document the active left-lateral displacement on these faults (Lacassin et al., 1998; Wang et al., 1998). We interpret faults of both ages and senses of slip to be parts of two regional systems related to clockwise rotation about the eastern Himalayan syntaxis, the first beginning in early Cenozoic time, and the second beginning ca. 12 Ma (Wang et al., 1998).

The long, narrow, N-S–trending basin containing Quaternary and Recent strata that lies along the Nujiang River in the western part of the Baoshan Unit is a graben in its southern and widest part, bounded by active normal faults (Plate 1). The basin increases in width to the south, as does the probable magnitude of extension across the basin, and contains a number of small, west-tilted blocks rotated down from the higher mountains to the west. This basin was erroneously related to shortening by Wang et al. (1998), but this error was corrected in a recent paper by Wang et al. (2008). To the west of this basin the NE-striking Ruili fault, along the SW boundary between the Baoshan and Tengchong Units, shows active left-lateral displacement and the development of several pull-apart basins along its trace, with both Neogene and Quaternary deposits (Wang et al., 1998; Wang et al., 2008). These young basins and their bounding faults have their origin as the boundaries of crustal fragments within the clockwise-rotating crust around the eastern Himalayan syntaxis. These faults are similar to faults discussed above for the late Cenozoic Xianshuihe-Xiaojiang fault system. The Ruili fault accommodates the clockwise rotation with increasingly greater left-slip displacement and increasingly greater east-west extension to the east, where it forms the southern end of the Nujiang River graben (see also Wang et al., 2008). Other such examples of similar structural relations occur within the Tengchong Unit to the northwest (see below).

Whether a majority of faults within the Baoshan Unit are related to the young rotation around the syntaxis is unclear, because they do not have obvious features related to active displacement, and their ages are unknown. Some of the faults could be related to the early Cenozoic deformation, such as the NW-striking faults within the northern part of the Baoshan Unit; however, the ages of most of these faults remain unknown.

**Boundaries of the Baoshan Unit**

Because the Baoshan Unit contains rocks that are unlike those of its neighboring units it has been considered to be a continental fragment that was accreted to other major fragments within Indochina and southern China during Mesozoic time (e.g., see Metcalfe, 1996, 1998, 2006, and references therein), its boundaries have been considered to be suture zones where possible oceanic crust was consumed. Thus the boundaries of the Baoshan Unit are regionally important not only for the assembly of continental fragments of SE Asia, but also because the boundaries have been reactivated to form some of the major fault zones in this part of China. Where its bounding shear zones come together at its northern end, the Baoshan Unit pinches out (Plate 1), and correlative rocks have not been found in Tibet to the north. The termination of the Baoshan Unit thus marks the northern termination of the Sibumasu crustal fragment of Indochina.

The western boundary of the Baoshan Unit is formed by the N-S–trending Gaoligong Shan shear zone, a zone of mylonitized granite and metasedimentary rocks recently studied by Akciz (2004; Akciz et al., 2008) and Lin et al. (2009). This shear zone is ~10–15 km wide and consists of two mylonitic parts, an orthogneiss section to the west, and a paragneiss section to the east, both of which consist of rocks metamorphosed to amphibolite and greenschist grade. Most of the foliation in these metamorphic rocks dips steeply ($\geq 60°$) both to the west and east, showing a well-developed mylonitic foliation that contains a prominent subhorizontal lineation, interpreted as a stretching lineation, formed by the elongation of quartz, feldspar, and biotite. Shear criteria, such as asymmetric tails on deformed feldspar porphyroclasts, S-C surfaces, asymmetric boudins, and centimeter- to decimeter-scale asymmetric folds observed in horizontal sections perpendicular to the foliation and parallel to the lineation, indicate a consistent right-lateral sense of shear (Akciz, 2004; Akciz et al., 2008; Wang et al., 2006). Syntectonic pegmatite dikes and leucogranite bodies generally form concordant veins, layers, and lens-shaped bodies, and many of these bodies contain a foliation formed by micas, indicating that they have been affected by the same ductile deformation as their host mylonitic rocks. Geochronological work indicates that this shear zone was active during early and late Cenozoic time and was one of the faults forming the western boundary of the extruding Indochina crustal fragment (Fig. 11-2; Akciz, 2004; Akciz et al., 2008; Wang et al., 2006; see below).

Southward, the Gaoligong Shan shear zone and its mylonitic fabric bend to the SW (Plate 1) and continue into northern Burma (Myanmar). Along the SW-trending part of the shear zone the mylonitic rocks are poorly exposed. They have not been well studied and are shown as mylonitic metamorphosed rocks on Plate 1 (m). They are mostly mylonitic, granitic rocks, but with rare exposures of mylonitic metasediments. These rocks are interpreted to be the direct continuation of the Gaoligong Shan shear zone rocks farther north. The mylonitic granitic rocks in the shear zone belong mostly to the poorly dated Mesozoic and early Cenozoic plutons of the southward continuation of the Gangdese batholith of the adjacent Tengchong Unit and its equivalent in the Lhasa Unit to the north. This area is covered by heavy vegetation, and outcrops are poor. Chinese maps show that the mylonitic rocks strike NE and dip both NW and SE. More work is needed in this area to understand this part of the shear zone.

The southern boundary of the mylonitic rocks where they trend NE is a north-dipping thrust fault. Footwall rocks contain a NW-dipping section of Paleozoic and Mesozoic rocks that range from Cambrian to Cretaceous(?) (Plate 1). The pre-Cretaceous rocks are similar to those in the Baoshan Unit, except they lack the characteristic upper Carboniferous Woniusi basalt, probably owing to an unconformity or a low-angle normal fault (Wang et al., 2008). Cretaceous(?) rocks at the top of the section contain poorly diagnostic plant fossils, and within these rocks, and at the thrust contact, are bodies of serpentinized mafic and ultramafic rocks. Also directly below the thrust fault in its western part is a narrow belt of Upper Triassic rocks that are also in fault contact with Paleozoic rocks. Triassic rocks are not present at the contact between Middle Jurassic and Permian rocks in the Baoshan sequence to the south, and may represent a fault sliver that has traveled some distance. From the discussion above, this boundary has been the site of Cenozoic strike-slip displacement that is presently active and which overprints structures related to its earlier history. The fault zone is the locus of several large, late Cenozoic pull-apart basins (Plate 1), recently studied by Wang et al. (2008).

The western boundary of the Baoshan Unit marks a major contact between two crustal units, the Baoshan and Tengchong Units (see below), with very different Paleozoic and Mesozoic sequences (Metcalfe et al., 1996, 1998, 2006, and others). We support the suture-zone interpretation for this boundary, and the evidence will be presented below. The only direct evidence that it is a suture is the presence of the serpentinized bodies in the Cretaceous(?) rocks along the southwesternmost extent of the boundary in China, and the contrast in stratigraphic units between the Baoshan and Tengchong Units. Its northward continuation is probably the Bangong suture, but the connection between the Gaoligong and Bangong areas is not clear and will be discussed in more detail below.

Our regional correlations place the mylonitic granitic rocks in the western part of the Gaoligong shear zone to be part of the southward continuation of the Gangdese batholith belt of Tibet, which makes up the eastern part of the Tengchong Unit (Akciz, 2004). In the north, NS–striking mylonitic metasedimentary rocks form part of the shear zone along its eastern side. These metasedimentary rocks do not contain rocks that can be correlated with the Baoshan section, but rather they correlate best with upper Paleozoic rocks of the Lhasa Unit to the north, into which they can be traced (see below). Both the granitic and sedimentary rocks were intruded and faulted before the major shearing took place. Along this segment of the shear zone there are no rocks of oceanic affinity, and the suture—the Nujiang suture—is cryptic, and it lies along the shear zone east of the Gaoligong mylonitic belt, although the mylonitic belt contains significant post-accretion shear. Where cryptic north of where the Baoshan Unit pinches out, the suture marked by the Gaoligong shear zone is juxtaposed against the Chong Shan shear zone, the shear zone that marks the east side of the Baoshan Unit farther south (see below); thus to the north the two shear zones are adjacent to one another.

Along most of the north-south trace of the western boundary, the eastern limit of mylonitic rocks is sharply juxtaposed against unmetamorphosed strata of the Baoshan Unit, and the boundary is marked by a young brittle fault (Plate 1). It dips vertically and is very straight, but it shows no sign that it has an active strike-slip component, as small tributaries to the Nujiang River show no evidence of lateral displacement. Thus, in this segment the boundary is a young fault that has overprinted the older mylonitic fabrics related to the Gaoligong shear zone.

In the north the eastern boundary of the Baoshan Unit is marked by the Chong Shan shear zone, a belt of mylonitic rocks of mixed protoliths. This shear zone separates the Baoshan and Lanping-Simao tectonic Units where Cenozoic shearing has been superposed on a late Paleozoic–Mesozoic suture between these two units. The suture is also cryptic, lacking any evidence of oceanic rocks, and having been highly modified by Cenozoic shearing and large displacement. This shear zone consists of mylonitized metaigneous and metasedimentary rocks that reach kyanite grade, and it was intruded by leucogranite dikes and sills. Foliation in these rocks dips steeply both east and west and contains a subhorizontal stretching lineation (Akciz et al., 2008). Kinematic indicators are abundant and indicate dominantly left-lateral shear, but locally interlayered rocks show both left- and right-lateral shear where the Chong Shan and Gaoligong shear zones are juxtaposed (Plate 1). At the south end of the mylonitic belt, lower Paleozoic rocks of the Baoshan section increase in metamorphic grade toward the shear zone and become mylonitic and are part of the protolith of the shear zone. However, other parts of the shear zone contain rocks unlike those in the Baoshan Unit, and bodies of granite of late Paleozoic–early Mesozoic age are present locally within the shear zone. The age and composition of these granites are similar to those of rocks in the Lin Cang Unit farther south. Other rocks within the shear zone may have protoliths from metamorphic rocks within the Lin Cang and Changning-Menglian Units to the south, where these units appear to be traceable into the Chong Shan shear zone along strike (Plate 1). The contact between the shear zone and rocks of the Lanping-Simao Unit is a sharp fault contact, a part of the Lancang fault zone, with brittle-ductile and brittle deformation. This contact is the result of late-stage modification of the Chong Shan shear zone against the Lanping-Simao low-grade to unmetamorphosed sedimentary rocks, a juxtaposition discussed in more detail below. The contact is clear, as the shear-zone rocks are strongly mylonitic and contain granites, whereas the Lanping-Simao Unit consists of lower grade, mainly meta-sandstone and meta-siltstone.

Where the Lancang River makes a westward bend across the shear zone the strongly mylonitic rocks of the Chong Shan end, and to the south the boundary between the Baoshan and Lanping-Simao Units is marked by the brittle character of the Lancang fault zone of Cenozoic age between upper Paleozoic rocks of the Baoshan Unit and Lanping-Simao Mesozoic red beds. South of where the NNE–striking Wanding fault intersects this boundary fault zone the position of the boundary becomes less clear (Fig. 12-3). We propose that the southern continuation of the Cenozoic zone of deformation probably lies along the western boundary of the Lanping-Simao Unit (see above). However, this boundary to the south also marks the major boundary between the Lanping-Simao and Changning-Menglian and Lin Cang Units (here often referred to as the southern continuation of the Lancang suture) but not the eastern boundary of the Baoshan Unit. These foliated metasedimentary and metaigneous rocks belong to a northward-narrowing belt of the Changning-Menglian and Lin Cang Units (see below and Plate 1).

The southern continuation of the eastern boundary of the Baoshan Unit diverges from the boundary fault with the Lanping-

Simao Unit and trends more southward or south-southwestward, but it is very irregular, and the original boundary has been strongly modified by younger faults. The Changning-Menglian and Lin Cang Units broaden to the south between the Baoshan and Lanping-Simao Units (Plate 1). South of the Wanding fault are two possible positions for the eastern boundary of the Baoshan Unit: one lies east of the unmetamorphosed rocks of the Baoshan Unit, and the other lies east of the metamorphic rocks (Pzl2) that we regard as the eastern part of the Baoshan Unit. Whereas the fault between the metamorphosed and unmetamorphosed rocks is an important fault, we take the eastern boundary of the metamorphosed rocks to be the main boundary of the Baoshan Unit. These foliated metaclastic rocks rarely contain fossils. Unlike the rest of the Baoshan Unit, where the stratigraphic section is generally complete from the latest Neoproterozoic to the Middle Jurassic, the metamorphic rocks probably range from latest Neoproterozoic to Silurian(?), but also with some fault-bounded pieces of Devonian and Permian rocks from the adjacent tectonic unit to the east. These infaulted pieces may be fault slivers, or perhaps windows, but no mapping has been undertaken to evaluate their position. The rocks are unconformably overlain by unmetamorphosed Upper Triassic or, more widely, Middle Jurassic strata, which without other constraints, suggests that the deformation and metamorphism are pre– to syn–Late Triassic. Most workers regard the deformation to be Triassic, based on the Triassic closure of the oceanic track in the Changning-Menglian belt to the east.

The eastern boundary of the Baoshan Unit can be followed by the juxtaposition of the metamorphosed lower Paleozoic strata (Pzl2, Plate 1) and sedimentary rocks of very different character that belong the Changning-Menglian belt to the east (see below). An unusual feature of the eastern boundary faults is that they are marked by a belt of Middle Jurassic strata that unconformably overlies the boundary fault. In fact, many of the major fault boundaries that juxtapose different rocks are overlain by belts of Middle Jurassic sedimentary rocks (Plate 1; see also Fig. 12-3). Additionally, the Jurassic rocks are cut by younger faults, so that the original nature of the boundary fault has been disrupted and obscured. Many faults cut the boundary obliquely, offsetting it in complex ways. The faults are post–Middle Jurassic, and in a few places the faults cut Paleogene rocks and alkalic plutons assigned a Paleogene age. Locally the boundary fault is marked by Quaternary basins that are part of the late Cenozoic strike-slip system. All of these relations suggest that the boundary is a major crustal zone that has been reactivated many times. Thus to decipher its original structure when it served as the suture between the Baoshan and Changning-Menglian Units will require much detailed work.

The low-grade metamorphism of the eastern belt of upper Neoproterozoic–lower Paleozoic rocks of the Baoshan Unit suggests that they were tectonically buried more deeply and deformed more ductilely than the unmetamorphosed part of the Baoshan Unit to the west. A poorly constrained tectonic scenario might suggest that they were part of the eastern passive margin of the Baoshan Unit, which became part of a short-lived active Andean margin in Late Triassic time when underthrust from the east by the Changning-Menglian Unit. The Upper Triassic volcanic and plutonic rocks within the eastern part of the Baoshan Unit would be the magmatic expression for this short lived Andean margin. The more deeply buried and hotter rocks along this active margin were back thrusted to the west over the unmetamorphosed Baoshan Unit rocks near the end of the Andean margin development. Unfortunately the nature of the boundary faults and internal structure of this belt of metamorphic rocks is poorly known and does not shed much light on how this belt fits into any tectonic setting. The boundary faults are hard to interpret, because since they formed they have been greatly modified. The faults that bound this belt strike generally north-northeast in the south but bend into a north-northwest strike to the north. Thus they make the same convex east bend as do the structures in the Baoshan Unit to the west. At their south end they are offset left-laterally near the Burma (Myanmar) border by the active Heihe fault. South of this fault the eastern boundary fault is completely covered by Middle Jurassic sedimentary rocks. As elsewhere in all the tectonic units from the Baoshan Unit to the east, the Middle Jurassic rocks are folded, commonly strongly so, but not metamorphosed.

The Baoshan Unit strata end abruptly to the north between the Gaoligong and Chong Shan shear zones along a NNE–striking fault (Plate 1). North of this fault the high-grade metasedimentary part of the Gaoligong shear zone first appears and continues northward, where it becomes juxtaposed with the Chong Shan shear zone. The two shear zones can be separated in the field by their lithological contrast (Akciz et al., 2008). In this area, interlayered mylonitic rocks show both right- and left-lateral shear, a relationship that indicates that the two shear zones were active contemporaneously, with active shearing being partitioned across the boundary between them. Such a relation is important regionally when trying to assess how the major crustal fragments behaved during southeastward extrusion (see below).

Southward the two shear zones diverge, and the Baoshan Unit appears between them. First, Mesozoic rocks of the Baoshan Unit were faulted against the mylonitic rocks of the Gaoligong Shan shear zone along a younger brittle-ductile and brittle fault that is near vertical, and farther south, lower Paleozoic (Cambrian?) strata of the Baoshan Unit are faulted against mylonitic rocks of the Gaoligong Shan shear zone. The sedimentary rocks of the Baoshan Unit are largely unmetamorphosed, and the younger fault juxtaposes rocks of very different metamorphic grade and modifies the original suture zone, and it remains cryptic. Farther south, where the Gaoligong Shan shear zone bends to the southwest and becomes wider, the relationship between the two tectonic units becomes less clear, largely because of poor outcrops and because the boundary has been modified by Cenozoic strike-slip faults. Baoshan Unit rocks occur south of the active NE-striking Cenozoic Ruili fault zone, but along this fault not only has there been left-lateral displacement, but the fault in places dips northward, and mylonitic rocks have been thrust southward over the Baoshan Unit strata. Rocks in the hanging wall of the thrust contain mainly granitic mylonite and rare mylonite from metasedimentary protoliths (for a recent discussion of this contact, see Wang et al., 2008).

Northward, the metasedimentary rocks of the Gaoligong shear zone become less metamorphosed and pass into a broad sequence of dark-gray to black slates and foliated sandstone that contains interbeds of conglomerate. These rocks are compositionally similar to upper Paleozoic rocks farther north in the Lhasa Unit. The continuations of the sutures between the Baoshan and Tengchong Units, and of the Chong Shan shear zone, pass into Tibet and remain cryptic for at least 100 km, becoming less mylonitic, and the locations of the sutures are less clear (see below).

### Age of Shearing in the Boundary Shear Zones

Ages determined for the sheared mylonitic rocks by $^{40}Ar/^{39}Ar$ methods show a range from $32.1 \pm 0.3$ to $19.41 \pm 0.42$ Ma by Wang et al. (2006) and Akciz (2004), but a somewhat younger age, $14.07 \pm 0.5$–$17.9 \pm 0.5$ Ma, by Lin et al. (2009), and by U-Pb on monazite from 24.5 to 16.5–17.5 Ma by Akciz (2004) (Fig. 11-2). The argon ages are cooling ages, but they show that the shear zone was active at least by early Oligocene time. However, some of the data obtained by Akciz (2004) suggested that some of the deformation within the shear zone may have begun to develop in latest Cretaceous time and earliest Paleogene time. Akciz (2004) obtained U-Pb ages of 65–79 Ma from syn-kinematic leucogranites from the Gaoligong shear zone, which suggested that lateral shearing was occurring during the late stages of the magmatism. These ages are pre-collisional, but they may imply that shearing along the eastern boundary of the Gangdese batholith was initiated in pre-collisional time. Akciz also obtained an age of 18.5 Ma from non-mylonitic leucogranite dikes that cut all the fabrics in the shear zone, indicating that shearing had ceased by that time. Ackiz obtained $^{40}Ar/^{39}Ar$ ages of 23–19 Ma, and U-Pb ages from monazite of 26–16.5 Ma, from mylonitic rocks in the north where the two shear zones merged. Wang et al. (2006) obtained $^{40}Ar/^{39}Ar$ ages of 32–28 Ma from hornblende and biotite from the Gaoligong mylonitic rocks. Work by Socquet and Pubellier (2005) obtained similar ages for the continuation of the Gaoligong shear zone in Burma (Myanmar). Unfortunately, even though there are some data to show that the shear zones were active in Cenozoic time, their detailed histories remain incomplete, particularly when they were first initiated.

## MAJOR PROBLEMS FOR THE BAOSHAN UNIT

The Gaoligong shear zone separates a major batholithic complex on the west from the Baoshan Unit on the east that is devoid of major plutons except at its southern end, where two poorly dated plutons intruded early Paleozoic rocks (Plate 1). In this respect it is similar to the Lancang shear zone, which bounds the Lin Cang Unit (see below). Why these shear zones should form such sharp boundaries to major batholithic terranes is hard to explain unless they formed prominent lithospheric boundaries during intrusion, or formed major shear zones that removed parts of the magmatic rocks during or after intrusion. Part of the problem is the lack of data for the ages, and the lack of detailed study of the Gangdese batholith east of the Eastern Himalayan syntaxis. The evolution of the magmatism is virtually unknown.

The two bounding shear zones have at least four characteristics in common: (1) they are subvertical zones of mylonitic rocks, (2) they have subhorizontal stretching lineations indicating major horizontal shear, (3) they commonly follow older sutures or terrane boundaries, and (4) they form belts of high-grade rock within regions of low-grade to non-metamorphic rocks. The last characteristic is one that has not received much attention. This point will be emphasized in a regional context below, but suffice it to say here that we interpret this relation to suggest that the mylonitic rocks were developed in regions of elevated crustal temperatures and were extruded upward during their displacement, so they have a component of vertical motion that was overwhelmed by large-scale horizontal motion. The dynamics of these and other shear zones in the area of Plate 1 will be discussed in more detail below.

# CHAPTER 13

## *Changning-Menglian Unit*

The Changning-Menglian Unit forms a belt 60–75 km wide in the southern part of Plate 1 (on DVD accompanying this volume and in the GSA Data Repository[1]), which lies between the Baoshan and Lin Cang Units, and, when traced to the north, it narrows and possibly pinches out (Fig. 1-1B). It is one of the most poorly understood and complex map units shown on Plate 1, because it consists of rocks within and adjacent to a suture zone, and it lies within an area of dense vegetation with generally poor outcrops. It is hard to make a specific tectonic discussion of the internal parts of this unit, but new, unpublished mapping (Chengdu Institute of Geology and Mineral Resources, 2002) across the central part of this unit has developed new subdivisions of the stratigraphy and tectonic elements that can serve as a basis for understanding the continuation of these rocks to the north and south. Some workers have included the metamorphic rocks here regarded as the eastern part of the Baoshan Unit as part of the Changning-Menglian Unit, but for reasons discussed below we do not.

The Changning-Menglian Unit contains rocks that range in age from Precambrian(?) to Permian and which are unconformably overlain by Triassic rocks of different ages, placing an upper limit on the age of deformation within this complex suture zone. Major rock units are separated by fault zones that define rock assemblages interpreted to represent significant Paleozoic paleogeographic elements from passive and active margins to oceanic domains. Its eastern and western parts may represent the tectonic transitions from early passive to active margin development, with rocks in the eastern part metamorphosed from low to high grade partly in high-pressure, low-temperature (HP-LT) facies, interpreted to be part of a subduction complex (Zhang et al., 1993). Centrally located within the metamorphic rocks are rare exposures of serpentinized ultramafic rocks, often considered remnants of an ophiolite sequence. Jurassic and, locally, Upper Triassic marine and non-marine strata overlap most of the major tectonic boundaries, indicating that suturing was complete by Late Triassic time; however, locally, Lower and Middle Triassic rocks rest unconformably on the deformed rocks, suggesting that closing of the oceanic terrane was diachronous.

## STRATIGRAPHY OF THE CHANGNING-MENGLIAN UNIT

New studies by Chinese geologists (unpublished map, Chengdu Institute of Geology and Mineral Resources, 2002), in the central part of the Changning-Menglian Unit, divide the area into three subunits, separated by major(?) faults, based on the interpretation of their stratigraphy (Fig. 13-1). This provides a new basis for subdivision within the Changning-Menglian Unit, and we have tentatively attempted to extend these subunits both to the north and south, but much further study is necessary to develop a better understanding of these rocks.

The westernmost subunit, Subunit I (Fig. 13-1, C-M I; Pzu1 on Plate 1), consists mainly of fine-grained clastic rocks, with both graptolites and plant fossils of Early to Late Devonian age, that interfinger with siliceous, fine-grained clastics and chert. Deposition appears to have been in shallow- to deep-water environments. These lithologies are unconformably overlain locally by a thick section of Lower Carboniferous basalt (Pzub, Plate 1) and fossiliferous Carboniferous limestone. In the upper part of the section these rocks interfinger with a Carboniferous-Permian basalt and clastic sediments with conodonts and chert of Permian age. A Permian unit consisting of fossiliferous limestone rests unconformably on both Carboniferous and Devonian strata. This section is interpreted to be a partly distal passive margin, at times the site of rifting, grading into deeper water sections with volcanic islands. It is capped by a carbonate bank and fringing carbonate deposits that grade into deep-water (oceanic?) sediments. This Upper Paleozoic sequence, with its thick Lower Carboniferous basalt and abundant chert of Devonian and Permian ages, contrasts greatly with the Baoshan Unit section to the west (see above).

Subunit II forms a belt of faulted outcrops in the middle part of the Changning-Menglian Unit, which pinches out to the north (Pzu2, Plate 1) and is characterized by some rocks that are hard to interpret. It consists of a basal unit, ~600 m thick, of fine-grained metaquartzite with a foliation that becomes more intense eastward (Fig. 13-1). The rocks appear bedded, but as the bedding is parallel to the foliation it is unclear if it is true bedding or a thoroughly transposed fabric. The uniformity of the lithology across the unit is striking, and it is not clear if the lithology is very fine grained, quartz-rich clastic rocks or perhaps even chert. These rocks are unfossiliferous and are assigned a Devonian–Carboniferous age. They are overlain by more varied sandstone and shale with some fossiliferous limestone that contains Permian fossils. These are the only two rock types within Subunit II and are distinct from strata in Subunit I by the lack of mafic volcanic rocks and their uniform lithology. We have used these characteristics to continue this subunit to the south, where it becomes wider and complexly faulted, together with mafic, volcanic-bearing strata that we assign to Subunit I. This complex

---
[1] GSA Data Repository Item 2012297, Plates 1–4, is available at www.geosociety.org/pubs/ft2012.htm, or on request from editing@geosociety.org or Documents Secretary, GSA, P.O. Box 9140, Boulder, CO 80301-9140, USA.

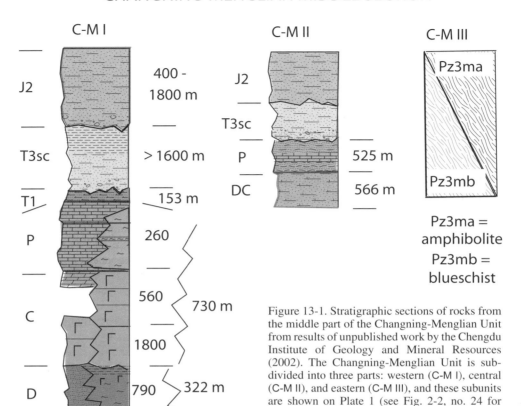

Figure 13-1. Stratigraphic sections of rocks from the middle part of the Changning-Menglian Unit from results of unpublished work by the Chengdu Institute of Geology and Mineral Resources (2002). The Changning-Menglian Unit is subdivided into three parts: western (C-M I), central (C-M II), and eastern (C-M III), and these subunits are shown on Plate 1 (see Fig. 2-2, no. 24 for location).

interleaving of these two rock units may in fact reflect the original distributions and paleogeography more than the faulted contacts that are shown on Plate 1.

The higher grade rocks along this unit's eastern boundary are, subunit III, both faulted against and intruded by Permian–Triassic plutons of the Lin Cang Unit. On most maps these higher grade metamorphic rocks are interpreted to have a Precambrian protolith and to belong to the older continental crust that is the host for the Lin Cang plutonic belt.

Some of the major unresolved questions regarding these rocks are (1) the age of the HP-LT metamorphism, (2) the existence and nature of any Precambrian protoliths and their structural interleaving, and (3) their relation to the Lin Cang plutonic belt to the east and their general tectonic setting.

In carrying the subunits used in the new mapping (Chengdu Institute, 2002) to the north and south, the following criteria were used: Subunit I is characterized by the abundance of basalt (Pzub) that is generally lacking in Subunit II. Subunit III is identified by higher grades of metamorphism, and it lies directly west of the Lin Cang batholith. This seems like an adequate characterization in the south; however, the extrapolation to the north is much less certain. The extrapolations of the subunits can only be regarded as most tentative, and their complexly faulted boundaries in the south may not be unrealistic, particularly as the original paleogeography of these rocks was probably one of mostly irregular and interfingering rock assemblages within a complex oceanic terrane. However, it is clear that the subunits are exposed in a wide belt in the south and that they narrow greatly to the north. Subunit I (Pzu1) appears to be the only subunit that can be extended to the north, but it is very narrow. Subunit II appears to pinch out to the north, and individual facies of Subunit III are combined in one metamorphic unit to the north (m, Plate 1). Viewed at the broader scale, not only do the individual subunits of the Changning-Menglian unit narrow and pinch out to the north, but the major tectonic units on both sides also do the same and curve from N-S to NW trends as they approach the Lancang fault zone (Plate 1).

## STRUCTURE OF THE CHANGNING-MENGLIAN UNIT

The contacts between the different belts of rocks that define the subunits trend generally N-S and follow the trends of the adjacent tectonic units, but they are hard to define not only because of poor outcrops but also because of extensive, younger deformational overprinting. In general the main structures that first bring the Changning-Menglian Unit together with tectonic units on either side appear to be late Paleozoic to Triassic. Most of the understanding of the rocks in the Changning-Menglian Unit comes from the new mapping in the central part of the area. The faults marking the western boundary with the Baoshan Unit metamorphic rocks have been so highly modified that it is difficult to interpret; in places the faults are shown as a west-dipping thrust fault that cuts obliquely across N-S- to NW-trending, curved fold axes in its footwall and generally curving N-S–trending folds in its hanging wall. However, the faults also cut Middle Jurassic rocks, and in one place they cut a small alkalic pluton assigned a Paleogene age; thus these contacts offer little evidence for their original late Paleozoic or early Mesozoic history. Farther north the boundary thrust dips to the N or NE along another arcuate fault trace, placing the rocks of Subunit I (Pzu1) above the Baoshan Unit rocks (Pzl and Pzu), suggestive of a west- or southwest-vergent thrust displacement. In these places the faults also cut Middle Jurassic rocks and thus are younger faults. Of interest, the Lower Carboniferous basalt commonly occurs at the base of the thrust that juxtaposes Subunit I (Pzu1 and Pzub) against other units to the east, suggesting that the basalt controlled the localization of the thrust. Subunit II (Pzu2) to the east does not contain the Carboniferous basalt and contains folds that trend N-S to NNW, similar to folds in the hanging wall of the boundary thrust. These folds, in both the hanging wall and footwall, are mapped as having been truncated by the thrust, suggesting a protracted deformational history. The rocks in Subunit II become more foliated eastward, and the general trend of the foliation is shown on Plate 1 by the dashed lines.

From this newly mapped area the stratigraphic and structural units can only be tentatively projected, generally to the north and south. In carrying the subunits to the north and south, the following criteria were used. Subunit I is characterized by the abundance of basalt that is generally lacking in Subunit II, and Subunit III is identified by its higher grades of metamorphism and position directly west of the Lin Cang batholith. To the south the belt of Changning-Menglian rocks is offset left-laterally ~25 km by the NW-striking Cenozoic Heihe fault, which extends southeastward across the Lin Cang batholith into the western part of the Lanping-Simao Unit. South of this fault is a broad area of Devonian to Permian rocks that have not been mapped recently but which do have stratigraphic similarities to the rocks in the newly mapped central area (Fig. 13-2). The Devonian rocks are up to 2800 m thick (although thicknesses are unreliable) and are overlain by a Carboniferous succession that is several thousand meters thick and consists of a thick, lower clastic sequence overlain by basalt, and an upper carbonate-dominated section. Above are Lower and Upper Permian limestones, overlain by clastic Upper Permian strata, locally with coal at the top. These strata are similar to those of Subunit I. However, here, unlike the central section described above, Carboniferous basalt extends across the entire belt in places. We have placed the basalt-bearing rocks into Subunit I (Pzu1) and the rocks without basalt into Subunit II (Pzu2) on Plate 1. This makes for an extremely irregular and faulted boundary, and one that is probably paleogeographically realistic.

To the east the rocks become metamorphic and contain both blueschist and Barrovian metamorphic rocks in fault and intrusive relations with the Lin Cang plutonic belt, with the higher grade rocks adjacent to the plutons. However, the metamorphic facies of these rocks have not been separated (Pzm3), but they have the characteristics of Subunit III. These rocks are shown on Chinese maps as having Precambrian protoliths, although they contain thick metabasalts, which might indicate that some have an upper Paleozoic protolith.

Using the same criteria mentioned above, Subunit I (Pzu1) appears to be the only subunit that can be extended very far to the north, but it becomes very narrow. Subunit II (Pzu2) appears to pinch out to the north, and individual facies of Subunit III are combined in one metamorphic unit to the north (m, Plate 1). Toward the east, rocks in Subunit III become higher grade, and along their eastern boundary they are both faulted against and intruded by Permian–Triassic plutons of the Lin Cang Unit. On most maps these higher grade metamorphic rocks are interpreted to have a Precambrian protolith and to belong to an older continental crust that is the host for the Lin Cang plutonic belt (Zhang et al., 1993; Heppe, 2004).

Farther north the upper Paleozoic rocks of Subunit I can be traced in a narrow and fault-disrupted belt northward, and the Carboniferous basalt is present both along its western and eastern sides as well as within the central part of this belt. The upper Paleozoic rocks continue north as an increasingly narrower belt only a few kilometers wide that pinches and swells along NNW–striking faults, appearing to end at the Wanding fault. In this northernmost area, the basalt forms a narrow belt along the east side, and the Paleozoic rocks are intruded by two plutons, one assigned a Cretaceous age ($\gamma_5^3$), and the second a Jurassic age ($\gamma_5^2$), but neither pluton has been dated. The metamorphic rocks east of the sedimentary and low-grade metasedimentary belt of Subunit I are continuous with the blueschist and Barrovian rocks mapped as Subunit III to the south in the central belt, but here they are not separated and are shown as an undifferentiated unit (m, Plate 1), and the general trend of their foliation is shown by dashed lines. Northward, these metamorphic rocks are separated from another belt of lower grade metasedimentary rocks to the NE that forms a large pendant within the Lin Cang batholith (Pzm). To the northwest these two belts of metamorphic rocks merge, and an arbitrary contact has been drawn between them, because the eastern belt of metamorphic rocks is regarded as having a Paleozoic protolith, whereas the western belt is assigned a pre-Ordovician age. Farther NW the higher grade rocks pinch out, and the metamorphosed metasedimentary rocks continue northwest, with faults on both sides, and these rocks continue north and strike into the metamorphic rocks of the Chong Shan mylonitic shear zone (Pzm and mch), which contains rocks of probable lower Paleozoic protolith (Pzm; Akciz et al., 2008; see below).

Although not subdivided in detail, the Changning-Menglian Unit can be well defined as rocks lying between the metamorphic rocks of the Baoshan Unit on the west and the Lin Cang

## CHANGNING - MENGLIAN
## (SOUTHERN SECTION)

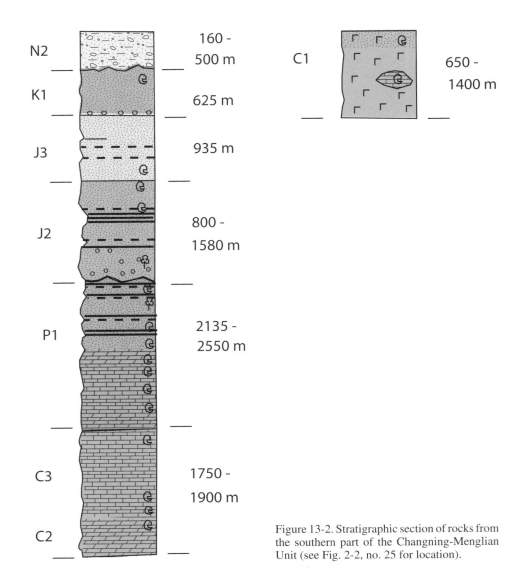

Figure 13-2. Stratigraphic section of rocks from the southern part of the Changning-Menglian Unit (see Fig. 2-2, no. 25 for location).

plutonic Unit on the east. On Plate 1 these rocks, with a general N-S trend to the north, show a striking regional bend to the NW and a dramatic narrowing, as does the Lin Cang Unit to the east. The Lancang fault zone and structures within the Lanping-Simao Unit also show a similar bend to the northwest. Much of this major bending and narrowing is probably Cenozoic in age and will be discussed in detail in its regional context below. These extrapolations of the subunits within the Changning-Menglian Unit can only be regarded as tentative. Their complexly faulted boundaries in the south may not be realistic, particularly as the original paleogeography in such rocks would imply originally highly irregular and interfingering boundaries; however, the general configuration of the tectonic unit is valid.

The structural interleaving of blueschist and Barrovian metamorphic rocks in Subunit III are suggestive of an original thrust or accretionary marginal setting, and many of the rock units are bounded by wide shear zones that generally strike N-S and which may have considerable strike-slip, but the detailed structural work to support such a setting has not been done. Some of the thrusting could be post-accretionary reworking. The interpretation by Chinese geologists (unpublished map, Chengdu Institute of Geology and Mineral Resources, 2002) shows the rocks in Subunit III to have been originally formed in an east-dipping subduction zone and related to the Lin Cang batholith to the east. This supports the interpretation of Zhang et al. (1993), who regarded the Changning-Menglian Unit and units to the east as parts of a paired metamorphic-magmatic unit related to eastward subduction, an interpretation not shared by Heppe et al. (2007; see below).

The details of the Paleozoic geology of the Changning-Menglian Unit is only poorly known, so its tectonic setting can only be generalized. The Paleozoic rocks are mainly Devonian to Permian in age (except for the metamorphic rocks on its eastern side, whose protolith age remains unknown but probably contains some Precambrian rocks), and they consist of both shallow- and deep-water facies, which can sometimes be shown to be lateral equivalents and are associated with thick basalts of Carboniferous age. Within Subunits I and II no rocks older than Devonian crop out, which might suggest this was the time of opening of the deep-water basin in which they were deposited. The general setting suggests a possible oceanic setting, which is supported by the presence of rare ultramafic rocks that occur only rarely in the middle of Subunit II (Pzu2), in the southern part of Subunit I (Pzu1), and in the northern part of Subunit I (see Zhang et al., 1985) of the Changning-Menglian Unit (Plate 1; e.g., see Zhong, 2000). The rare outcrops in the central part of the unit were studied recently by Jian et al. (2009a; Fig. 7-2), where a metagabbro yielded a U-Pb zircon age of 267.1 ± 3.1 Ma and had a chemistry suggestive of a supra-subduction zone ophiolite. The mafic-ultramafic rocks in the southern part of the unit have not been studied, but the study by Zhang et al. (1985) in the north gives a description of a dismembered ophiolite. Thin continental crust in some parts of the limit may be possible, but none is exposed. The interfingering of shallow- and deep-water rocks and their association with basalt suggests oceanic islands. The deformation and metamorphism of the rocks is as young as Late Permian and may have extended into the Early and even the early Late Triassic in some places and may have been diachronous (see below), but the deformation and metamorphism were completed by Late Triassic time. The progression from blueschists and higher grade rocks along the east side of the Changning-Menglian Unit suggest eastward underthrusting of the unit, consistent with its relation to the Permian–Triassic age of plutonism in the Lin Cang Unit to the east (see Zhang et al., 1993).

The Upper Triassic volcanic rocks and rare plutons in the eastern part of the Baoshan Unit to the west (see below) suggest that short-lived, Late Triassic westward subduction may also have occurred. However, as already discussed, the western boundary of the Changning-Menglian Unit has been so modified by post-Triassic deformation that the nature of the original boundary has been obscured.

All the rocks in the Changning-Menglian Unit, and even units to the east and west, are unconformably overlain by Upper Triassic sedimentary, largely non-marine rocks (see below). Triassic rocks provide an upper limit to the closure of the possible suture within the Changning-Menglian Unit. In many places, particularly in the north, Middle and Upper Triassic rocks (T2 and T3, Plate 1; Plate 3 [see footnote 1]) unconformably overlie all older rock units and are unmetamorphosed. The Middle Triassic rocks, however, contain volcanic rocks, which suggests that volcanism continued into the Middle Triassic. These rocks consist of ~1100 m of conglomerate and sandstone with interbedded felsic and intermediate volcanic rocks of the Middle Triassic (Fig. 13-3). In the central part of the Changning-Menglian Unit, Upper Triassic non-marine strata (T3sc) overlie the strongly foliated upper Paleozoic rocks very close to the outcrops that contain the ultramafic rocks. In one small area in the southern part of the unit, 150 m of fossiliferous Lower Triassic (T1, Plate 1) sandstone and conglomerate, with some intermediate volcanic rocks, are present unconformably overlying folded Permian limestone (Plates 1 and 3). They are unmetamorphosed but deformed. The extent of these rocks, which unconformably overlie deformed rocks related to closure of the Changning-Menglian oceanic terrane, is so limited that their regional relations remain unclear. The relation of the Triassic strata to the older rocks indicates that the deformation and associated HP-LT metamorphism had ceased by at least Middle Triassic time over much of the northern part of the area, and by the Late Triassic in the central part of the area (and possibly by Early Triassic time at least locally in the southern area), although Triassic rocks are absent from much of the unit. These variations in relation to the possible time of closure of the Changning-Menglian suture suggest that the closure was probably diachronous across the unit as it is now exposed, so that relations in one area need not necessarily apply to the entire unit.

In Middle Jurassic time a very different structural setting is present. Middle Jurassic rocks are abundant and lie unconformably on all older rocks over much of the Changning-Menglian Unit and the units to both the east and west (Plates 1 and 3). Characteristically they form elongated, arcuate basins that trend generally N-S and contain thick sections of up to 1500 m of gray, brown, and red non-marine and marine conglomerate, sandstone, and finer clastic rocks. The rocks are poorly dated by plant fossils. They are strongly folded in most places, with inclined, even overturned, axial surfaces, and in tightly folded areas they contain cleavage that extends into the underlying rocks. The basins commonly trend parallel to the underlying structures, particularly major fault zones, where they cover critical relationships. These Jurassic strata are similar to rock units that extend across the eastern part of the Baoshan Unit to the west and the Lin Cang Unit to the east, indicating widespread basin development across a broad region. They are not stratigraphically uniform laterally but appear to have been deposited within individual basins whose setting has not been well studied. Deposits in the two basins at the southern end of the Changning-Menglian Unit and one basin in the adjacent Lin Cang Unit continue into the Lower Cretaceous (K1, Plate 1). Their significant thickness changes suggest that they may have been at least partly deposited in extensional grabens. Middle Jurassic into Cretaceous time might have been a time of extension, followed by younger shortening across a broad region.

The time of the youngest shortening within the Changning-Menglian Unit is not well established. The next strata younger than Jurassic or Early Cretaceous are generally Neogene, both Miocene and Pliocene, and of limited distribution. Cretaceous and Paleogene strata are absent within the most of the unit except for Paleogene strata along narrow fault zones in the northwest at the contact with the Baoshan Unit and Cretaceous rocks in the far south of the Changning-Menglian Unit (Plate 1). Paleogene strata in adjacent units do show folding in both pre- and post-Paleogene time, but how much of the folding of Mesozoic rocks is Cenozoic across the Changning-Menglian Unit is unclear and needs careful examination.

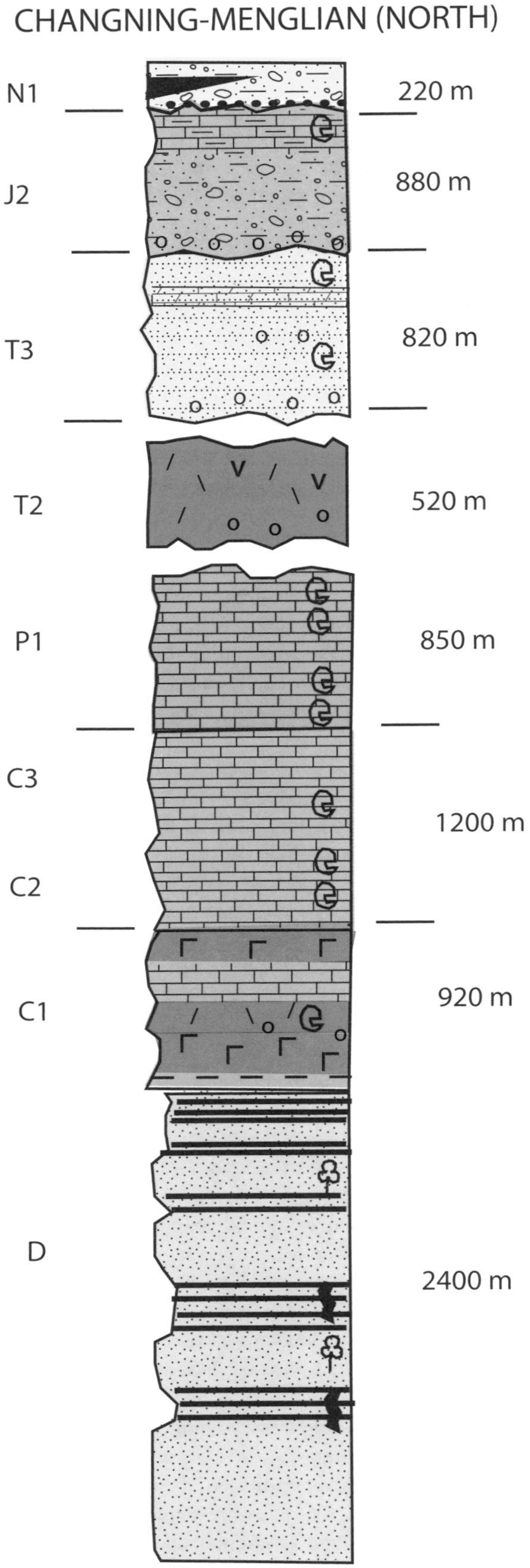

Neogene strata were deposited in generally equidimensional basins that are present across all the Changning-Menglian Unit. They consist of tan, gray, and yellow conglomerate, and sandstone, mudstone, and important units of coal. In places they are >1500 m thick. Most of the basins contain fossils, indicating they are Miocene in age, but a few smaller basins contain fossils of Pliocene age. The strata rest unconformably on all older rocks and are also folded in places but not as strongly as the older rock (Plate 3).

**Strike-Slip Faults within the Changning-Menglian Unit**

All the rocks in the Changning-Menglian Unit are cut by numerous young faults of Cenozoic age. The NE-striking Nan Tinghe and Menglian faults and the NW-striking Heihe fault are most well known (Plate 1 and Fig. 2-10). They cut through and displace all the units in the Changning-Menglian Unit and are known to have both an early and a late Cenozoic history (Lacassin et al., 1998). The early Cenozoic faults are right lateral and the late Cenozoic faults are left lateral. Abundant smaller faults displace the folded Jurassic strata, and probably many of these faults are Cenozoic in age; what role they play in the overall deformation of the region will be discussed below.

**Major Problems for the Changning-Menglian Unit**

The major problem in this area (Plate 1) is the relationship between the Paleozoic and early Mesozoic geology of the Changning-Menglian Unit and its relation to the batholithic belt in the Lin Cang Unit to the east. This key problem will be addressed at the end of the discussion of the Lin Cang Unit, below.

Some additional major questions regarding the Changning-Menglian Unit that remain unresolved are as follows: Was there an oceanic crust beneath major parts of the Changning-Menglian Unit, the age of the HP-LT metamorphism, how the structural interleaving of blueschist and Barrovian metamorphic rocks occurred, and the existence and nature of any Precambrian rocks that may be present within Subunit III and which also underlie the Lin Cang batholith. The age of final closure of the Changning-Menglian Unit may have been diachronous, but this remains unclear.

Figure 13-3. Stratigraphic section of rocks from the northern part of the Changning-Menglian Unit (see Fig. 2-2, no. 26 for location).

CHAPTER 14

# Lin Cang Unit

The Lin Cang Unit consists of a major batholithic complex lying between the Changning-Menglian and Lanping-Simao tectonic units (Fig. 1-1B). This unit might better be considered to be the eastern part of a larger combined Changning-Menglian–Lin Cang Unit, but its relation to other units remains unclear. Jian et al. (2009b) interpreted the Lin Cang as basement rocks for the Lanping-Simao Unit. Even though the interpretation that the plutonic complex is a magmatic arc related to the eastward subduction within the Changning-Menglian Unit remains likely, it is regarded as unproven by some workers, who have considered the Lin Cang batholith to be related to westward subduction along its eastern side at the Lancang "suture"—our Lancang fault zone (Fan and Zhang, 1994; Scrotese and Golonka, 1992). The ages of the plutons within the Lin Cang batholith are generally regarded to be Permian–Triassic, but good geochronological data are lacking. Older Rb-Sr ages reported by Chen (1987) show a wide range from 348 to 236 Ma. U-Pb ages from zircons have yielded a tighter age range, between 254 and 212 Ma (Liu et al., 1989), but these ages are from only limited sampling of the multi-intrusion plutonic complex. A recent analysis of plutonic rocks from the Lin Cang batholith by Heppe (2004) shows that these rocks are mainly S-type granites with high initial $^{87}Sr/^{86}Sr$ ratios, showing little evidence for derivation from mantle rocks. Heppe interprets the plutons as largely post-orogenic and formed mainly by melting of older Proterozoic continental crust.

The evidence for the presence of an older continental basement within this complex comes from the geochronological and geochemical analyses of metamorphic rocks of unknown protolith (m, Plate 1 [available on DVD accompanying this volume and in the GSA Data Repository[1]]) and metasedimentary units of possible Paleozoic age (Pzm units, Plate 1). These rock units occur as screens between plutons along shear zones and within the high-temperature metamorphic rocks along the west side of the batholith that are here placed arbitrarily into the easternmost part (Subunit III) of the Changning-Menglian belt. The metamorphic rocks of largely sedimentary and volcanic origin grade into or are structurally interleaved with blueschists that form the western wall rocks for the batholith, and because of their blueschist character were placed into an accretionary complex of the Changning-Menglian Unit (Subunit III). Within the batholith the metamorphic rocks occur as pendants, and also within shear zones with mineral assemblages that range from lower amphibolite to lower granulite assemblages (Zhang et al., 1993). These rocks are regarded as Precambrian on the basis of different criteria: (1) a whole-rock Sm-Nd isochron age of 1437 ± 17 Ma (Zhai et al., 1990b); (2) Sm-Nd model ages of granitic gneiss and amphiboles between 1.7 and 1.9 Ga, reported by Heppe (2004); (3) the suggestion by Zhang et al. (1993) that these rocks were part of a Mesoproterozoic volcanic arc; and (4) the correlation of these rocks by Cong and Zhai (2000) with basement rocks in the Yangtze Unit.

The protoliths for the western wall rocks, including the blueschists, here assigned to the Changning-Menglian Unit, are regarded also as Precambrian on the recent unpublished map from the Chengdu Institute (2002), referred to earlier, and within the central part of the Changning-Menglian Unit. The strong late Paleozoic–Triassic overprint on these rocks has made it difficult to unequivocally demonstrate the older age. The belt of metamorphic rocks within the northwestern part of the Lin Cang batholith (Pzm, Plate 1) is shown on some Chinese maps as having a pre-Ordovician protolith and even a suggested Cambrian age for some of the rocks. However, the belt to the east within the batholith is shown as having a probable Paleozoic protolith. From his geochemical work, Heppe (2004) suggested that the plutons of the Lin Cang Unit were products of melting of an older Precambrian basement. Needless to say, there is considerable work to be done on these rocks to show the nature of the crust in this region and the origin of the batholith.

The batholith in general consists of coarse-grained granodiorite to granite, and some diorite plutons and generally lacks an obvious fabric. Heppe (2004) performed detailed analyses of some of the plutons, showing high initial $^{87}Sr/^{86}Sr$ ratios from 0.7455 and 0.7545 and $^{143}Nd/^{144}Nd$ values of 0.511768 and 0.511893. Sm-Nd model ages yielded a 1.9–2.1 Ga age. These analyses show a dominance of S-type chemistries, and Heppe (2004) concluded that the plutons are a post-orogenic assemblage formed from melting of a Proterozoic continental crust. He argued that there is no geochemical evidence for the existence of an island arc protolith.

In the northeastern part of the Lin Cang Unit are critical areas where the plutons of the batholith intruded fossiliferous volcanic and sedimentary rocks of Middle and Late Triassic age (Mz1 and Mz2b, Plate 1). This is the only area where such relationships can be observed. There are no other places where Triassic rocks are not in fault contact with the Lin Cang plutons. The volcanic rocks comprise a bimodal assemblage of basalt and felsic volcanics and could be regarded as the volcanic equivalents of the plutons of the Lin Cang batholith. Heppe (2004) suggested

---
[1]GSA Data Repository Item 2012297, Plates 1–4, is available at www.geosociety.org/pubs/ft2012.htm, or on request from editing@geosociety.org or Documents Secretary, GSA, P.O. Box 9140, Boulder, CO 80301-9140, USA.

a bimodal character for all the volcanic rocks that lie along and just east of the Lin Cang plutonic belt (Mz1 and Mz2b, Plate 1), but west of the Lancang fault zone, which defines the eastern boundary of the Lin Cang Unit. The recent unpublished map of the central area of the batholith demonstrates that some of the volcanic and interbedded sedimentary rocks originally assigned a Late Triassic age are as young as Early Jurassic and that their bimodal character continues into rocks of this age. Some of the Lower Jurassic sections are reported to be >4000 m thick. These Triassic and Lower Jurassic volcanic rocks form a possible, but complicated, connection of the Lin Cang Unit with the Lanping-Simao Unit to the east and are relevant to the nature of the eastern boundary of the Lin Cang Unit, as discussed above.

The presence of the volcanic rocks extending into the Early Jurassic suggests that magmatism was active through the Triassic into the Early Jurassic. The same Triassic volcanism may be the source for the Middle Triassic volcanic rocks in the northern part of the Changning-Menglian Unit (T2 and T3, Plate 1), where they rest unconformably on blueschist facies rocks. These relations apparently contradict other geological data. The unconformity at the base of the Middle and Upper Triassic rocks above blueschists of the Changning-Menglian Unit (see above, and Plate 2 [see footnote 1]) provides an apparent upper limit on the age of the blueschist and suggests that subduction was completed by Middle and perhaps locally by Early Triassic time (see above). However, the Triassic rocks in the Changning-Menglian Unit have limited distribution and may not provide an upper limit for all rocks within the unit; so this may suggest that subduction continued into the Late Triassic, a point discussed above in the section on the Lanping-Simao Unit. Thus precise ages for the blueschist in the Changning-Menglian Unit, for the plutons of the Lin Cang Unit, and their relations remain major unresolved problems for the region.

Middle Jurassic conglomerate, sandstone, and shale lie unconformably above the plutons (Plates 1 and 3 [see footnote 1]) and provide a certain upper limit for most of the batholith, but in a few places rare small bodies of granite also intrude the Jurassic strata and are considered to be Cretaceous ($\gamma_5^3$) or even Paleogene ($\varepsilon_6$) (Plate 1). The Middle Jurassic strata are the basal and most abundant part of a section of Middle and Upper Jurassic rocks (J2 and J3) that extend locally into the Cretaceous (K1) in the Lin Cang Unit (Plate 1). These Jurassic strata are widespread across all the units from the western Lanping-Simao Unit to the eastern Baoshan Unit (J2, Plates 1 and 3). Within the Lin Cang Unit and other adjacent units these strata locally reach thicknesses of >1500 m. These rocks are dominantly red, brown, tan, and yellow non-marine rocks, but locally they contain marine interbeds. The strata are strongly folded in places, along with the underlying granite, and in some places contain a well-developed axial planar cleavage that commonly extends into the underlying granites. Along the margins of some of the generally N-S linear belts the underlying granite has been thrust over the Jurassic rocks. The local distribution of these rocks suggests that their basins of deposition were also local, a feature we suggest indicates an extensional origin for these basins, an interpretation that requires detailed study.

The new unpublished map by the Chengdu Institute (2002) shows the plutons of the Lin Cang batholith to contain numerous shear zones that parallel the N-S trend of the unit as a whole but that also cut at a low angle across it to the NNE. These shear zones thus parallel pluton boundaries but also locally cut across them. The map also shows that some of these shear zones displace the unconformity at the base of the overlying Middle Jurassic sedimentary rocks. Thus the age of these shear zones is at least partly post–Middle Jurassic, but how young they might be is uncertain. The shear zones are cut by NE- and NW-striking brittle faults that show mostly left-lateral displacement and are probably of late Cenozoic age as part of the regional strike-slip systems of that age in the map area.

In the southeasternmost part of the Lin Cang Unit, Paleogene (Eocene?) sedimentary rocks interbedded with basalt unconformably overlie the plutons (Plate 1). This is the only place where Paleogene strata are present within the Lin Cang Unit. These strata are folded along N-S–trending axes. They are unconformably overlain by unfolded Neogene strata, showing that at least in this area Paleogene deformation affected the eastern part of the Lin Cang Unit.

The folds in the Eocene and Jurassic rocks are interpreted to be early Cenozoic in age on the basis of regional evidence. The Lanping-Simao Unit to the east contains a continuous stratigraphic section that ranges from Early Jurassic to early Cenozoic in age. Although some folds of Cretaceous age are present in the southeastern part of the Lanping-Simao Unit, folding of most of the Lanping-Simao rocks is early Cenozoic in age. On this basis we interpret the folding of the Jurassic and Eocene rocks within the Lin Cang Unit also to be early Cenozoic in age. A similar conclusion was reached for Jurassic and Cretaceous rocks in tectonic units to the west (see above). The age of folding, however, is poorly constrained and remains to be established as a critical piece of data for a better understanding of the tectonics of this part of the map area.

Similar to the other tectonic units to the west, the northern part of the Lin Cang batholith curves and plunges to the NW, beyond which the unit is represented by a pendant of low-grade Paleozoic rocks (Pzm, Plate 1) that may continue into the Chong Shan shear zone to the NW (Akciz et al., 2008). In our interpretation the curvature of the batholith is due to early Cenozoic deformation related to clockwise rotation of crustal fragments around the eastern Himalayan syntaxis, accommodated by shear along the Chong Shan shear zone and Lancang fault zone (Wang and Burchfiel, 1997; Akciz et al., 2008). The Lin Cang Unit and all the tectonic units to the west bend westward and end against the Cong Shan shear zone and the Lancang shear zone: the latter is its possible southeastern continuation (Plate 1). No offset continuation of these belts is obvious northeast of the shear zone, a relationship that requires more explanation (see below). We will suggest that regionally the few elongated bodies of late Paleozoic intrusives to the north, and the rocks they intrude that continue far to the northwest, may be fragments of the Lin Cang batholith caught in the major Chong Shan and Lancang shear zones postulated above, and marked by tectonic lenses of which the Chong Shan shear zone is a part.

As pointed out above, the Lin Cang Unit and other units to the west are cut by the NE-trending Nan Tinghe fault, which lies within the axial plane of a first-order, convex-east bend in the Lin Cang Unit. This bend is interpreted as a fold with a vertical axis that also involves the Lanping-Simao Unit, and its relation to movement on the Nan Tinghe fault and other structures to the north suggests the structure to be an early Cenozoic feature (discussed above).

Neogene basins (N1, Plates 1 and 3) lie unconformably on the plutons in the central and southeastern parts of the Lin Cang Unit. They contain fossiliferous Miocene conglomerate, sandstone, and mudstone. They are generally equidimensional to NW elongate basins and are moderately folded along NW and N-S trends. Their axial directions and the arcuate nature of the folds we interpret to be related to the clockwise rotational system active during late Cenozoic to Recent time. These basins are not in contact with major faults, so they do not provide any limits on the ages of the young faults.

## TECTONIC INTERPRETATION OF THE CHANGNING-MENGLIAN AND LIN CANG UNITS

Our favored interpretation for the relations of the Lin Cang and Changning-Menglian Units is only preliminary because of the complexity, poor outcrops, and lack of recent detailed studies, except for the unpublished work from the Chengdu Institute of Geology and Mineral Resources (2002) on which some of our interpretation is based.

The Changning-Menglian Unit we regard as the site of, first, an opening of an early late(?) Paleozoic oceanic basin (Paleotethys), followed by its closing by eastward subduction during Permian–Triassic time. The oldest rocks within the Changning Menglian Unit are Devonian deep-water strata, suggesting that the opening of an oceanic realm would be at least that old. Unfortunately even though there are exposures of mafic and ultramafic rocks associated with Devonian-Carboniferous rocks within the unit, the only age dates so far reported are Permian (267.1 ± 3 Ma). Closure of the oceanic realm is recorded by the eastward progressive increase of high-pressure, low-temperature (HP-LT) metamorphic rocks within the eastern part of the Changning Menglian Unit adjacent to the Lin Cang magmatic rocks. Ages for these rocks are given by older Rb-Sr data of 260–240 Ma (Zhou and Lin, 1982). The HP-LT rocks are interleaved with rock metamorphosed to greenschist and amphibolite facies, suggesting later or postsubduction deformation within these metamorphic units. The Lin Cang magmatic belt lies to the east and has yielded ages of plutons of Permian and Triassic age (254–212 Ma; Liu et al., 1989) consistent with a subduction-related scenario during Permian-Triassic time. However, Heppe (2004) showed that some of the plutons are S-type intrusives formed as a result of melting of older crustal rocks, although the plutons he describes may be only the shallow-level parts of a magmatic arc, as they intruded the Middle and Upper Triassic volcanic rocks still preserved as pendants and as overlying some plutons. Subduction may have been protracted, beginning in Permian time and ending by Late Triassic time, but the spatial and temporal evolution of the subduction process remains largely unclear. Unconformably overlapping Triassic strata on subduction-related rocks consist of rare Lower Triassic rocks with some volcanics, Middle Triassic strata also with volcanics, and more widespread Upper Triassic rocks. These Triassic units, however, do not occur always in the same area and do not cover all parts of the Changning Menglian Unit, so it remains possible that closure and final deformation within the unit was diachronous. The Lin Cang Unit represents a probable Andean magmatic arc, the late magmatism of which was post-tectonic and related to melting within a hot, thickened crust. Extension within the arc may have begun by latest Triassic or Early Jurassic time, but extension was more regionally expressed by Middle Jurassic time, following cessation of subduction activity by the development of the numerous basins of thick non-marine and locally marine strata. Middle Jurassic extension spread widely across the area between the Lanping-Simao Unit westward into the eastern Baoshan Unit. This extension was not accompanied by major magmatic activity, as is common in regional extensional tectonism, a feature discussed in more detail below.

In our interpretation the geology of the Changning-Menglian and Lin Cang Units can best be interpreted as parts of a paired subduction-magmatic complex (Zhang et al., 1993), an interpretation not accepted by Heppe et al. (2007), who regarded the Baoshan, Changning-Menglian, and Lin Cang Units all to be part of a single Gondwana continental terrane. However, the presence of fossils of both late Paleozoic Gondwana and Cathaysian realms, blueschists that become progressively higher grade eastward, and rare mafic and ultramafic rocks indicate to us that the Changning-Menglian Unit is the site of an oceanic realm (Paleotethys) that closed by latest Triassic time. In this tentative interpretation the Lin Cang Unit was a magmatic complex related to subduction and was intrusive into a continental fragment that belongs to the pre-Mesozoic collage underlying the Lanping-Simao Unit (see above).

Regionally the belt of Permo-Triassic plutonism, the Lin Cang Unit, can be continued into Thailand through the Doi Inthanon metamorphic belt (MacDonald et al., 1993), with some zones of structural offset, making it a >1000-km-long metamorphic magmatic belt. However, the biggest problem is that if the subduction is on the west side of the magmatic belt in China, interpretations of the rocks farther south place the magmatic belt on its east side in Thailand. How this difference is resolved remains unknown.

There is some evidence for a short-lived westward subduction that formed plutons and volcanic rocks in the southern and eastern Baoshan Unit to the west (Plate 1), but these rocks are not as voluminous as similar rocks in the Lin Cang batholith and probably of shorter duration.

We argue below that rocks of the Lin Cang Unit are present farther north, where they may be represented in the large lenses that separate the Qiangtang I, II, and III units from the Lhasa Unit, an interpretation that has great implications for the regional tectonics (see below). The evidence that rocks in the lenses to the north may be disrupted parts of the Lin Cang Unit is suggested by the presence of Triassic plutons intruding metamorphic rocks, and unconformably overlain by Upper Triassic volcanic rocks along the western boundaries of the Qiangtang I and II units and the Lanping-Simao Unit. These lenses may be the only remnants of the Lin Cang Unit after the main body of the unit was extruded to the southeast in early Cenozoic time (see below). If this is correct, the bending to the west, shown when the Changning-Menglian Unit and units both to the east and west are traced northward, must be a Cenozoic deformation. In fact, the end of the Nan Tinghe fault at a large, vertical axis fold in the Wuliang Shan (WS in Fig. 2-10) we interpret to the be result of late Cenozoic, post-extrusion deformation. Heppe et al. (2007) reported several cooling ages from the Wuliang Shan metamorphic rocks, the youngest of which is 17 Ma, which they regard to be related to the exhumation in the Wuliang Shan, which supports a late Cenozoic deformation.

**BOUNDARIES OF THE LIN CANG UNIT**

The entire eastern boundary of the Lin Cang Unit is marked by the Lancang fault zone, described above, with the Lanping-Simao rocks to the east. The plutonic belt is ~20–50 km wide, and its contact is intrusive into its wall rocks, which lie to the west. However, along the Lancang fault zone the contact with the plutonic rocks is sharp, and there are no rocks that belong to the Lin Cang batholith east of the fault, although its probable Mesozoic volcanic cover may extend to the east. The Lancang fault, or fault zone, changes character along strike from a mylonite zone to the north, where it forms the eastern boundary of the Chong Shan shear zone (see below), to a brittle-ductile to brittle fault zone of probable complex origin to the south. In its northernmost part the contact is a sharply defined fault with metamorphic rocks of the northernmost Lin Cang belt to the south, and deformed and variably metamorphosed Mesozoic and older rocks of the Lanping-Simao Unit to the north. In this segment it makes a convex south trace, and then a convex north trace to the east, where volcanic and sedimentary rocks of Triassic age belonging to the northern end of the Lin Cang Unit lie to the south. The boundary farther south is more complex and is hard to interpret. Where the fault zone turns south, covered unconformably by a small area of Paleogene strata, Triassic volcanic rocks are present to the east and west, divided by a narrow, N-S–trending belt of sheared Permian rocks with a few small bodies of serpentine and intruded by small stocks of probable early Mesozoic age. The Permian rocks form a discontinuous narrow belt of faulted rocks that end at the Heihe fault. Farther south the fault separates Mesozoic sedimentary and volcanic rocks from plutonic rocks, and near the Laotian border a large body of poorly dated upper Paleozoic rocks lies to the east of the fault zone. In one place the Permian rocks are intruded by a small gabbroic pluton with small bodies of serpentinite interpreted to be an ophiolite by Liu et al. (1993), and a zoned mafic-ultramafic complex by Jian et al. (2009a, 2009b). This is one of three similar bodies that occur within the Lanping-Simao Unit along the Lancang shear zone; the other two are the Banpo and Gicha mafic-ultramafic bodies (see above) of late Paleozoic age. The eastern boundary of the Lin Cang Unit is reasonably well marked by the Lancang fault zone, whose significance will be discussed in detail in a later section.

The eastern boundary fault zone is marked by both a narrow brittle fault zone, or in its N-S stretch, a 3–5-km-wide belt of strongly sheared rocks of Permian age and granitic rocks of the Lin Cang batholith, which are locally mylonitic. We interpret the Lancang fault to be a regional fault zone that played a major role in the early Cenozoic extrusion tectonics of SE China. However, it may not be the only fault zone to have accommodated large displacements during extrusion. In some places within the batholith, displacement appears to have been accommodated by several parallel and closely spaced faults; some of these faults may have absorbed some of the displacement of the Chong Shan shear zone. Other, similar faults may lie within the Changning-Menglian Unit or even in the eastern Baoshan Unit to the west, but few of these faults have been studied. During early Cenozoic time the Lancang and other parallel faults were probably straighter or gently arcuate, convex to the east, and have been folded by younger deformation as recorded in the Lin Cang Unit and the Wuliang Shan in the Lanping Simao Unit by movement related to the left-lateral Nan Tinghe fault (see above), and displaced by other late Cenozoic left-lateral, strike-slip faults to the south (Plate 1 and Fig. 2-10). These left-lateral faults are the youngest deformational structures to have affected the Lin Cang Unit; they were initiated ca. 8–10 Ma and are active today (Fig. 2-10). The faults are the same NW- and NE-trending faults that pass through the Changning-Menglian Unit into the Lin Cang Unit. The relations of these brittle faults to the more ductile shear zones within the Lin Cang Unit are of regional importance for both the early and late Cenozoic history of the region, and require considerably greater future study.

**EXPOSURE OF THE LIN CANG BATHOLITH**

It is evident that the Lin Cang batholith was exposed before the deposition of Middle Jurassic coarse-clastic rocks. The batholitic rocks also intruded its volcanic cover of Triassic age in the north. Pendants of the host rocks for the batholith are rare, present only in the north where the Paleozoic(?) metamorphic rocks (Pzm) are exposed within the batholith. This indicates that the batholith was at shallow depth during the Triassic and was exposed sometime between Late Triassic and Middle Jurassic time. This was the time in which thick sections of Mesozoic sedimentary rocks were deposited to the east in the Lanping-Simao Unit above their Paleozoic basement. Intrusive rocks are rare within the Lanping-Simao Unit. This indicates that the eastern boundary fault of the batholith, the Lancang fault zone, probably had important relative displacement, lowering the Lanping-Simao in pre–Late Triassic time and forming the sharp eastern boundary to the batholith at that time. Later displacement by faults of the Lancang fault zone cut the Triassic, Jurassic (and Cretaceous?), and Paleogene rocks in the south, indicating that the fault zone was reactivated by one or more periods of younger movement. These relations are unlike those along the western boundary of the Lin Cang batholith, where its western contacts are intrusive into its high-grade wall rocks and the contact is unconformably covered by Middle Jurassic clastic strata. However, the western boundary is similar to the eastern boundary in that very few small intrusive bodies lie to the west of the boundary.

# CHAPTER 15

# *Ailao Shan Unit*

The Ailao Shan Unit is a northwest-trending, 350–km-long narrow belt of metamorphic rocks (m) that lies between the Lanping-Simao and Yangtze Units in the southern part of the area of Plate 1 (available on DVD accompanying this volume and in the GSA Data Repository as Item 2012297). It is fault bounded on both sides. In the south is an area underlain by an unmetamorphosed section of Paleozoic and Triassic rocks that we include in the Ailao Shan Unit. This unit has figured prominently in the interpretations of early Cenozoic extrusion of crustal fragments from central Tibet eastward into southeastern Asia (e.g., see Tapponnier et al., 1982; Leloup et al., 1995), an interpretation that will be discussed in detail in the synthesis chapter below.

## ROCKS AND STRUCTURES OF THE AILAO SHAN AND RELATED UNITS

The Ailao Shan Unit consists of greenschist- to amphibolite-grade metamorphic rocks (m, Plate 1) that end in the northwest between the two bounding fault zones, gradually widening southward to >20 km before continuing into Vietnam. The rocks are strongly foliated and variably mylonitized in the northern and central parts with foliation dipping from moderately to steeply northeast and locally overturned to dip southwest along the northeastern side. To the southeast the width of foliated and mylonitized rocks becomes narrower and mainly along the northeastern side of the unit, and near Vietnam the mylonitic rocks are only 1–2 km wide, and to the west the metamorphic rocks are intruded by abundant plutons (Plate 1 and Fig. 15-1). The metamorphic and mylonitic rocks are bounded on the west side by a fault that generally dips moderately to steeply to the northeast, and in most places the metamorphic rocks lie above the belt of mélange at the northeastern side of the Lanping-Simao Unit shown as Paleozoic (Pzu with vertical pattern) in the north and by mélange shown as Mesozoic (Mz1 with vertical pattern) in the south. The mélange in the footwall to the west is mostly metamorphosed to low grade and commonly with weak to strong foliation (see the serial cross sections in J. Zhang et al., 2006b). In the south there is a thick section of Paleozoic strata capped by Triassic rocks that we include in the Ailao Shan (Plate 1), because they are intruded by the plutonic complex that also intrudes the metamorphic rocks (Plate 1 and Fig. 15-1).

The age of the metamorphic and mylonitic event is usually reported to be between 27 and 34 Ma, and ending at ca. 17 Ma (Gilley et al., 2003) (Fig. 11-2); however, older ages are reported from the metamorphic rocks where matrix monazite grains from south Mosha have yielded ages of 74.1 and 52.9, but these are

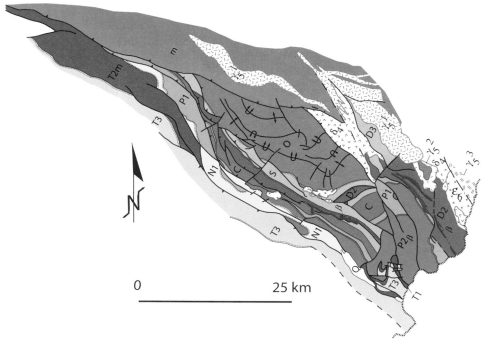

Figure 15-1. Geological map of the Paleozoic and Mesozoic rocks at the southeastern end of the Ailao Shan (see Plate 1 for location). These rocks are intruded by poorly dated plutons that also intrude the metamorphic rocks of the Ailao Shan (m, Plate 1), indicating that all these rocks were combined with the belt of metamorphic rocks that made up the bulk of the Ailao Shan prior to its Cenozoic deformation and metamorphism. Unit designations: β—basalt; m—metamorphic rocks; O—Ordovician; S—Silurian; D2—Devonian; D3—Upper Devonian; C—Carboniferous; N1—Miocene; P1—Lower Permian; P2$_β$—Upper Permian Emei Shan basalt; Q—Quaternary; T1—Lower Triassic; T2m—Triassic mélange; T3—Upper Triassic.

discounted as incompletely reset detrital grains (Gilley et al., 2003). Recently, J. Zhang et al. (2006b) suggested that shearing was partitioned across the Ailao Shan such that it may have occurred at different times at different places, and they reported ages that range from 58 to 56 Ma from LS-tectonites, 27–22 Ma from mylonites that formed near the brittle-ductile transition, and ages of 13–12 Ma on low-grade phyllites that lie in the footwall of the western boundary fault (which they call the *Ailao Shan fault*). In a recent paper on the timing of left-lateral shear along the Ailao Shan, Searle et al. (2010) concluded that the shearing occurred between 31.9 and 24.2 Ma, and at 21.7 Ma, with exhumation having continued until ca. 17 Ma (however, they call the deformed belt the *Red River shear zone*, not the Ailao Shan). Because of the variable nature of the foliation and mylonitization across the Ailao Shan the possibility that the deformation is strongly diachronous, and possibly of more than one age, still needs to be investigated more thoroughly.

In the southeastern part of the Ailao Shan, poorly dated plutons become increasingly abundant and make up nearly half the width of the belt (Plate 1 and Fig. 15-1). These plutons range from diabase to alkalic granite and are largely undeformed and not mylonitized. These plutons are shown to be from Triassic to Cenozoic in age, but many of them are undated, and for those that are dated the techniques used are old and not reliable. The one exception is an alkali granite within the plutonic complex near Jinping that has yielded U-Pb ages that range from 35 ± 3 Ma to 38.6 Ma (Zhang and Schärer, 1999: $\varepsilon_6$ in Fig. 15-1). It is a single, mainly undeformed pluton within the complex. Importantly, plutons that are interpreted to be late Paleozoic ($\delta_4$) to Jurassic ($\gamma_5^2$) in age intrude Devonian rocks that are part of a Paleozoic sequence that lies along the southwest side of the metamorphic rocks and will figure prominently in the interpretation presented below.

Along the entire length of the Ailao Shan metamorphic belt are smaller bodies of amphibole-bearing intrusions of alkaline trend and less abundant leucogranitic layers. These rocks were studied by Zhang and Schärer (1999), who showed that they range in age from 22 to 33 Ma, with the larger body at Jinping being slightly older (see above). These authors concluded that the alkaline intrusives were generated by lithospheric melting that required elevated temperatures at shallow depth, and these rocks intruded not only the Ailao Shan metamorphic rocks but also unmetamorphosed rocks outside of the Ailao Shan belt. The leucogranites, on the other hand, show abundant evidence for crustal contamination and anatectic melting and are confined to the Ailao Shan belt. Zhang and Schärer (1999) concluded that during the shearing the temperature within the lithosphere was significantly elevated not only in the area of the Ailao Shan shear zone but also across a broad area that extended beyond the Ailao Shan. They rejected both subduction and lithospheric thickening as models for the melting and suggested that a broad, significant thermal anomaly below this region was required to explain the melting. We agree and will develop another hypothesis in the synthesis chapter below.

The Ailao Shan rocks are separated from the northwest continuation of the Day Nui Con Voi mylonitic rocks (m, Plate 1) that lie a few kilometers to their east in the southeastern part of the area of Plate 1 by narrow belts of Triassic strata that are both metamorphosed and unmetamorphosed. The two groups of mylonitic rocks cannot be traced directly into one another (Plate 1), as shown on some maps. The metamorphism and mylonitization for the Day Nui Con Voi rocks in Vietnam, like the Ailao Shan, have dominantly left-lateral shear fabrics and have yielded a range of metamorphic ages from 220 to 44 Ma. However, younger ages have been reported for rim inclusions that range from 33 to 21 Ma, but older ages (107 Ma) also were reported (Gilley et al., 2003). It is generally accepted that the younger ages that are similar to those from the Ailao Shan date the main deformation, but for the same reasons given above for the Ailao Shan, such an interpretation needs greater investigation. Older ages of ca. 250 Ma to as young as 220 Ma are reported within northern Vietnam, not only for rocks of the Day Nui Con Voi but also for rocks over a large area on both sides of the mylonitic rocks of the Day Nui Con Voi, and are considered to date a major early Mesozoic regional deformational event, the Indosinian event (Carter et al., 2001; Lepvrier et al., 2004). Although these ages have not turned up across the border within China to the northwest, they do occur farther northwest in the Dian Cang Shan west of Erhai Lake (Searle et al., 2010; Plate 1) and will figure importantly in a reconstruction below. The metamorphic rocks in the Dian Cang Shan have been considered a continuation of the major Ailao Shan left-lateral shear zone (e.g., Tapponnier et al., 1986; Leloup et al., 1995; Searle et al., 2010).

**Day Nui Con Voi Structure in China**

The mylonitic rocks of the Day Nui Con Voi in China form a long, narrow, antiformal structure that plunges to the northwest (m, Plate 1) east of the Ailao Shan (see the southeastern part of Plate 1). Where the Day Nui Con Voi plunges, metamorphic Triassic rocks wrap around the plunging nose and become less metamorphosed away from the contact, and these rocks contain fossils (see Burchfiel et al., 2008a). Everywhere else in the region near to the Day Nui Con Voi structure there is a section of Yangtze-type upper Paleozoic rocks, beginning with the Emei Shan basalt and overlying Triassic strata (Plate 1). North of the Triassic rocks the Neoproterozoic and Paleozoic rocks that underlie them contain unconformities (Plate 3) but are typical for Yangtze Paleozoic strata, but these rocks are missing at the contact with the Day Nui Con Voi metamorphic and mylonitic rocks. The NW-striking Nanxihe fault locally bounds the northeastern side of the Day Nui Con Voi, and east of this fault is a mostly complete section of very thick Neoproterozoic phyllitic strata and Cambrian through Ordovician strata that are not typical for the Yangtze section; the unconformably overlying Devonian and younger rocks, however, are typical Yangtze strata (see above). The Nanxihe fault thus appears to have a large pre-Devonian displacement, and possibly also a pre-Triassic displacement, because it does not displace the Triassic rocks along its northern trace by the same magnitude required to displace the pre-Devonian rocks. Cenozoic displacement on the Nanxihe fault is shown by the fault slivers of Paleogene and Neogene rocks along it, and along its northern trace it bounds the Mengzi basin, where it appears to be active. All the rocks east of the Nanxihe fault below the Emei Shan unit are not present at the contact of the Day Nui Con Voi metamorphic and mylonitic rocks and the overlying Triassic rocks. These relations have been interpreted by Burchfiel et al. (2008a) to indicate that the contact between the mylonitic and Triassic rocks is a low-angle normal fault that cut out the Neoproterozoic and Paleozoic sections and was later folded into the Day Nui Con Voi antiform. These relationships place the high-grade rocks of the Day Nui Con Voi beneath rocks of the Yangtze Unit. The core rocks of both the Day Nui Con Voi and the Ailao Shan Unit consist of rocks that cannot logically have Neoproterozoic and Paleozoic rocks of the Yangtze Unit as their protoliths, as they lack the abundant carbonate rocks characteristic of these units. We suggest that most of the gneiss and schist is probably Precambrian, but plutons that intrude these rocks are probably Mesozoic and Cenozoic in age, but remain undated. Our preliminary study of zircons from mylonitic granite within the Day Nui Con Voi shows that nearly all the zircons yield U-Pb ages of 830–850 Ma. Such ages are common from the basement of the Yangtze Unit, supporting the interpretation that these granitic rocks were derived from Yangtze basement rocks that underlie the Day Nui Con Voi, and that may also be the protolith for the Ailao Shan metamorphic rocks.

**Displacement along the Ailao Shan**

The data presented above are used to interpret that the Day Nui Con Voi and the Ailao Shan to have been cored by basement rocks of the Yangtze Unit that had been heated, partially melted, and strongly deformed during early Cenozoic extrusion (Burchfiel et al., 2008a), an interpretation that will be discussed in the regional analysis presented below. If these rocks are the basement for the Yangtze Unit, it indicates that early Cenozoic shearing was at least partly or wholly within the Yangtze craton, although some shearing may also have occurred in the mélange belt to the west. This raises the question of where did these units come from, and how were they emplaced?

At the southern end of the Ailao Shan along its southwest side in China is a nearly complete section of rocks that range from Ordovician to Triassic (Plate 1 and Fig. 15-1). Nowhere are these strata in depositional contact with the Ailao Shan rocks. In most places they lie in the footwall of a thrust that dips north and carries the Ailao Shan metamorphic rocks above them. They were, however, intruded by plutons that tie these rocks to the Ailao Shan metamorphic rocks at the time of intrusion. The ages of the plutons are not well constrained, with the exception of the 35 Ma alkali granite near Jinping, mentioned above, and dated by Zhang and Schärer (1999). Chinese maps assign ages of late Paleozoic to Jurassic to these plutons, but they need new radiometric work, as they hold one possible key to the lateral displacement of the Ailao Shan. If these plutons are that old, they tie together the Paleozoic rocks to the metamorphic rocks of the Ailao Shan long before Cenozoic extrusion and deformation. Thus, the paleogeography of the Paleozoic section provides important information on their possible source before displacement. Most of the strata are similar to the Yangtze Unit, but two of the important parts of this section, a thick Ordovician section of fine-grained clastic strata, and a very thick, >4000 m, Emei Shan basalt, are unlike the strata of the Yangtze Unit to their northeast. These two rock units are similar to the rocks in the Transitional Unit near Erhai Lake ~300 km to the northwest, where there is also a complex of plutonic rocks assigned ages from Paleozoic to Cenozoic (see above), similar to the complex in the southern Ailao Shan. These relations were pointed out by others (e.g., Wu, 1993; Chung et al., 1997). Recently, Searle et al. (2010) reported Triassic ages (ca. 239–243 Ma) from zircons from schists and gneisses in the Dian Cang Shan west of Erhai Lake. These are similar to ages from rocks in northern Vietnam and also support the interpretation of a major early Mesozoic deformation in the rocks surrounding Erhai Lake. These schists and gneisses lie within the structurally lowest part of the major east-vergent antiformal structure we proposed above. Our interpretation of the Ailao Shan and Day Nui Con Voi rocks is that they represent a mobile lower crust that developed during transpressional extrusion where left-shear lateral motion was dominant, but which was associated with a vertical component that extruded these rocks obliquely upward along a major ductile shear zone. This would explain the thrust fault at the base and the normal fault at the top, both with dominant horizontal motion, an interpretation we present for other high-grade shear zones in the region (see below). This could account for some of the displacement during the early Cenozoic: the southeastward extrusion of these rocks, accompanied by a component of southwestward thrusting relative to the Yangtze Unit, but probably not all of it. A discussion of this interpretation will be detailed below.

Along the eastern and western sides of the Ailao Shan belt, younger deformation has modified the metamorphic and structural features of the belt. Along the eastern side the late Cenozoic brittle Red River fault zone locally forms the boundary between the metamorphic and mylonitic rocks of the Ailao Shan and the unmetamorphosed rocks of the Yangtze Unit, but in many places the boundary lies a few kilometers east of the metamorphic rocks. The Red River fault in China is an active right-lateral fault in contrast to the left-lateral shear zone of early Cenozoic age in the Ailao Shan mylonitic rocks, a relation pointed out by Tapponnier et al. (1990). It is important that the rocks directly east of the Ailao Shan metamorphic rocks are unmetamorphosed all along the Ailao Shan except at its southernmost end. A narrow belt of Paleogene and Neogene strata lie along the east side of the Ailao Shan in its middle part (Plate 1), and these strata are folded along axes generally parallel to the Ailao Shan and are cut by active strands of parallel faults of the Red River fault zone (Schoenbohm et al., 2005). There is abundant evidence for active right-lateral displacement along the strands of the fault zone, as shown by right-lateral displacement of streams and stream valleys where they cross the faults, but surprisingly this fault zone has been seismically quiet. Near the southern end of the Ailao Shan these N-W–striking faults bound slivers of Triassic strata that separate the Ailao Shan from the Day Nui Con Voi rocks, and these faults also bound narrow slivers of metamorphic and mylonitic gneiss that probably were slivered from a broader metamorphic area that originally lay between the Ailao Shan and Day Nui Con Voi. These slivers are most abundant where the Red River faults pass into Vietnam (Plate 1).

At the north end of the Ailao Shan the metamorphic rocks are truncated at a low angle by the active Red River fault zone where it offsets the Red River by ~6 km (Allen et al., 1984; Wang et al., 1998; Schoenbohm, 2006b; Plate 1). At the north end of the Ailao Shan are marbles mixed with the mylonitic rocks that might be remnants of the Neoproterozoic rocks of the Yangtze Unit, unconformably overlain by Pliocene strata that are cut by the Red River fault zone. Along the east side of the northwesternmost part of the Ailao Shan are slivers of mélange with Permian clasts and some granite slivers that are interpreted to have been displaced right-laterally from rocks normally along the west side of the Ailao Shan (Pzm). These offset pieces indicate a magnitude of 27–54 km of late Cenozoic right slip on the Red River fault zone (Wang et al., 1998; Schoenbohm et al., 2006b). Northwest from the end of the Ailao Shan metamorphic rocks the Red River faults form the boundary between the Yangtze and Lanping-Simao Units. These rocks were described recently by Schoenbohm et al. (2006b), who showed that the active trace of the Red River fault zone can be followed to the north through the Midu basin to the south end of Erhai Lake. This part of the fault zone is active, and its relation to the process of early Cenozoic extrusion will be discussed below.

On the southwestern side the Ailao Shan metamorphic rocks are thrust on moderate to steeply NE-dipping faults above rocks of the mélange belt at the eastern side of the Lanping-Simao Unit (Pzu with vertical pattern). The mélange belt is largely a latest Paleozoic and early Mesozoic structural belt that is unconformably overlapped by Upper Triassic strata, but in places the overlap strata are also cut by thrust faults that place the mélange above the Upper Triassic strata, indicating that it has been involved in post-Triassic deformation. We interpret the younger activity along this thrust to be related to the folding during Paleogene time of the Lanping-Simao thrust belt to the southwest. Here, much of the shortening in the thrust belt was taken up below the Ailao Shan metamorphic and mylonitic belt in a transpressional environment during extrusion. The latest movement on this oblique thrust outlasted the shortening, as it obliquely truncates the more northerly trending folds in the Lanping-Simao Unit (Plate 1). This relation suggests that there may have been considerable Cenozoic strike-slip on these faults and within the mélange belt along the western side of the Ailao Shan.

Thus the geology in the Ailao Shan and the Day Nui Con Voi is hypothesized to include a highly mobile, partially melted, reactivated basement of the Yangtze Unit that was overlain by a Neoproterozoic to Mesozoic cover. During early Cenozoic extrusion in a transpressive environment it was sheared left-laterally, and mobile crustal rocks were extruded upward from the lower crust, which was bounded by a thrust zone at the base and a normal fault at the top (Burchfiel et al., 2008a). The Paleozoic rocks and plutons at the southern end of the Ailao Shan appear to be best correlated with the rocks near Erhai Lake, supporting a left-lateral displacement of ~300–350 km. But we suggest this to have been a minimum displacement, because there must have been additional displacement along the southwestern side of the Ailao Shan belt as well (see below).

But there are problems with this hypothesis. If the Paleozoic rocks at the south end of the Ailao Shan are repositioned by reversing the left-lateral shear to Erhai Lake, it would imply that the long belt of metamorphic rocks to their north within the Ailao Shan would be positioned farther north and contain the shear related to their southward movement. The metamorphic and mylonitic rocks within the shear zone thus would have distributed deformation accommodating a ~300 km of movement along both the fault zones at the shear-zone margins as well as within the sheared mylonitic rocks. The rocks in the displaced section of the Ailao Shan continue into the Upper Triassic; thus their displacement should be no older than Late Triassic, and whereas some of the displacement could have been early Mesozoic, most of it would appear to have been early Cenozoic. The early Mesozoic shear may lie within the mélange belt to the west of the Ailao Shan and could have been continued northward into the boundary between the Jinsha accretionary complex and the Yidun Unit, but this does not resolve the problems of how much younger modification of this boundary has occurred (see above). A common interpretation places the large, early Cenozoic displacement along the faults that bound the narrow sliver of mylonitic rocks of the Yuelong Shan (Leloup et al., 1995, 2001), which we interpret to be the main fault that separates the Lanping-Simao Unit and the Qiangtang I unit. These units have no exposed basement rocks, with the exception of the metamorphic rocks in the Yuelong Shan, and the associated Paleozoic rocks are unlike any rocks in the section at the southwestern side of the Ailao Shan, particularly the thick Emei Shan basalt, thus suggesting that there is additional shear and displacement within the mélange belt west of the Ailao Shan.

All the displacements and shear zones within this belt may have been responsible for removing the southwestern margin of the Yangtze Unit. If so, the displaced rocks should be found somewhere to the southeast, a problem which is discussed in the synthesis chapter (Chapter 16) below.

## CHAPTER 16

# *Synthesis*

## INTRODUCTION

The region of Southeastern Tibet consists of fragments of mostly continental rocks from both the Gondwanan and Cathaysian paleogeographic realms, separated by oceanic regions of variable but unknown widths that were accreted in Mesozoic and Cenozoic time to form the present-day underpinnings of the Tibetan Plateau and adjacent areas. During and following accretion the fragments were deformed and modified by intracontinental processes that continue until today. Although the discussions in the preceding chapters have included the geology from Precambrian to Recent, the focus here is to synthesize the evolution of this complex region from late Paleozoic and mainly Mesozoic and Cenozoic time to the present. The older history will be considered where rocks and structures formed in pre–late Paleozoic time have influenced younger geologic development.

As pointed out in preceding chapters, there are a great many uncertainties in the geology of this complex region, thus uncertainties in developing a complete and coherent synthesis remain. Such uncertainties will be highlighted below. A set of figures attempting to show a time-space sequence of events and what is known and what is interpreted is presented (Figs. 16-4 through 16-7 and Fig. 16-10). These figures have been difficult to construct, as we attempt to show the regional distribution of structural and magmatic events, but each figure requires considerable extrapolation of events beyond the areas where they are constrained by data. For example, the age of folds and faults within a region can be determined in only limited areas where unconformable relations date them, and their distribution can be extrapolated by assuming that parallel structures are of the same age. However, in making such extrapolations, contradictions arise as pointed out in the discussions above. Also, in a region where the structures tend to have a general parallelism regardless of age, this implies that parallel structures may have more than one age, older structures having been reactivated during younger events. All these considerations make the following discussion only a tentative blueprint for research to bring more reality into future syntheses.

## SUTURES AND CONTINENTAL FRAGMENTS

The major continental fragments that make up the region shown on Plates 1 and 2 (available on DVD accompanying this volume and in the GSA Data Repository[1]) can be identified by remnants of mafic and ultramafic rocks that separate them and were formed in different oceanic and tectonic settings, and by cryptic sutures where these rocks are missing by the processes of accretion and later tectonism (Fig. 16-1). In almost all cases the mafic and ultramafic rocks occur as discontinuous pieces or remnants within mélange or along shear zones. Such remnants are rarely large enough to contain a full ophiolite sequence, and their tectonic setting has been interpreted from geochemical analyses. Geochronological dating of these remnants within a single zone often yields a range of ages. All these relations indicate that these remnants are only highly limited samples of what were oceanic regions, thus giving only a general and incomplete range for the ages of the oceanic realms. Most of these belts of oceanic fragments are commonly reconstructed in what might be considered an accordion model in which the zones are simply opened and closed normal to the trend of the zone, but many, if not all, of these zones have been subject to considerable strike-slip displacement that may be of large magnitude, and the fragments may be from disparate parts of the oceanic realms.

Within the area of Plates 1 and 2 are seven and possibly eight zones where fragments of oceanic rocks are present within both narrow and broad zones that separate tectonic elements that are both continental fragments and collage units (Fig. 16-1). These oceanic realms include the Jinsha accretionary complex, the mélange belt along the eastern side of the Lanping-Simao Unit, the Changning-Menglian Unit, the Bangong-Nujiang zone, the Yarlung-Zangbo shear zone in the Syntaxial Unit, the Longmen Shan belt, the subduction zone at the eastern margin of the Yidun Unit, and possibly the Lancang belt (Fig. 16-1). There are segments along these zones where no fragments of oceanic rocks are present, and these are interpreted to be cryptic sutures and include segments of the Longmen Shan belt, the Changning-Menglian Unit, the Bangong-Nujiang (Gaoligong) zone, and the Chong Shan shear zone. Other zones where remnants of oceanic rocks are either rare or not present, interpreted but not proven to have been the sites of oceanic regions that include the Parlung-Jiali fault zone, the Lancang fault zone, shear zones within the Lhasa Unit, and shear zones within and along the east side of the Yidun Unit.

In addition, in what tectonic setting these oceanic fragments formed has been determined from geochemical studies that also have yielded a range of tectonic settings. At present these data are too few within the region of Plate 1 to say more than that there was oceanic crust that separated large first-order continental crustal fragments. Also, some of the zones with mafic and ultramafic rocks are narrow, discontinuous, and separate rocks

---

[1]GSA Data Repository Item 2012297, Plates 1–4, is available at www.geosociety.org/pubs/ft2012.htm, or on request from editing@geosociety.org or Documents Secretary, GSA, P.O. Box 9140, Boulder, CO 80301-9140, USA.

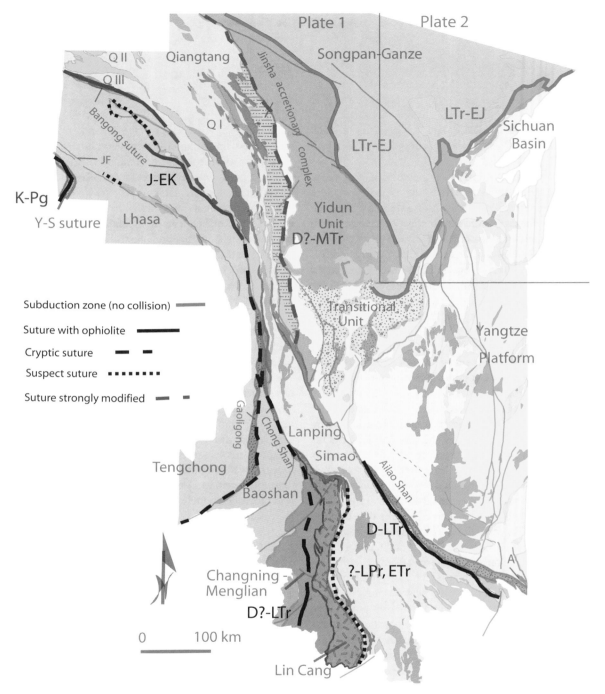

Figure 16-1. Locations of the major sutures and the age ranges for the oceanic terranes they represent. Solid lines indicate sutures marked by mafic and ultramafic rocks, and dashed lines indicate where the sutures are cryptic. The dotted line marks the Lancang shear zone, which remains a suspect suture. The faded background represents the tectonic units discussed in the text and also shown in Figure 1-1B. JF—Jiali fault; Y-S—Yarlung-Zangbo suture.

of similar character. Such zones may be repeated by strike-slip faults or be the result of fragmentation of crust by local extension in pull-apart basins or within intra-arc rifts. In all cases there is no zone that has not been disrupted by younger tectonic events.

Taking these zones, which might also be called sutures, with all their caveats, seven(?) continental fragments and the zones that separate them can be identified generally from west to east, although some fragments may not extend across the

## Lancang (Lancangjiang) Shear Zone

The Lancang shear zone is one of the most difficult structural belts to understand and has been interpreted by some workers as a suture zone (e.g., most recently by Hennig et al., 2009), who studied a volcanic-plutonic complex at Nanlianshan near the southern end of the Lancang shear zone and the nearby Jinghong granodiorite (Fig. 16-3). A gabbro associated with basaltic andesites at Nanlianshan yielded a U-Pb zircon age of 292 ± 1 Ma, and its geochemistry is indicative of formation at a spreading ridge. Small granodiorite intrusions nearby at Jinghong yielded U-Pb zircon ages of 284 ± 1 Ma and 282 ± 1 Ma; geochemically they have an arc-type signature. These Early Permian ages are older than those of the more prominent Middle and Upper Triassic belt of volcanic rocks that occur almost continuously along the western side of the Lanping-Simao Unit in the south and across the boundary between the Lanping-Simao and Lin Cang Units in the north (Plate 1; see also Fig. 7-5). These rocks lie along a complex structural belt of late Paleozoic rocks that are commonly sheared with blocks in a mélange-like matrix that marks the general boundary between the Lanping-Simao and Lin Cang Units. This same zone to the north contains the Banpo zoned ultramafic body, which has yielded ages of 285.6 ± 1.7 Ma and 285.8 ± 3 Ma, and with a geochemical signature of an arc intrusive body (Jian et al., 2009a; Fig. 7-2). Farther north near the north end of the Lin Cang plutonic complex, several sheared serpentinized blocks are present within a sheared zone of sedimentary rocks that have yielded Permian fossils from enclosed blocks (Plate 1). These rocks also lie along the prominent belt of upper Paleozoic rocks at the western border of the Lanping-Simao Unit.

There is only limited evidence for an oceanic area in the Lancang shear zone. One of the characteristic features of this zone is the sharp tectonic boundary it makes with the Lin Cang batholithic belt to the west. The narrow shear zone is marked by the upper Paleozoic rocks that are mapped as unconformably overlain by Upper Triassic volcanic and sedimentary rocks, and similar Middle Triassic rocks that are less common can be interpreted to lie above it regionally. The Triassic rocks are not sheared like the upper Paleozoic rocks, suggesting that most of the shearing is pre–Middle Triassic. Younger post-Triassic faulting has affected the belt of upper Paleozoic rocks, but the evidence suggests that most of the shearing is pre–Middle Triassic. Thus, if a late Paleozoic oceanic area was present, the closure is best interpreted as pre–Middle Triassic. In our interpretation the

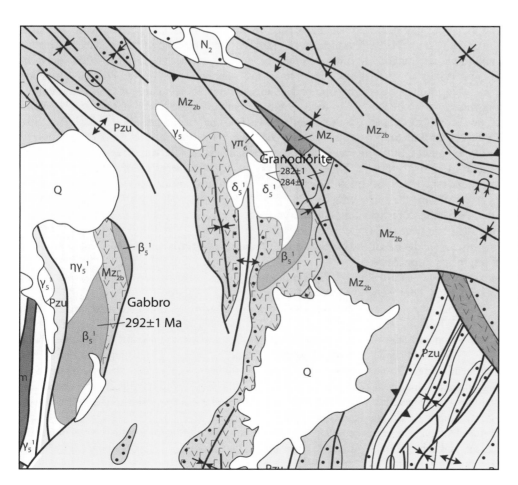

Figure 16-3. The dated gabbro and granodiorite at Nanlianshan along the Lancang fault zone in the southwestern Lanping-Simao (from Hennig et al., 2009). The gabbro in this area yielded a zircon age of 292 ± 1 Ma, and zircons from the granodiorite yielded ages of 284 ± 1 Ma and 282 ± 1 Ma (fig. is from Plate 1).

Paleozoic rocks that are exposed in the erosional windows below Mesozoic strata of the Lanping-Simao Unit are part of a collage of faulted fragments accreted together prior to the subsidence and deposition of Upper Triassic strata, and that Lower(?) and Middle Triassic strata may have been either deposited during earlier extensional faulting or are part of the accreted collage formation. This interpretation is supported more strongly by relations in the Qiangtang I region to the north, but it is also our interpretation for the pre-Mesozoic rocks that underlie the Lanping-Simao Unit as well. The collage contains upper Paleozoic rocks that in some sections contain volcanic rocks, and the collage may represent fragments of both volcanic and non-volcanic units disrupted by faults that created small areas of oceanic rocks separating them rather than larger oceanic regions. All of these different rock assemblages were assembled into a collage during late Paleozoic convergence.

The belt of sheared upper Paleozoic strata along the Lancang shear zone, which forms the boundary between the Lin Cang and Lanping-Simao Units, is unusually sharp for reasons that are not clear, and it is similar to the sharp truncation of the batholiths in the eastern Tengchong Unit, sharply truncated by the Gaoligong shear zone. The Lin Cang plutons are poorly dated, and new U-Pb zircon dates reported by Hennig et al. (2009) of $229 \pm 3$ Ma, $230 \pm 4$ Ma, and $231 \pm 6$ Ma are within the range of ages reported by Liu et al. (1989) of 254–212 Ma. The plutons are mostly Triassic, consistent with their intrusive relations into the Triassic volcanic cover at the north end of the Lin Cang Unit (Hennig et al., 2009) and are younger than the ages of mafic and ultramafic magmatism within the narrow belt of sheared upper Paleozoic rocks that mark the Lancang shear zone. Thus we interpret the volcanic and plutonic rocks within the Lancang shear zone to be the result of convergence or oblique convergence during formation of the pre-Mesozoic collage, and possibly related in time to the subduction mélange along the eastern side of the Lanping-Simao Unit and to the formation of the Lin Cang plutons by subduction within the Changning-Menglian suture to the west. Following this interpretation would suggest that the host rocks for the Lin Cang plutonic belt, which includes probable Precambrian rocks, is also a fragment within the pre-Mesozoic collage, an interpretation that needs to be tested. Thus, in our interpretation, the major oceanic realms that separated continental fragments were within the Changning-Menglian and Ailao Shan suture zones, and the other rocks suggestive of oceanic environments were more local and part of the collage development.

The Lancang shear zone continues to the north from the north end of the Lin Cang Unit, where it forms the boundary between the Lanping-Siamo and other units to the west. For ~150 km north of the most northerly plutons of the Lin Cang Unit the fault zone marks the eastern boundary of the Chong Shan mylonitic belt, where mylonitic rocks are juxtaposed with lower grade metasedimentary rocks of the Lanping-Simao Unit, and where the Chong Shan shear zone marks the cryptic Changning-Menglian suture. Farther north a belt of Middle Triassic felsic and intermediate volcanic (Mz1a) and sedimentary (Mz1s) rocks crop out west of the Lanping-Simao red beds and continues as a belt of variable width northward with faults on both sides. We place these volcanic rocks within the Lanping-Simao Unit; however, they are unique to the unit, as no other similar rocks are present in the unit except within this belt. It is possible to consider the faults on both sides of this volcanic belt as belonging to the Lancang fault zone. West of the volcanic rocks the mylonite belt of the Chong Shan continues north with large lens-shaped bodies of plutonic rocks that are shown on Chinese maps as Jurassic and Cretaceous, but without dates. Akciz et al. (2008) suggested that these plutons could be Permo-Triassic and correlative with the Lin Cang plutons on the basis of their compositional similarities to slivers of plutonic rocks at the north end of the Lin Cang Unit, where they became sheared into the Chong Shan belt. Thus in this area the Lancang fault zone is juxtaposed with the Chong Shan belt that marks the cryptic Changning-Menglian suture. The faults on both sides of the volcanic belt continue north into an area where the fault zone we interpret to be the boundary between the Lanping-Simao Unit and the Qiangtang I unit also strikes into the Lancang fault zone. In this area, several major fault zones that mark cryptic sutures converge, but it is not clear how they interact. This is the area where the Bangong-Nujiang and Changning-Menglian cryptic sutures appear to meet, but how they continue through this area remains unknown. North of this area the large lens-shaped bodies of Paleozoic, metamorphic, and Triassic volcanic rocks begin that we interpret to mark a broad shear zone related to the early Cenozoic extrusion. Some workers have called this zone of anastomosing faults the *Lancang fault zone*. It is a critical area, but one that is poorly known. However, no rocks along the anastomosing faults are indicative of marking an oceanic suture.

To the south, within the northern part of the Lanping-Simao Unit, is the large Gicha zoned mafic-ultramafic body (Fig. 7-6), which is associated with numerous small bodies of serpentinized rocks and which contains rocks that yield a range of ages from $280 \pm 3$ Ma to $306.2 \pm 4.6$ Ma, interpreted by Jian et al. (2009b) as originating in a backarc rift. In our interpretation this body is entirely within the Lanping-Simao Unit and is one of the mafic-ultramafic complexes that are associated with collage formation.

## Jinsha Suture (Jinsha Accretionary Complex)

The Jinsha suture in our interpretation consists of some lower Paleozoic but mostly upper Paleozoic and Lower and Middle Triassic rocks that form an accretionary complex that is well exposed beneath the extensive thick succession of Upper Triassic to Paleogene strata of the Qiangtang I unit. The complex is an assemblage of fragments of crustal and oceanic rocks, commonly mixed with zones of mélange, that formed during closure of an oceanic realm of probable late Paleozoic and early Mesozoic age and was strongly modified by deformation of Late Cretaceous and early and late Cenozoic ages. The Jinsha complex is in fault contact with the Yidun Unit to the east, but the fault boundary truncates elements of the complex, so it cannot be followed south of the Jinchuan basin; thus the position of a major ocean is not obvious. The nature of the complex probably also gives a clue as to what the assemblage of upper Paleozoic rocks beneath the Lanping-Simao Unit to the south resembles.

The discussion of the Jinsha accretionary complex presented above shows that it is a complexly faulted assemblage of more or less foliated arc volcanic rocks and carbonate and terrigenous clastic strata separated by narrow zones of sheared rocks, some of which are mélange, both containing sheared and dismembered fragments of volcanic, variably serpentinized mafic and ultramafic bodies. In some places the metamorphism reaches kyanite-sillimanite grade that has overprinted the rocks to the degree that the protoliths are unknown. In many places plutons have intruded the complex and are themselves sheared, forming part of the dismembered assemblage. Examination of Plate 1 shows several belts of mélange and zones with serpentinized mafic and ultramafic bodies; where a large oceanic realm was located is hard to determine. Probably before later modification, the Jinsha complex was separated from the Yidun Unit by a large oceanic region the evidence for which is now obscured by younger faults.

An example of the difficulty in reconstructing the oceanic terrane comes from the study of the Susong ophiolite, presented by Jian et al. (2009a, 2009b; Fig. 6-1), where they show the ophiolite to consist of sheared fragments of gabbro, anorthosite, plagiogranites, and trondhjemites within mélange that yielded ages that range from 340 Ma to 282–285 Ma, a time span of 60 m.y. The geochemical signatures of these rocks are interpreted to have formed at a spreading ocean ridge and at a continental rift margin. Within the same mélange belt amphibolite xenoliths have been dated at between 401 and 451 Ma (Jian, 2009a, 2009b) and which are interpreted to be related to an early rifting episode, suggesting that at least some part of the oceanic realm began to open in Early Devonian time. The assemblage of rocks in the Jinsha accretionary complex and their sheared contacts we interpret to have formed from continental and arc environments that were subjected to intra-arc rifting and shearing over a long period of time during which the rocks were diachronously rifted, sheared, and accreted. Much of the accretionary assembly was completed by the late Paleozoic, but it continued to a lesser degree into the Early and Middle Triassic. The best exposure of the Jinsha accretionary complex is along the east side of the Qiangtang I unit, but it extends across most of the width of the unit in the south and may underlie much of the unit in the north, where diverse Paleozoic rocks are exposed within erosional windows through the more uniform Mesozoic strata of the Qiangtang I unit.

**Songpan Ganze Suture**

In our interpretation the contact between the Yidun and Songpan Ganze Units was a southwest-dipping subduction zone that was active during Late Triassic time; however, its surface expression has been modified by Cenozoic faulting that has reversed the sense of thrusting along this margin. In addition, the dominance of Triassic flysch in both units and their lithologic similarities suggest that these two flysch terranes were connected and were part of a single, large region of flysch deposition prior to initiation of westward subduction during part of Late Triassic time and was expressed by magmatism within the Yidun Unit, and a belt of mélange along the subduction zone that was elevated at times as an outer arc high (see above). In this interpretation the Yidun and Songpan Ganze Units were always part of a single realm with a basement of either oceanic or thinned continental crust that was subducted. The flysch within both the Songpan Ganze and Yidun Units was greatly shortened, but at times the Yidun arc was subjected to intra-arc extension. In this sense there was no collision between separate tectonic elements within the Songpan Ganze and Yidun Units within the area of Plate 1.

**Longmen Shan Suture**

Like the Songpan Ganze suture, the boundary between the Yangtze and Songpan Ganze Units in the Longmen Shan was a contact between two paleogeographic realms that were always adjacent to one another and not the site of an accretion of independent tectonic units. The Songpan Ganze flysch can be shown to overlie Yangtze-type Paleozoic rocks along its eastern side (Plate 2). The contact between the two units was shortened during Late Triassic–Early Jurassic time (e.g., Mattauer et al., 1992; Burchfiel et al., 1995) along a left-lateral transpressional boundary, but the amount of westward subduction was limited. The Songpan Ganze subduction zone and the Longmen Shan boundary faults converge to the south within the area of Plates 1 and 2, but how these two zones interact in the area to the south where they converge is poorly understood.

## PALEOGEOGRAPHIC ORIGIN OF THE MAJOR TECTONIC UNITS

The suture zones shown in Plates 1 and 2 outline the major tectonic fragments of largely continental origin that were accreted during Mesozoic and Cenozoic time. Known fossil and stratigraphic evidence indicates that during late Paleozoic time they belonged to two different paleogeographic realms: a southern Gondwanan realm and a more northerly Cathaysian realm. The rocks that lie to the west of the central part of the Changning-Menglian Unit in the south (Plate 1) and west of the Bangong-Nujiang suture farther north, belonging to the Lhasa, Tengchong, and Baoshan Units, contain late Paleozoic glacial-marine sedimentary rocks, and Carboniferous and Lower Permian fauna and flora, evidence for a Gondwanan origin (Li et al., 1985; Zhong, 1998; Zhang et al., 2002; Metcalfe, 2006; Jin et al., 2011). To the east the rocks have yielded fossils of Cathaysian origin, although in central Tibet, west of the area of Plate 1, there are localities with a mixed biota (Sengor, 1990; Li et al., 1985); their boundaries, sometimes called the *Shaunghu suture*, lie within the Paleozoic rocks in a large anticlinorium of older rocks surrounded by Mesozoic strata in the Qiangtang Unit. These are boundaries of Paleozoic rocks that mark an accretion of these realms in late Paleozoic or Early Triassic time, later overlain by

Mesozoic rocks of the Qiangtang Unit (see Zhang et al., 2002); the eastward continuation of these boundaries into the area of Plate 1 are largely covered by younger rocks. To the east (Plate 1) the boundary between these two realms lies within the central part of the Changning-Menglian Unit, and, from discussion of this unit above, this boundary would probably lie within the cryptic suture within the Chong Shan shear zone. It might lie within the western part of the Qiangtang Unit (sensu lato) as used here, but it has not been identified.

In the south the boundary between the two realms was described near Menglian in the southern part of the Changning-Menglian Unit by Wu et al. (1995), who reported a mixed Gondwanan and Cathaysian biota they interpreted to be within a structurally interleaved sequence of rocks. They interpreted the rocks in this area to be a complex mixture of oceanic island and mid-oceanic-ridge basalts. We agree with their interpretation that this belt of rocks is a structurally complex zone of fragments from many different parts of a broad oceanic realm whose inception was at least Early Devonian and not closed until Late Triassic time. Most importantly, the biota in this area places the Lin Cang magmatic arc, its wall rocks (Pz3ma and Pz3mb) and probably part of subunit Pzu2 (Plate 1) into the Cathaysian realm and probably also within the collage of Paleozoic fragments.

Within the Baoshan and Tengchong Units to the west the paleontological and stratigraphic evidence indicates that it is of Gondwanan origin (Wang, 1983; Cao, 1986a, 1986b; Wu et al., 1995; Metcalfe, 2006). However, there is no evidence to indicate how much separation there may have been between the two units. The strata of the Baoshan Unit has a Paleozoic stratigraphy that can be followed across the unit, although somewhat disrupted in its eastern part, and is a single, coherent, continental unit. Unfortunately the Tengchong Unit is so poorly known because of metamorphism, plutonism, and lack of detailed study, so it is not clear if it is a single unit. It is the direct continuation of the Lhasa Unit to the north and may have some of the same potentially disrupted character.

All the areas east of the Changning-Menglian suture and its northward continuation contain Cathaysian biota, but important belts of late Paleozoic ophiolitic mélange within the eastern Lanping-Simao Unit and Qiangtang I unit that separate them from the Yangtze, Transitional, and Yidun Units. Both mélange belts contain evidence of oceanic regions of Devonian to Permian age that closed in Early–Middle Triassic time (Jinsha accretionary complex) and Late Triassic time (eastern Lanping-Simao Unit). How wide these oceanic areas were remains unknown. Because the rocks west of and for the Jinsha complex within, these two mélange belts contain Cathaysian biota similar to that of the Yangtze Unit, and they are often regarded as being of Yangtze origin. However, as the oceanic rocks within the mélange belts indicate, they must have been separated from the unit to the east. Also, the stratigraphy of the late and some early Paleozoic rocks within the collage is unlike the units to the east. In the Lanping-Simao Unit these rocks are only exposed in erosional windows below Mesozoic strata, and their stratigraphies have differences between exposures, locally with the presence of volcanics rocks within similar age strata to nonvolcanic strata in the Yangtze Unit. In the Qiangtang I unit the Paleozoic rocks commonly contain mafic and intermediate volcanic rocks, some in great abundance, although they lack the characteristic Emei Shan basalt, as do the rocks in the Lanping-Simao Unit, and they do not correlate well stratigraphically with rocks in the Zhongza (Yangtze) subunit. In our interpretation the rocks of the Jinsha accretionary complex yield the clue to the pre-Mesozoic nature of the crust for both the Lanping-Simao Unit and the Qiangtang I unit. As the name implies, the pre-Mesozoic rocks represent a collage of continental fragments, some with arc or extensional volcanic and intrusive rocks separated by collapsed narrow(?) oceanic regions that include intra-arc rift, oceanic, and ocean island environments, which, during accretion, formed zones of mélange or shear. Thus, in our interpretation, this collage formed a basement on which strata of Early to Late Triassic age were deposited (Fig. 16-2). In this interpretation the basement for the Lanping-Simao Unit and the Qiangtang I unit was not a coherent paleogeographic unit of continuous rocks similar to rocks that make up the pre-Mesozoic rocks and other major tectonic units such as the Yangtze, Transitional, Baoshan, Tengchong, and Lhasa Units. Although, as discussed above, even some of these tectonic units did undergo limited fragmentation by rifting and strike-slip faulting, accompanied locally by mafic volcanism, but of a magnitude far less than that in the pre-Mesozoic rocks of the Lanping-Simao Unit and the Qiangtang I unit. How fragmentary the Paleozoic crust is below the Qiangtang II and III units remains uncertain. Thus the definitions of the Lanping-Simao Unit and the Qiangtang I unit are entirely based on the more continuous Mesozoic strata that overlie the Paleozoic collage. The nature of this basement probably may partly explain why this area subsided to form the thick sequence of Mesozoic strata above it. It appears that most of the collage consists of fragments of Cathaysian origin (Figs. 16-1 and 16-2), although within central Tibet this is not the case, and the collage contains both Gondwanan and Cathaysian components. Models that regard these two units as coherent Paleozoic continental fragments need reevaluation and are considered in the post-Paleozoic tectonic evolution of the region presented below.

## SEQUENTIAL EVOLUTION FROM LATEST PALEOZOIC TO RECENT

We make an attempt here to synthesize the age and distribution of the different deformations within the areas of Plates 1 and 2. This has been very difficult for the following reasons: (1) structures and associated metamorphism and magmatic activity are poorly constrained by good data; (2) structures can be dated locally, but the distribution of structures of the same age is hard to determine where constraints are lacking; (3) the parallelism of structures to dated structures is used, but it is known that structures are often reactivated and may have formed at more than one time; and (4) parallelism of structures in different tectonic units now adjacent to one another may not have been close originally because of large lateral displacements.

The distribution of deformational events within Plate 1 can be seen on Plate 3, the unconformity map, where unconformities that overlie deformed rocks give a generalized view of the pre-unconformity extent of the deformation. For example, the Late Triassic uplift of the Kungdian high within the Yangtze Unit is well shown by the distribution of the T3 unconformity, and the pre–Middle Jurassic deformation of the units in the western Lanping-Simao Unit and units to its west is shown by the distribution of the J2 unconformity. Plate 2 also shows some of the pitfalls in trying to show the distribution of some events, such as the critical early Cenozoic deformation related to extrusion where the Paleogene unconformities are widely separated, but based on parallelism of structures, this deformation is interpreted to be widespread across the area of Plate 1 (related to deformations of early Cenozoic time, see below).

**Late Paleozoic to Early and Middle Triassic Time**

During late Paleozoic time all tectonic units showed very different paleogeographies and tectonic events that emphasized their separation by major oceanic sutures. For example, near the end of the Paleozoic there was a great outpouring of Emei Shan basalt within a terrestrial environment across most of the southern Yangtze and Transitional Units, reaching 5+ km in the southern Transitional Unit (Plate 1). To the north the basalt was extruded in a marine environment in the western Transitional and southern Yidun and Songpan Ganze Units. The Emei Shan basalt and underlying rocks extend to the west and southwest, where they abruptly end against the faults at the boundary with the Jinsha accretionary complex and the Ailao Shan Unit. Their continuations to the west are not recognized. The rocks that make up the late Paleozoic collage are shown in Figure 16-2 and include the rocks that form the lens-shaped bodies separating the Qiangtang I from the Qiangtang II and III subunits; because they may be parts of the collage, their Paleozoic affinities remain unknown. It is also necessary to consider the strong Mesozoic and Cenozoic structural overprints on all rocks within all the time-space figures within this section because of possible large-scale post-accretionary movements that would have altered their original relative positions, but such reconstructions are not attempted. The collage was assembled initially in late Paleozoic to Middle and Late Triassic time, and the deformations recorded in these rocks are both late Paleozoic and early Mesozoic (Plate 3). Thus the collage formed a possibly unique assemblage of rocks and cannot be considered as a simple accretion of a continental terrane to adjacent units. These rocks are probably the most poorly understood group of late Paleozoic and early Mesozoic rocks within the area of Plate 1. They form the basement for more uniform sequences of Mesozoic strata that define such units as the Lanping-Simao, Qiangtang (sensu lato), Lin Cang, and Changning-Menglian Units. All other units are apparently more coherent tectonic units that can be considered to be continental fragments, although there are still some problems as to the singular nature of some units, such as the Lhasa, Tengchong, Yidun, and Songpan Ganze Units.

**Late Triassic–Early Jurassic Time**

Late Triassic and Early Jurassic stratigraphic sequences within the Lanping-Simao, Qiangtang (sensu lato), Yidun, Songpan Ganze, Transitional, and Yangtze Units can be followed broadly across all these units, and they form a seal for the collage units that lie below in the Lanping-Simao and Qiangtang Units. In contrast, the early Mesozoic rocks to the west are coarse clastic deposits that are more restricted areally, and some may have formed within isolated basins. These rocks rest unconformably above rocks in the Lin Cang to eastern Baoshan Units (Plate 3), and also in the Lhasa and Tengchong Units where Mesozoic strata are less uniform laterally and commonly are only locally present. For example, in the western Baoshan Unit there are disconformities that separate Triassic and Jurassic strata that are mainly marine, with local basalt in the Jurassic, and these strata show a major contrast to the broad depositional environment of dominantly red beds in the Jurassic rocks to the east. The differences between the early Mesozoic units within the northern and eastern parts of the Lhasa Unit are very different from the rocks in the Qiangtang units to the east, suggesting that they were separated from each other even though rocks in the Bangong suture indicate that the suture may not have opened until Late Triassic or Middle Jurassic time, an ambiguity that requires further study.

In Middle and Late Triassic time the Songpan Ganze and Yidun areas rapidly subsided to become the site of thick flysch deposition that may have locally extended into the Early Jurassic. Deformation began during Late Triassic time by southwestward subduction along the present boundary between the Sonpang Ganze and Yidun Units, and first, mafic, followed by intermediate and felsic volcanic rocks, were extruded and interbedded with the flysch of the Yidun Unit. Mélange units and a forearc high formed along this subduction boundary (Plates 1 and 3), but flysch deposition continued in both areas. In latest Triassic and Early Jurassic time the Yidun and Songpan Ganze Units were strongly folded, and the Yidun Unit was intruded by large Triassic plutons that cut across most of the structures, and plutons of Late Triassic and Jurassic age intruded the Songpan Ganze folds (Fig. 16-4). Most of the plutons are largely unfoliated.

Along the eastern side of the Songpan Ganze and Transitional Units the Longmen Shan transpressional thrust belt telescoped rocks that formed the transitional boundary between the Songpan Ganze and Yangtze Units, creating a sharp structural boundary between the two stratigraphic sequences in the north. To the south the thrust faults of the Longmen Shan diverge, and transitional stratigraphy is preserved between the Transitional Unit and the Songpan Ganze and Yidun Units, although foreshortening occurred within these units. The easternmost thrust faults of the Longmen Shan belt appear to have continued southward along the boundary between the Transitional and Yangtze Units. The left-shear transpression of the Longmen Shan thrust belt is

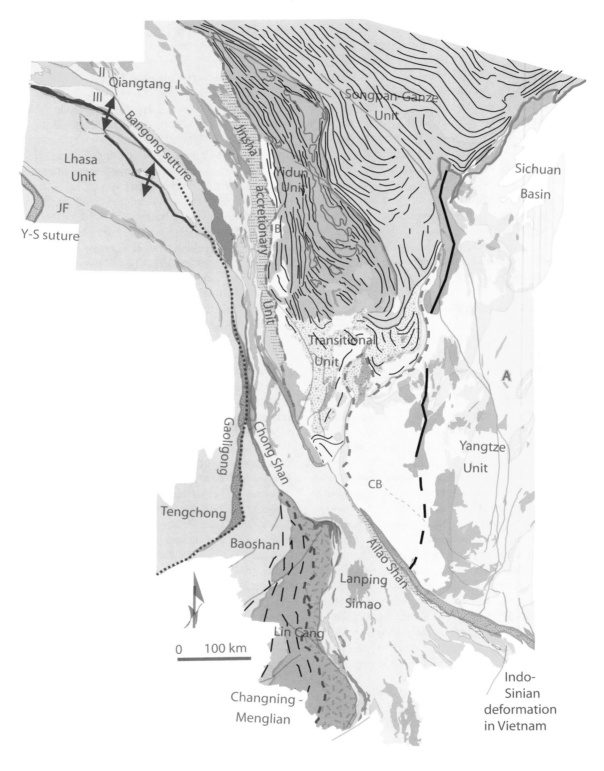

expressed in the folds in the eastern side of the Songpan Ganze and Transitional Units, where the west to northwest trends characteristic for the folds and faults in the Songpan Ganze and Yidun Units change direction, curving to the northeast (Fig. 16-4; Plates 1 and 2). The transpressional belt was developed as South China (Yangtze Unit) moved relatively northward, closing the oceanic realm to the north to form the Qingling belt during Late Triassic to Early Jurassic time.

Within the Yangtze Unit, a northeast-trending foredeep was formed east of the Longmen Shan thrust belt and was filled with

Figure 16-4. Deformation during Late Triassic to Early Jurassic time consists of mainly folds and faults with northwest to north trends within the Songpan Ganze, Yidun, and Transitional Units. Folds are upright and tightly appressed, with fanned cleavage. Along the east side the folds and faults curve into a northeastward trend and become a left-lateral transpressional thrust belt in the Longmen Shan, where the Songpan Ganze rocks are thrust eastward above those of the Yangtze Unit. The eastern boundary of the thrust belt is shown by a thick red line, but to the south the boundary is interpreted and shown as a dashed red line. The thrust belt is marked by thick foredeep deposits of Late Triassic to Cretaceous age east of the thrust belt (Fig. 2-11). Plutons of Late Triassic age are shown within the Yidun Unit, but numerous plutons within the Songpan Ganze Unit of late Triassic and Jurassic age are not shown, but they are shown on Plate 2. The axis of the north-south Kungdian high within the Yangtze Unit is shown by a heavy black line, but also has an older history of uplift. The folds in the southernmost part of the South China fold belts in the Yangtze Unit may be of this age, but they are poorly dated and are shown as dashed lines. The Yangtze, Transitional, and Yidun Units were together at this time and have not been disrupted significantly by younger displacements, but tectonic units west of their western and southwestern boundary (dashed red line) have been subjected to major modification since Late Triassic and Early Jurassic time. The green line with divergent arrows shows the beginning of the opening of the Bangong ocean. To the south the line is dashed where it is now the Bangong-Nujiang cryptic suture, but would mark the position where the tectonic units on both sides would have been separated by an oceanic realm. The formation of these structures may overlap in age with the latest deformation within the region shown in Figure 16-2 within the Changning-Menglian Unit (thin dashed lines), but they are probably mostly younger. This deformational event, referred to as the *Indosinian event*, is well known in Vietnam and along the eastern part of South China, but how these structures continue into the area of Plate 1 is uncertain. B—Batang; CB—Chuxiong basin; JF—Jiali fault.

Upper Triassic to Cretaceous non-marine strata derived from the thrust belt. The foredeep subsidence is superposed across the north-south trend of the slightly older Kungdian high, reversing its earlier uplift history. The foredeep, thrust belt, and Kungdian high can be followed to the south, where they end abruptly against the Ailao Shan Unit. The Kungdian high, a north-south–trending element in the western Yangtze Unit, was elevated during early Late Triassic time. Along this high all the Paleozoic rocks, including the thick Emei Shan basalt, were removed, so that the first deposits to cover it were uppermost Triassic strata of the foredeep in its northern part. Along the northwest end of the high, a major northwest-trending belt of Triassic plutons lies east of the thrust belt and was intruded within the continental crust of the Yangtze Unit, a relationship that remains largely unexplained.

The Songpan Ganze and Yidun Units show transitions to the south, in both stratigraphy and structure, into the Transitional Unit. There does not appear to be any evidence that these units continued farther south than where they are today. The folds within the Songpan Ganze and Yidun Units are mostly upright near isoclinal folds and thus are the result of at least 100% shortening, although no quantitative estimate presently can be made. These folds also appear to be terminated along their base by a subhorizontal, ductile décollement shear zone (Harrowfield and Wilson, 2005). If these two units end to the east against the Longmen Shan thrust belt, and to the south into the Transitional Unit, the restoration of the Songpan Ganze–Yidun flysch basin requires that the units to the south rotated counterclockwise by a significant but unknown amount. Thus the present wedge shape of the flysch basin is nothing like its original shape, owing to strong folding, and was a basin with a much broader westward opening.

The formation of the Upper Triassic flysch basin and its subsequent deformation and magmatic activity contrast greatly with the contemporaneous deposition of shallow-marine and non-marine strata to the west in the Qiangtang I unit and the Lanping-Simao Unit. These two units are now in fault contact with the Yidun, Transitional, and Yangtze Units, but they do not show a transition into them, either stratigraphically or structurally, during Late Triassic to Early Jurassic time. The Yidun Unit contains only rare Jurassic strata, and the Transitional Unit has none, whereas the Lanping-Simao Unit and Qiangtang I unit exhibit continuous deposition from Late Triassic to at least Late Jurassic and Cretaceous time, and were not folded until Late Cretaceous and/or Paleogene time.

At about the same time as the deformation and sedimentation were occurring in the Songpan-Ganze and Yidun Units, subduction was ending along the mélange belt at the eastern side of the Lanping-Simao Unit and within the Jinsha accretionary complex to the west, where Upper Triassic strata are unconformable on mélange and the accretionary complex. In our interpretation the ophiolitic mélange belt or belts were formed by west-dipping subduction during late Paleozoic and Early and Middle Triassic time; thus structural contacts with rocks to the east must be younger. The Yangtze Unit to the east does not show deformation at this time, and the mélange is separated from it by the metamorphic rocks of the Ailao Shan Unit, interpreted to be a left-lateral mylonitic shear zone displaced in the early Cenozoic. Contacts on both sides of the Ailao Shan shear zone are faults, and presently the Ailao Shan rocks are thrusted to the southwest above the mélange zone. What lay along the southwest side of the Yidun, Transitional, and Yangtze Units at this time is unclear, as the boundary has been tectonically modified by younger deformation, and there is no evidence for a southwest-facing margin of this age. Thus these relationships indicate that none of these three units were in their present positions in Late Triassic time.

During this time interval, deformation related to the closing of the oceanic realm was concluding within the Changning-Menglian Unit. The time of closure of this oceanic terrane is poorly known, and evidence suggests that it was diachronous. In different places, Lower, Middle, and Upper Triassic strata overlie the deformed rocks of the suture zone. Plutons and volcanic rocks of Triassic to perhaps Early Jurassic age, probably derived from eastward subduction within the Changning-Menglian belt, are present to the east in the Lin Cang and western Lanping-Simao Units, suggesting that subduction did not cease until Late Triassic time. The deformation appears to be unconnected to the activity in the units to the north (Fig. 16-4).

## Middle Jurassic–Early Cretaceous Time

The time of opening of the Bangong-Nujiang oceanic area is uncertain. The fossil evidence from two ophiolitic mélange belts that form part of the Bangong suture suggests that the oceanic realm was open by Middle Jurassic time. The geology of the two ophiolite-bearing belts supports the interpretation that they formed in a continental margin environment and are mainly part of the rifted Lhasa side of the ocean that opened (similar to the Galica and Newfoundland banks in the modern Atlantic). Upper Triassic strata of the Lhasa Unit are shallow-marine carbonate rocks, but in the Early Jurassic, and particularly in the Middle and Late Jurassic, thick flysch-like strata were deposited, suggesting subsidence at a Lhasa northern passive margin. Within two narrow belts in the metamorphic rocks of the Lhasa basement, between the two Jurassic mélange belts, are sheared ultramafic and mafic rocks locally associated with Paleozoic rocks and rarely Triassic rocks. These two belts might indicate local or regional attempts at stretching of the Lhasa crust, but the evidence is of limited extent and exposure, so their origin remains uncertain (Fig. 16-4).

On the north side of the Bangong suture the Qiangtang III unit contains little evidence for development of a continental margin in Middle and Late Jurassic time, although the Jurassic strata do become marine in contrast to the marine–non-marine deposition in the Qiangtang II unit to the north. If there were Jurassic strata of an offshore to deep-water, continental-margin facies, they are not present now. Marine Triassic strata in the Qiangtang III unit are very different from those in the Lhasa Unit south of the suture. Within the eastern part of the Qiangtang III unit is a north-south–trending fault across which Upper Triassic rocks thicken abruptly toward the suture, which might indicate that extension began at that time within the Qiangtang III unit, a date older than Middle Jurassic, as suggested by the ophiolitic rocks within the suture zone. However, from the Qiangtang III unit there is little direct evidence for the age of opening of the Bangong ocean, and these rocks are probably separated from those in the Lhasa Unit by younger lateral displacements.

Evidence for the beginning of closure of the Bangong suture is supported by the plutons of Jurassic and Early Cretaceous age within the northern Lhasa Unit, which are lacking within the Qiangtang III unit to the north and support southwestward subduction (Fig. 16-5). Final closure is well marked by Upper Cretaceous red beds that unconformably overlie the Bangong mélange belts. The Lower Jurassic to Lower Cretaceous clastic strata of the northern Lhasa Unit forms an Andean fold-thrust belt with a general, but not uniform, southwest vergence (Plates 1 and 4 [available on DVD accompanying this volume and in the GSA Data Repository; see footnote 1]), and it is intruded by plutons of Jurassic(?) and mainly Early Cretaceous age (Fig. 16-5). The structures are unconformably overlain by Upper Cretaceous red beds, indicating that the folding is Early Cretaceous. The opening and closure of the Bangong oceanic terrane was apparently short lived. The Upper Cretaceous red beds unconformably overlie both the northern mélange belt and the fold-thrust belt, but they have also been folded and cut by faults, indicating that the two mélange belts have been deformed in post-Cretaceous time, furnishing evidence for original relations that might bear on the opening and closing of the oceanic realm and its post-closure disruption.

No Upper Cretaceous strata lie unconformably on the fold-thrust belt in the Qiangtang III unit, which involves only Triassic and Jurassic rocks. These structures are poorly dated because of the lack of plutons and Cretaceous strata. The structures are generally south-vergent and could be regarded as part of the same folding event as that south of the suture, a reasonable but unsubstantiated interpretation. These structures can also be interpreted to be the result of the final collisional deformation at the Bangong suture zone. Unfortunately, all the structures of the Qiangtang II and III units, and perhaps even parts of the Qiangtang I unit, are

Figure 16-5. Deformations of Middle Jurassic to Early Cretaceous are poorly constrained in many areas of Plate 1. Areas where they are best constrained are in the northern Lhasa and probably Yangtze Units. Much of the deformation in the South China fold belts as seen within Plate 1 is not well constrained. The ages of the South China fold belts are shown in two colors to emphasize that, when correlated with folding events in Southeastern China, the more southerly structures could be Late Triassic–Early Jurassic (blue), but to the north they can be correlated with Late Jurassic–Early Cretaceous deformation (black). One pluton that intrudes the anomalously thick section of lower Paleozoic rocks in the south is shown as Jurassic, and to the west is a plutonic complex, shown to contain both Cretaceous and early Cenozoic plutons, which also intrudes deformed strata, but these plutons are poorly dated. The structures within the South China fold belts can be traced westward into the eastern margin of the Kungdian high, where they appear to end, a relation that suggests there should be major left shear along the poorly defined western part of this folded belt. In the northern part of the Lhasa Unit, folds in the Jurassic Lower Cretaceous clastic strata are unconformably overlain by Upper Cretaceous strata, which are shown in thick (folds) and thin (faults) black lines and fall within this time period. In the southern Lhasa Unit, structures are intruded by Early Cretaceous and Jurassic(?) plutons (shown as white with red outlines). The fault that bounds the base of the Lhasa Unit at the syntaxis marks the pre-collisional underthrusting along what is now the Yarlung-Zangbo suture. The structures within the Tengchong Unit are shown to be Jurassic–Early Cretaceous on the basis of correlation with those structures in the Lhasa Unit to the north, but there was also extensive reworking of the Tengchong rocks in Late Cretaceous and Paleogene time (Figs. 16-6 and 16-7). Structures within the Baoshan, Changning-Menglian, and Lin Cang Units are poorly dated and are determined mostly to be post–Middle Jurassic and pre-Paleogene (red); thus, depending upon interpretations, they could be Late Jurassic, as they fold Middle Jurassic rocks, and even as young as Cretaceous. In rare locations within the western Yangtze and southern Transitional Units, structures of Late Jurassic and Early Cretaceous ages are determined. Two large plutons within the Yidun Unit have yielded Cretaceous ages, but related structures are not recognized, and the origin of these plutons remains unclear. Y-S—Yarlung-Zangbo suture; B—Batang; JF—Jiali fault.

poorly dated, and it must be entertained that Cretaceous deformation may have been widespread north of the Bangong suture; it is clear for some areas that deformation was pre-Paleogene, but by how much remains an open question (see below; Fig. 16-6).

Plutons of Jurassic and mainly Cretaceous age extend across the entire Lhasa Unit. The plutons in the west are related to northward subduction along the Yarlung-Zangbo suture, and they form the eastern part of the Andean Gangdese plutonic belt, but the plutons in the north are more likely related to southward subduction along the Bangong suture zone. The Paleozoic and rare Mesozoic rocks that form the wall rocks for the plutonic belts are strongly folded and variably metamorphosed. Thus the plutons

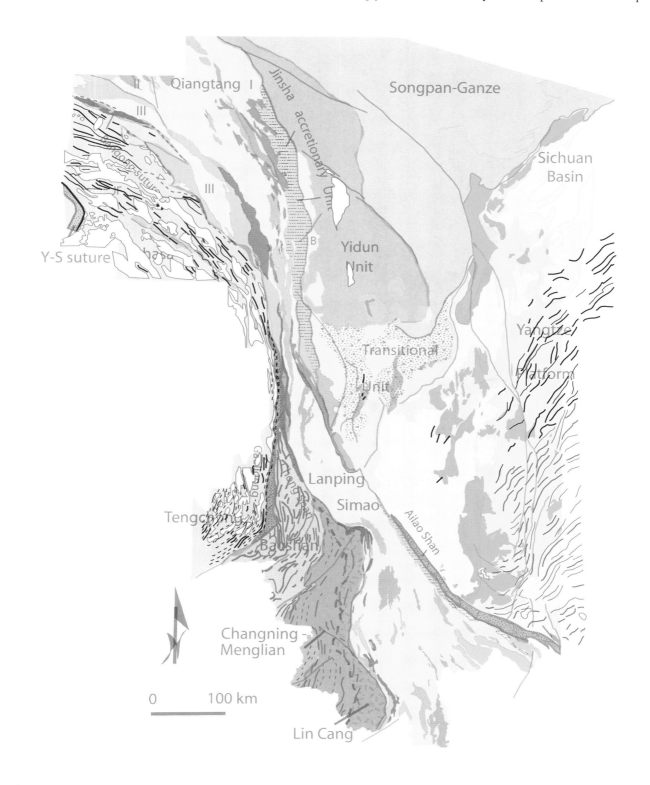

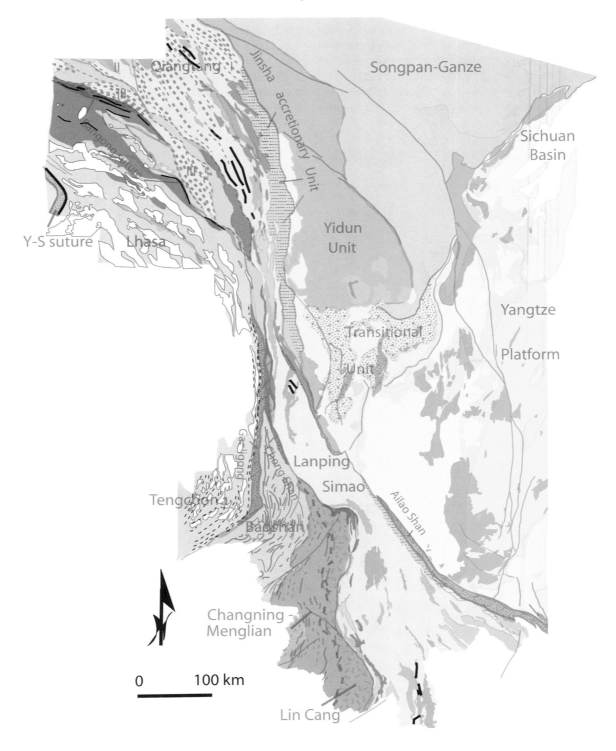

and deformation within the Lhasa Unit are related to subduction from both sides of the unit, but subduction from its south side outlasted that from its north side, and it continued into Paleogene time, as witnessed by the abundance of Late Cretaceous and Paleogene plutons.

## Post–Middle Jurassic, Pre-Paleogene Deformation within the Tengchong, Baoshan, Changning-Menglian, and Lin Cang Units

Late Jurassic–Early Cretaceous folding is evident within both the Tengchong and Baoshan Units, but the general lack of

Figure 16-6. An attempt is made to show where Late Cretaceous structures are known or suspected, as this time period forms the transition from pre-India–Eurasia collision to post-collisional intracontinental deformation. Unfortunately there are not many places where structures can be constrained to be of Late Cretaceous or earliest Cenozoic age. In the northern part of the Lhasa Unit, Upper Cretaceous rocks are folded and are unconformably overlain by Paleogene strata in a few places (heavy black lines). Within the Gangdese belt in south-central Tibet, Late Cretaceous deformation is well constrained, so it can be assumed that similar deformation is present within this eastern continuation of the Gangdese belt in both the Lhasa and Tengchong Units, but well-determined ages of individual structures are rare. Few plutons are dated in these units, and a few are Late Cretaceous. All the plutons in the Lhasa and Tengchong Units are shown, but how many are Late Cretaceous is unknown. From determined Late Cretaceous structures along the Bangong suture it can be interpreted that the suture was a zone of weakness; post-closure deformation was concentrated there, and along with the deformation within the Gangdese batholith, deformation may have extended to rocks more broadly to the north in the Qiangtang Unit (sensu lato). However, north of the Bangong suture there are no dated Cretaceous strata, except in the southeastern part of the Qiangtang I unit, and only poorly dated Paleogene strata unconformably overlie faults and folds that involve Jurassic strata, which could be of Late Cretaceous age. These folds and faults can be traced southward into the southern part of the Qiangtang I unit, where strata as young as Late Cretaceous are folded; thus it is possible that there was widespread Late Cretaceous deformation across much of the Qiangtang Unit (shown with faded red dashed lines). Paleogene strata are present but are not widespread, and they are deformed. Thus it is clear that there is deformation of both pre-Paleogene and post-Paleogene age within the Qiangtang Unit, although the structures of both ages are mostly parallel, making it impossible to be sure of the age of individual structures. Thus, in both Figures 16-6 and 16-7 the folds and faults are shown, and only where the lines are thick are they of determined age. In the Baoshan, Changning-Menglian, and Lin Cang Units, Late Cretaceous structures may be present, but they cannot be definitely proven and are shown as faded red lines. These structures can be determined to be only post–Middle Jurassic and pre-Paleogene or pre-Neogene. Y-S—Yarlung-Zangbo suture.

post–Middle Jurassic strata and age data for foliated and crosscutting plutons in these two units makes it difficult to determine. Deformation of this age is shown in Figure 16-5; however, this is based on the continuation of structures from the Lhasa Unit south into the Tengchong Unit. Within the Baoshan Unit, conglomerate of Middle Jurassic age is present only in the southeastern part of the unit, where it lies unconformably on deformed Triassic rocks, suggesting that at least some deformation occurred in pre–Middle Jurassic time. The conglomerate is not widespread, so the evidence as to how extensive structures of this age are is unclear. The conglomerate is also folded, some places strongly indicating post–Middle Jurassic deformation.

In the western part of the Baoshan Unit the stratigraphic sequence continues into the Middle Jurassic, and the rocks are strongly folded and faulted, but there is little evidence to indicate the upper age limit of the deformation. Folded Neoproterozoic strata that appear to be part of the deformed belt in the southwestern part of the Baoshan Unit are intruded by poorly dated plutons (Plate 1 and Fig. 16-5) that are shown on Chinese maps as both Paleozoic and Cretaceous (Plate 1), offering little support for the age of the folding. In the eastern part of the Baoshan Unit the structures are unconformably overlain by Paleogene strata, but this does not provide a constraint on the deformation in the western part of the belt. In Figures 16-5 and 16-6 these structures are shown as possible Late Jurassic or Cretaceous structures. The Middle Jurassic conglomerate within the southeastern part of the Baoshan Unit is also folded and is unconformably overlain by Paleogene strata, also supporting a poorly determined Late Jurassic to Cretaceous age. The Middle Jurassic conglomerate becomes more abundant to the east, where it unconformably overlies all older rocks as far east as the southeastern Lin Cang Unit. Below the unconformity the rocks are as young as Late Triassic in places, and the older rocks in the Baoshan Unit are metamorphosed. This pre–Middle Jurassic deformation we consider to be part of the final events in the late Paleozoic to Late Triassic collage accretion (Fig. 16-4). The Middle Jurassic conglomerates apparently were deposited locally in north-south–trending grabens and are weakly to strongly folded, indicating that the post–Middle Jurassic age for the folding was widespread. There is little evidence for an upper limit to this folding, however, because the area to the east mostly lacks Paleogene strata. Southeast of the Lin Cang Unit the Jurassic strata are shown to continue into Lower Cretaceous beds, so here at least the folding is Late Cretaceous or Paleogene. In the southwesternmost part of the Baoshan Unit a few outcrops of poorly dated Cretaceous rocks are mapped as unconformably overlying metamorphosed lower Paleozoic rocks, but their relations to the Middle Jurassic conglomerate is unknown, and they provide little evidence for an upper limit to the regional folding. Thus the regional folding within the Baoshan Unit and the units to the east can only be determined as post–Middle Jurassic and pre-Paleogene, and these structures are shown in Figures 16-4, 16-5, and 16-6.

The structures in the areas northeast of the Bangong-Nujiang suture and their continuation to the south can be interpreted as of a similar age to those west of the suture in the Lhasa and Tengchong Units because of structural parallelism. It might be logical to interpret these structures as deformation related to pre–Late Cretaceous closure along the suture. However, as we will discuss below, these areas of deformation on opposite sides of the suture may not have been in the relative positions they now occupy at that time. We propose a major lateral translation of the rocks north and east of the suture during early Cenozoic extrusion. This exemplifies the point that simple parallelism of structures cannot be used to determine their age (see below).

### South China Fold Belts

Within the northeast-trending folds and faults within the South China fold belts in the Yangtze Unit the ages of the structures are poorly determined, as shown on Plate 1. Some interpretations suggest that the folding began in the Late Triassic or

Early Jurassic and became progressively younger into the Early Cretaceous in the north. The rocks within these belts are mainly no younger than Triassic in the south, but they contain Jurassic rocks in their middle and northern parts and are unconformably overlain sporadically by Paleogene strata; thus the age limits for the deformation within Plate 1 is generally poorly determined.

Farther to the east within South China the fold-thrust belt has been better but incompletely studied. In this region the structures have been interpreted to have formed in a backarc Andean setting related to westward subduction beneath southeastern China, where it formed during Late Triassic–Early Jurassic time. The eastern side of the northward-moving South China continental element (Yangtze Unit) occurred at the same time as the transpressional Longmen Shan thrust belt formed its western side. The convergent structures in southeastern China have been determined to have begun in Late Triassic to Early Jurassic time and progressed either continuously or by at least two periods of deformation separated by extensional events, ending in the northwest in early Late Cretaceous time. However, the structures cannot be traced from the northeast in this deformed belt into the area of Plate 1 because they occur in arcuate patterns that are disrupted by northwest-trending faults and folds. Only the northwestern structures from the northeast, where they are dated as Early Cretaceous in age, can be traced into the area of Plates 2 and 1 (Fig. 2-11) and are the only structures that can be confirmed to be of Early Cretaceous age in Figure 16-5. In any case, the structures in the South China fold belts are related to different dynamic systems from either the Late Triassic–Early Jurassic structures in the Yidun, Songpan Ganze, and Transitional Units, or the Middle Jurassic and Early Cretaceous structures in all the tectonic units west of the Ailao Shan and Jinsha sutures.

**Late Cretaceous to Earliest Cenozoic Deformation**

Late Cretaceous to earliest Cenozoic time is the period just before the India-Eurasia collision, following which deformation within southeastern Tibet and adjacent regions became dominated by intracontinental convergent deformation. West of the area of Plate 1 an Andean magmatic belt, with formation of an associated backarc fold-thrust belt, began the shortening and thickening of the southern Tibetan crust. Plate 1 shows evidence for similar deformation, but much is more poorly constrained. Plutons occur across the entire width of the Lhasa Unit and are related to northward subduction from the Yarlung-Zangbo shear zone in the south and southward subduction from the Bangong suture in the north. Unfortunately Upper Cretaceous strata are rare within the Lhasa Unit for dating late Cretaceous deformation.

In the northern part of the Lhasa Unit, Upper Cretaceous rocks are folded and unconformably overlain by Paleogene strata in a few places (Fig. 16-6). Because within the Gangdese belt in south-central Tibet, Late Cretaceous deformation is well constrained, so it can be assumed that similar deformation is present within the eastern continuation of the Gangdese belt in both the Lhasa and Tengchong Units (Plate 1), but well-determined ages of individual structures are rare. Late Cretaceous structures can be determined along the Bangong suture and to its south (Fig. 16-6), where Upper Cretaceous red beds are folded and unconformably overlain by Paleogene red beds. This suggests that the suture was a zone of weakness and that post-closure deformation was concentrated there. It also can be suggested that along with the deformation within the Gangdese batholith, deformation also may have affected rocks more broadly to the north in the Qiangtang Unit (sensu lato). However, north of the Bangong suture there are no dated Cretaceous strata except in the southeastern part of the Qiangtang I unit, and only locally are there thick, but poorly dated unconformable Paleogene strata, where faults and folds involve Jurassic strata. A broad belt of folds and faults across the Qiangtang Unit (sensu lato) could be Late Cretaceous, because these folds and faults can be traced southward into the southern part of the Qiangtang I unit, where strata as young as Late Cretaceous are folded, and although there are only rare Paleogene strata that unconformably overlie them the Paleogene rocks in this area have a very different distribution and lie unconformably on rocks of many different ages (Plates 1 and 3). Thus it is possible that widespread Late Cretaceous deformation occurred across much of the Qiangtang Unit (Fig. 16-6).

Figure 16-7. Early Cenozoic structures extend across all the tectonic units within the southwest area of Plate 1. Some areas of Paleogene deformation are well defined (broad solid lines) where they involve Paleogene strata and where those strata contain dated syntectonic deposits, whereas in other areas the deformation is inferred by tracing parallel structures from known areas of early Cenozoic deformation (dashed lines). Less well age-defined folds and faults are shown in dashed red lines. Some of these structures can be constrained only to the pre–middle Paleogene, so they could be early Paleogene or Late Cretaceous, such as the folds in the northern Qiangtang Unit (sensu lato) or post–Middle Jurassic and pre-Paleogene in the Baoshan Unit, but how much of the structure away from areas of Paleogene strata is early Cenozoic remains unknown. The Baoshan, Changning-Menglian, and Lin Cang Units appear to be dominated by right-lateral, strike-slip faults, and the extent of other structural development is unclear except where rare Paleogene strata are present (heavy lines). The extent of Paleogene structures within the southern Yangtze Unit (Chuxiong basin [CB] and to the east and north) remains uncertain, as Paleogene strata are involved only locally. The continuation of dated early Cenozoic structures to the north into the region of the Sichuan basin becomes difficult, because the structures are parallel and merge with structures of clear late Cenozoic to currently active age. Major shear zones and faults are shown in dark red. The Ailao Shan and the Gaoligong and Chong Shan shear zones are broad mylonitic zones of early Oligocene (perhaps with earlier pre-mylonitic initiation) to early late Miocene age. Other short red-line segments are shear zones with evidence for early Cenozoic displacement. Heavy dashed red-line segments are faults of proven early Cenozoic displacement; the dashed line in the middle of the southern Lanping-Simao Unit is from a study by J.W. Geissman and B.C. Burchfiel (unpublished report). NF—Nantinghe fault; WS—Wuliang Shan; Y-S—Yarlung-Zangbo suture; JF—Jiali fault; B—Batang; A—Day Nui Con Voi. (Note: Transitional, Lanping-Simao, and Ailao Shan labels were darkened because they were lost in the pattern.)

The Paleogene strata are deformed; thus it is clear that there was deformation of both pre-Paleogene and post-Paleogene age within the Qiangtang Unit, although the structures of these ages are mostly parallel, making it impossible to be sure of the age of individual structures where appropriate strata are lacking. Thus in both Figures 16-6 and 16-7 the folds and faults are shown, and only where the lines are thick are they of determined age.

Southeast of the southern extent of the cryptic Bangong-Nujiang suture, some or most of the folds and faults within the Baoshan Unit (Fig. 16-6) may be of possible Late Cretaceous age, but the time of their deformation is poorly constrained. Triassic and Lower and Middle Jurassic strata are deformed and are unconformably overlain by the coarse-grained strata of J2 age in the east and south, and by Paleogene strata in the northeast. The

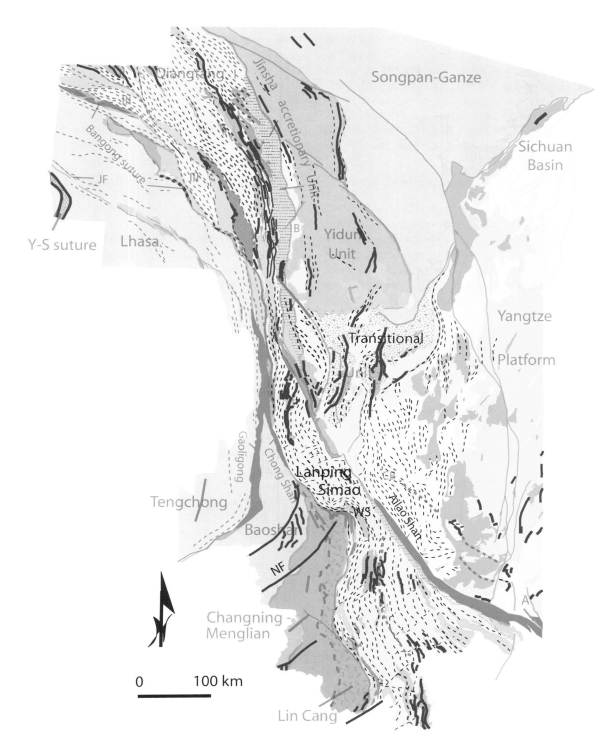

Middle Jurassic strata in the eastern part of the Baoshan Unit (J2, Plate 1) are in places strongly folded, but they and the Paleogene strata are less folded than the underlying rocks. The folding of these two stratigraphic units may be of the same post-Paleogene age, but the pre-Paleogene strata may also contain Cretaceous structures. If the entire Bangong-Nujiang suture was closed in Early Cretaceous time, as it was in the north, it may well have caused collisional deformation north and east of the suture, but this remains to be determined.

Note that the known and suspected Late Cretaceous deformation and magmatism are almost entirely related to activity along the Yarlung-Zangbo subduction zone, and the remainder of the area to the east shows little convergent activity (Fig. 16-6). During Late Cretaceous time, activity along the eastern margin of southeastern China became extensional following Early Cretaceous shortening in the South China fold belts, but this activity took place east of the area shown in Figure 16-6.

**Paleogene Time**

Following the India-Eurasia collision, intracontinental deformation related to continued northward convergence of India into Eurasia became the dominant tectonic system within the Tibet region, although the Western Pacific and Indonesian subduction systems played a major role east of the plateau (see below; cf. Figs. 16-6 and 16-7). Paleogene deformation (Plate 1) is widespread, but its true extent is uncertain (Fig. 16-7). This is the first time that deformation extended across most of the tectonic units that were assembled in pre-collisional time. The interpretation of structures of this age is critical, as these structures occurred during the possible large-scale southeastward extrusion of crust-lithosphere from eastern Tibet into Indochina.

The areas of Mesozoic red beds that overlay the Paleozoic collage appear to have been the locus of much of the deformation, although both the red beds and the deformation extend more broadly. The better constrained areas of deformation are within the Lanping-Simao Unit and the Qiangtang I unit. In the Lanping-Simao Unit, deposition locally is continuous from Upper Cretaceous into Paleogene strata, although there are local areas in the south where Late Cretaceous folds can be identified, and a broad area of disconformity is present within Upper Cretaceous strata (Plate 3). Several large areas of folded Paleogene strata, many of which appear to be syntectonically deposited, lie generally within the central part of the Lanping-Samao Unit. The folds are sinuous, forming two large, convex-west arcs, and where the folds follow the sinuous western boundary. However, the eastern boundary of the Lanping-Samoa Unit is linear, trending northwest, and the folds and faults strike it obliquely (Fig. 16-7), and where in our interpretation this obliquity requires major shear within the mélange belt in the south and along major faults in the north. The contact on the eastern side of the mélange belt adjacent to the Ailao Shan Unit is a steep to moderately east-dipping, west-vergent thrust contact carrying the mylonitic rocks of the Ailao Shan shear zone in its hanging wall. The eastern boundary of the Ailao Shan shear zone is marked by the brittle late Cenozoic to active right-lateral Red River fault zone. The Ailao Shan shear-zone rocks are thus fault bounded and discordant with respect to the folded belt in the Lanping-Simao Unit and the structures in the Yangtze Unit. At its south end (before entering northern Vietnam) the Ailao Shan metamorphic rocks and partially mylonitized rocks are tied to a Paleozoic sequence of rocks that contains thick Emei Shan basalt by poorly dated plutons of Mesozoic and Cenozoic age. The Paleozoic section is unlike rocks in the tectonic units on either side, but it has some similarity to rocks in the area of Erhai Lake, thus appearing to have been displaced by ~300 km to the south. It is also separated from the Lanping-Simao Unit to the west by left-lateral shear within the mélange belt, indicating that the Lanping-Simao Unit is also displaced by an unknown distance to the southeast relative to the Ailao Shan Unit.

In the southern part of the Qiangtang I unit the stratigraphic sequence is continuous from Upper Triassic into Upper Cretaceous strata before it was folded (Fig. 16-7). In some areas the section is shown to extend into the Paleogene, but dating of these younger strata is poor. In most of the area, younger Paleogene strata lie unconformably on the structures, but not where the Upper Cretaceous rocks are involved. Thus the age of folding, whether latest Cretaceous or earliest Paleogene, remains uncertain. Farther north the large NNW–trending Paleogene Gonjo basin is syndepositionally folded. East of the basin, folds in the Mesozoic rocks are parallel to the folds in the Gonjo basin strata and are thus most likely of Paleogene age. However, folds west of the basin strike obliquely into the basin and exhibit faulted contacts with the basin strata except at the northwest side of the basin, where the folds are unconformably overlain by Paleogene strata. However, the folded strata at the north end of the basin are Triassic, so the age of the folding is not well constrained. These folds regionally are parallel to folds farther south that involve Cretaceous strata, and on that basis they are tentatively assigned a Late Cretaceous age (Fig. 16-6). Farther to the north are several other basins with thick Paleogene strata that lie unconformably on Jurassic and Triassic rocks that are more strongly folded than the Paleogene strata. The ages of many of these folds are older than the Paleogene strata of the basins, but whether they are early Paleogene or older remains unclear (Figs. 16-6 and 16-7). The folds that affected the Paleogene strata can be traced to folds beyond the Paleogene basins that are parallel to folds in areas far beyond the basins, but their age is uncertain, and they are regarded as Late Cretaceous or Paleogene. This is true for the folds in the Qiangtang II and III units (Figs. 16-6 and 16-7).

Within the Yidun and Transitional Units there is clear evidence for synsedimentary Paleogene deformation within narrow basins that trend north-south (Fig. 16-7; cf. Plate 1). The best examples are in the east-vergent Lijiang thrust belt near Lijiang and in several narrow thrusted zones within the Yidun Unit (Fig. 16-7). Folds within the Chuxiong basin in the southwestern part of the Yangtze Unit involve mainly rocks as young as Late

Cretaceous, but they continue to the northwest into an area where they involve Paleogene strata. In the eastern part of the Chuxing basin local Paleogene strata are also involved in the folds, and north of the basin syntectonic Paleogene strata lie below the Jinhai-Qujiang thrust; thus it appears that the entire Chuxing basin may have been folded in Paleogene time, but how young the folds are remains uncertain.

Most of the evidence for Paleogene deformation is south of the Longmen Shan, but recent thermochronology fission-track data presented by Wang et al. (2012) (see Plate 2) suggest that the uplift of the Pengguan massif occurred in two episodes of accelerated cooling; the first from ~30–25 Ma and the second during the past ~10 m.y. This interpretation is important because it suggests Oligocene–early Miocene uplift in the region north of Xianshuihe fault. This area has limited outcrops of Paleogene rocks, and it has been interpreted that uplift of the Longmen Shan was a late Cenozoic feature beginning in the late Miocene and has been in response to eastward lower crustal flow from central Tibet (Royden et al., 2008). Their data suggest some uplift may have occurred earlier in the area of the Pengguan by an uncertain mechanism. Unfortunately, these data are basically a point source, and it remains unclear how much of the Longmen Shan was involved and when the eastern margin of the Tibet plateau began to form.

Farther to the east, within the western part of the South China fold belts, are narrow belts of Paleogene strata, some of which can be determined to be syntectonic; like folds in many other areas that involve Paleogene strata, however, widespread folding of this age remains unknown because of the limited extent of Paleogene strata. But at least it is clear locally that the north-south–trending structures overprint the older structural trends of the northeast-trending South China fold belts.

Contemporaneous with Paleogene folding are major strike-slip faults and shear zones. The broad mylonitic Ailao Shan, Gaoligong, and Chongshan shear zones are the best examples and can be shown to have been active from about early Oligocene to early late Miocene time. Dating of shear zone activity is based on thermochronological data, but if these shear zones were initiated by brittle deformation their inception could be older. Other fault zones of limited extent also can be shown to have most likely been active at this time within the Tengchong, Changning-Menglian, and Lanping-Simao Units. The Ailao Shan shear zone and short segments of faults to its north are left-shear, as is the Chongshan shear zone, but the Gaoligong shear zone is right-lateral. Along with the left shear within the mélange belt on the east side of the Lanping-Simao Unit, these faults and shear zones are supportive of the hypothesis first proposed by Tapponnier et al. (1982) for extrusion of crustal-lithospheric fragments from southeastern Tibet into Indochina. However, our interpretation differs from theirs because the crustal fragments were not rigid but strongly deformed and rotated; numerous fault zones were active during extrusion, and most importantly, extrusion was the result of the interaction of two adjacent dynamic systems, not just intracontinental India-Eurasia deformation (see below).

## Paleogene Extrusion Hypothesis

Key evidence for large-scale, early Cenozoic extrusion is provided by the paleomagnetic study of J.W. Geissman and B.C. Burchfiel (unpublished report) that showed that Paleogene and Upper Cretaceous rocks in the western part of the Lanping-Simao Unit were not only rotated 45°–120° clockwise (Fig. 7-8) but have been translated 500–1000 km to the south relative to South China on the basis of paleomagnetic inclination data similar to the translation reported by Tanaka et al. (2008). Considering that these rocks moved to the southeast, the displacement would have been 750–1500 km, a displacement at a maximum that would have placed rocks presently in the position of the Wuliang Shan (Plate 1) well into east-central Tibet. Such large displacements require major changes in present paleogeographic positions; for example, the southern part of the Lanping-Simao Unit would lie southwest of the Qiangtang I unit, and the other units between it and the Lhasa-Tengchong Units would lie at least in a similar position, but north of the Lhasa Unit. Otofuji et al. (2007) presented paleomagnetic evidence that during the Cretaceous there was a gap of >1200 km between the Lhasa and Qiangtang Units across the Bangong suture at 96° E that was closed by Cenozoic time. These authors suggest that at least parts of the Lanping-Siamo Unit filled this gap, but was moved southeastward during extrusion as the gap closed, an interpretation that is similar to ours (see below). With the large uncertainties in extrusion displacements, we have not shown quantitative amounts on paleogeographic reconstructions but only generalized reconstructions on maps (see below).

However, for such large changes of paleogeography the location and offset on major extrusional faults will determine where the extruded crustal units originated. Major extrusional faults can be identified or suggested by the geological data presented above (Fig. 16-8), and our present working hypothesis is outlined below, but it will require extensive testing, and some parts of the hypothesis have unresolved problems, as discussed below.

Most of the evidence for large extrusion-related faults is in the southern part of the area of Plate 1 (Fig. 16-8). The Ailao Shan and Gaoligong shear zones are usually taken as the major fault zones proposed as the boundaries for the early Cenozoic extruded crustal fragments from southeastern Tibet, but they are only two of many faults and shear zones that were active at that time (Fig. 16-8). We also interpret the Chong Shan, Lancang, and western Ailao Shan mélange belt to be shear zones that were major contributors to extrusion, and their hypothesized northward continuations are shown in Figure 16-8, where several large lens-shaped crustal fragments are outlined. These lenses are bounded by faults and consist mostly of metamorphic rocks intruded by poorly dated plutons, but a few have yielded Triassic ages that may be correlative with the Lin Cang batholith to the south. The Lin Cang batholith and its wall rocks have been strongly sheared left-laterally by the southern continuation of the Chong Shan shear zone, and at a later time by the southern continuation of the Lancang fault zone. In the northernmost lens, shown on Plate 1, Upper Triassic strata overlie the metamorphic and plutonic rocks,

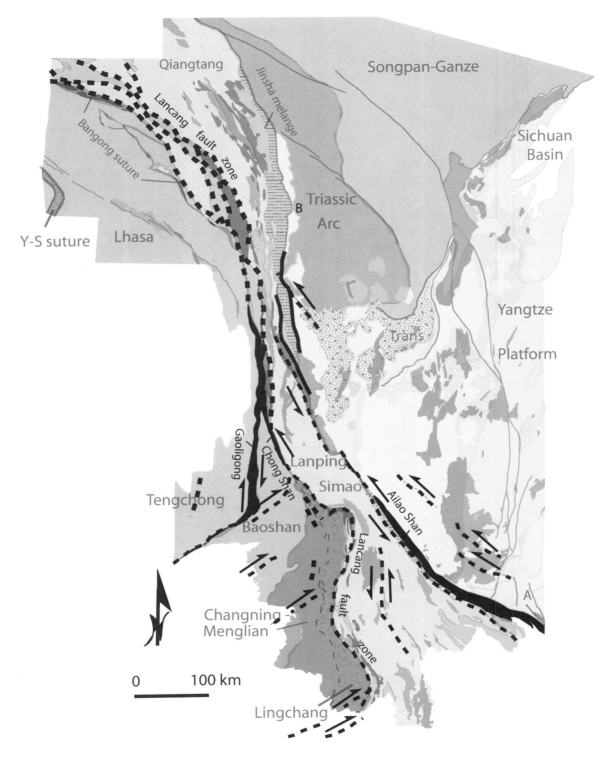

and their stratigraphy is unlike that in the adjacent Qiangtang II and III units. Other lenses to the south also contain metamorphic rocks intruded by plutons of both Paleozoic and unknown age, but they are overlain by Triassic sedimentary rocks in one place and by Triassic volcanic rocks in another (Plate 1). Triassic volcanic and hypabyssal rocks are part of another large fault-bounded lens (along the Lancang fault zone) with a very thick succession of felsic and intermediate rocks, largely ignimbrites and hypabyssal intrusives, that have been dated as Middle and Late Triassic, and these rocks are unlike any others within the adjacent Qiangtang I

Figure 16-8. Major faults and shear zones, known and interpreted to be related to the early Cenozoic extrusion hypothesis, as outlined in the text, are shown. Most of the best documented structures are in the south, where evidence has been presented for major left-lateral shear on the Ailao Shan and Chong Shan shear zones, and right-lateral shear on the Gaoligong shear zone (in black). The dashed line west of the Ailao Shan metamorphic and mylonitic belt shows the location of a late Paleozoic, Early Triassic mélange belt we interpret to have been reactivated as a major Paleogene shear zone. The Lancang shear zone at the western boundary of the Lanping-Simao Unit remains a poorly understood shear zone, but it may have been important during Paleogene extrusion (dashed line). The dashed left-lateral fault shown in the southern part of the Lanping-Simao Unit was identified by paleomagnetic data presented by J.W. Geissman and B.C. Burchfiel (unpublished report), and its significance remains unclear. These shear zones, when traced northward, are interpreted to merge into a broad zone of fault-bounded tectonic lenses (dashed lines) that are hypothesized to represent remnants of the extruded terranes to the south, and they mark the zone where the extruded crustal fragments originated. These lenses and faults have not received any major study. Short segments of faults, shown as dashed lines, with known strike-slip displacement, were active during early Cenozoic extrusion. A—Day Nui Con Voi; B—Batang; Y-S—Yarlung-Zangbo suture; Trans—Transitional Unit.

unit. Southward, these volcanic rocks form a narrow, discontinuous belt with faults on both sides; the western fault constitutes the eastern contact of the northern Chong Shan shear zone. We have interpreted the southern continuation of the Paleozoic and early Mesozoic rocks within the lenses to be part of the Paleozoic collage that lay east of the Changning-Menglian suture, but whether the rocks in the northern lenses are part of the collage remains to be determined. We include the Qiangtang III unit in the broad zone of lenses, but even though their Triassic strata have some similarities to the other Qiangtang units their Jurassic strata are dominated by marine limestone unlike the other Qiangtang units. How far to the southeast these Jurassic strata have been translated relative to the Lhasa Unit to their south is unknown, but as shown on Plate 1 their southern boundary fault is the one that bounds the Bangong suture on the north and becomes a cryptic suture boundary to the southeast. The shear zones along the Ailao Shan mélange belt, along with the Ailao Shan Unit itself, form the major northeastern boundary of the extruded fragments, but its northern continuation we interpret to be the fault that separates the Qiangtang I unit and the Lanping-Simao Unit (Fig. 16-8). The continuation of this fault would lie along the narrow zone of Mesozoic strata that separate two large belts of Paleozoic rocks. The northern one, with metamorphic rocks and mélange (Plate 1) belongs to the collage underlying the Qiangtang I unit, and the southern one is interpreted to belong to the Lanping-Simao collage. This fault zone has been modified along its entire length by late Cenozoic displacements. How all these major fault zones interact within the area where this fault strikes into the belt containing several of the other major north-south–striking faults is unknown.

In Figure 16-8 are shown other faults that were active during early Cenozoic time and were probably related to extrusion. Four left-lateral faults northeast of the Ailao Shan shear zone indicate that the left-lateral shear affected rocks in the Yangtze and Transitional crust east of the main boundary faults; the three southern ones were reactivated in late Cenozoic time as right-lateral faults. Several northeast-striking right-lateral faults south of the Lanping-Simao Unit were active during early Cenozoic time (Lacassin et al., 1998) and have been interpreted to be related to clockwise rotation of second-order continental fragments during extrusion (Wang and Burchfiel, 1997). Wang et al. (2006) reported shear of early Cenozoic age along faults in the Tengchong and Changning-Menglian Units, but the extent of these faults remains unknown. J.W. Geissman and B.C. Burchfiel (unpublished report) identified left-shear along a north-south zone within the southern part of the Lanping-Simao Unit. This fault was identified by paleomagnetic data, separating two areas, one with large clockwise rotation of rocks to the west, and the other with no rotation to the east. This unstudied fault may be a major fault that bounds large fragments within the Lanping-Simao Unit and is important in relating the crustal structure in southeastern China with crustal structure in northern Vietnam (see below).

Of all the suspected extrusion faults in the southern part of the area of Plate 1, the Lancang fault zone is the most enigmatic. It forms a sharp eastern boundary of the Lin Cang batholith, and only a few minor plutons lie to its east. Northward it bends westward and continues west, then it bends northward along the west side of the Lanping-Simao Unit. Many authors extend the fault northward along the eastern side of what we show in Figure 16-8 to be the belt of lens-shaped bodies that are interpreted to represent a broad shear zone of remnants within the area from which crust was extruded.

However, there are problems in making the Lancang fault zone a through-going structure with large, early Cenozoic displacement. Even though in the southern part (Plate 1) the Lancang fault zone forms a sharp boundary in pre-Mesozoic and early Mesozoic rocks, with plutons to the west and low-grade Paleozoic rocks to the east, the geological relations in the area where it makes a large northeastern convex bend may indicate limited early Cenozoic displacement: Triassic and Jurassic(?) volcanic rocks intruded by plutons at the north end of the Lin Cang batholith are cut by the Lancang shear zone, but they appear to be present on both sides of the fault. To the east of the shear zone these volcanic rocks lie on strata of the Lanping-Simao Unit, a relation that suggests very little post–early Mesozoic movement on the fault. Where the fault turns westward the volcanic rocks intruded by the Lin Cang plutons are sharply truncated, and no volcanic rocks are present to the north, even though the volcanic sequence contains a thick section of extruded volcanic rocks. It is possible that the Lancang fault zone at the north end of the batholith continues eastward and turns southward to lie within the Lanping-Simao Unit east of the volcanic rocks, but such an interpretation requires further field studies. It is clear, however, that the fault or shear zone along the east side of the Lin Cang batholith has a sinuous trace that we interpret to be the result of early and late Cenozoic folding around a vertical axis and related to the same deformation that produced the northeast-trending faults west of the Lancang fault zone. Many questions need answers about the Lancang fault zone such as, is it a single shear zone, what is its age, and is its main displacement pre-Mesozoic and/or early Cenozoic, or are there other unidentified branches of the shear zone that lie farther east? Although it has clear Cenozoic movement north of the Lin Cang Unit, the nature of its southern extent remains unclear.

In the northern part of the area of Plate 1, the evidence for large extrusion-related faults is less clear than in the south. One such fault bounds the north side of the Bangong suture. The contrast between the rocks on both sides of this fault suggests large lateral displacement: truncation of the ophiolitic mélange belts within the Lhasa Unit to the south, and truncation of Upper Triassic to Middle Jurassic dominantly marine clastics and limestone of the Qiangtang III unit to the north, where there is no significant evidence for continental margin deposits along its south side. In contrast, the northern part of the Lhasa Unit contains a thick Upper Triassic limestone unit, overlain by a thick Middle to Lower Cretaceous clastic and flysch-like sequence that unconformably overlies what we interpret to be an extended continental margin of the Lhasa Unit. The fault along the southern contact of the Qiangtang III unit marks the northern boundary of the Bangong mélange and continues to the southeast, truncating the mélange belt and all the rock units within the northern Lhasa continental margin. It continues south as a cryptic suture and appears to strike into the early Cenozoic mylonitic Gaoligong shear zone, one of the shear zones that is taken to be one of the major extrusion shear zones. Far to the south, where the Gaoligong shear zone has been modified by the younger northeast-striking Ruili fault are possible remnants of mafic and ultramafic rocks that may be relics of oceanic crust. Along the cryptic parts of the Bangong suture the Lhasa Unit and its continuation to the south, the Tengchong Unit, are in contact with the Qiangtang I unit and the Lanping-Simao and Baoshan Units to the east. One branch of the Bangong cryptic suture may merge with the Cenozoic Chongshan mylonitic shear zone and continue into the Changning-Menglian suture, where evidence indicates that this suture is much older than the Bangong suture (Fig. 16-1). How these age relations between the sutures is resolved remains to be worked out, but the continuation of the late Paleozoic to Triassic Changning-Menglian suture may continue northward, where it may lie within what we interpret to be a Paleozoic collage terrane, masked by Mesozoic strata and disrupted by Cenozoic faults so that its trace is now difficult to identify.

The recent studies by Lin et al. (2009) and Searle et al. (2007, 2010) suggested that the Jiali fault in the Lhasa Unit originally joined with the Ailao Shan shear zone during extrusion. Searle et al.'s connection is suggested by regional tectonic analysis. Lin et al.'s connection is suggested by $^{40}Ar/^{39}Ar$ ages they obtained for the Jiali fault that are similar to those from the Ailao Shan shear zone during the time from ca. 22–18 Ma, when they argued that both shear zones had left-lateral displacement. Only later, after 18 Ma, did the Jiali fault connect with the Gaoligong shear zone at ca. 13 Ma. We find the older connection with the Ailao Shan difficult to support, because if the Jiali and Ailao Shan shear zones had several hundred kilometers of displacement during extrusion, this would place the Lanping-Simao Unit within the Lhasa Unit. There is little paleogeographic evidence for this interpretation, as the Lanping-Simao Unit is composed dominantly of Mesozoic red beds and lacks the plutons so abundant within the Lhasa Unit. In addition we can find little evidence as to how the two shear zones would have traversed all the N-S–trending units and structures between them (Plate 1).

The western boundary of the Yidun and Transitional Units has been proposed to be a major Cenozoic fault zone, but its relation to the extrusion process remains unclear. This boundary is a faulted boundary, but evidence indicates it is not a single fault of one age. As discussed above, it is a boundary between two different terranes and is often called the *Jinsha suture*. The evolution of the two terranes on either side is very different, but when and how the two terranes were juxtaposed remains unclear. In our interpretation the Jinsha accretionary complex to its west is a Paleozoic and early Mesozoic collage that became the basement for the Qiangtang I unit in Middle to Late Triassic time. The Qiangtang I unit was not strongly deformed until late Cretaceous and early Cenozoic time. The Yidun and Transitional Units to the east were strongly deformed, metamorphosed, and intruded by plutons in Late Triassic time. Along a segment of the northern boundary is a mélange belt that is probably of Triassic age, but it is unclear whether the mélange is related to eastward or westward subduction. Even though in our interpretation the collage is related to westward subduction, the short segment of mélange in the north is probably also related to a short period of eastward subduction, but whether this is the time of accretion of the two terranes is not clear. Elsewhere along the boundary there is scant evidence for a subduction zone or suture.

Near Batang (B in Fig. 16-8) a large mafic pluton is mapped to intrude the boundary between the two terranes, and on weak evidence it is dated as Late Triassic. If this part of the boundary fault was intruded in Triassic time, it prohibits major strike-slip displacement in post–Late Triassic time, but the nature of the pluton and the boundary fault in this area requires additional careful study and dating before they can be interpreted properly. The central part of the boundary fault was offset left-laterally by the northwest-trending Zongdian fault in early Cenozoic time, and in the south the boundary fault both cuts and is overlain by Paleogene strata in the western part of the Jinchuan basin. Even if there is uncertainty in the relations, the Paleogene strata on both sides of the boundary faults are the same, restricting large strike-slip displacement. Also, east-west shortening of Paleogene age has affected the rocks on both sides of the boundary. Thus the evidence suggests, but remains to be tested, that this boundary has not been a major extrusion-fault boundary during early Cenozoic time.

Thus, in conclusion, we propose that the extrusion of crustal fragments did occur in early Cenozoic time from a domain north of the eastern Himalayan syntaxis. The extrusion occurred by southeastward movement along several shear zones, which in the north are marked by remnants of units now present as lenses within a broad zone of shearing, marking the cicatrix. The extrusion was accompanied by continual north-south shortening as India penetrated Eurasia so that the faults related to extrusion were transpressional in the area north and west of the syntaxis, but as the crustal fragments moved east of the syntaxis the regime became neutral or even transtensional. In this eastern part they merge into the area where the tectonics of southeastern Tibet are dominated by the India-Eurasia intracontinental shortening and pass into the tectonic regime of the Western Pacific–Indonesian subduction system, where the tectonics are dominated by trench rollback (Fig. 16-9). Further deformation of these lenses and shear zones occurred during late Cenozoic time under a different tectonic scheme (see below), and India continued to move northward relative to the crust east of the syntaxis. Restoration of the crustal fragments to their original positions would have broadened the area within the shear zone considerably in the north-south direction west of the syntaxis prior to extrusion, a geometry that helps to resolve the large space problem between India and Eurasia at the time of collision (Royden et al., 2008).

## Late Cenozoic Time

The tectonics in southeastern Tibet changed in late Cenozoic time. Deformation of late Cenozoic to Recent age is dominated by strike-slip faults related to rotation of crustal fragments around the eastern Himalayan syntaxis and shortening in the Longmen Shan and southern Sichuan basin area that has increasing right-lateral oblique slip to the northeast (Figs. 16-10 and 16-11). Active motion in these two regions is well shown by global positioning system (GPS) velocities (Fig. 16-11; Chen et al., 2000; Zhang et al., 2004; Gan et al., 2007) with respect to a fixed South China. The pattern of strike-slip faults within Figure 16-10 is dominated by major left-slip faults that strike northwest in the north but which curve progressively to the south to first, north-south strikes, and then terminate to the north or at the Red River fault. In the north, faults that bound the rotating crustal fragments are marked by narrow fault zones, but to the southeast they splay into more faults, distributed across a broad area and associated with numerous pull-apart basins. In the southeast the faults can be grouped into two systems: the Xianshuihe-Xiaojiang faults, with a total of ~60 km of left slip (Wang et al., 1998), and the Dali fault system to the east and west, respectively (Figs. 2-10 and 16-10). Both fault systems cut across preexisting structures within the major tectonic units,

Figure 16-9. These two dynamic systems are hypothesized to have controlled the Cenozoic tectonism of Southeast Asia: the intracontinental convergent system in the India-Eurasia collision zone, and the Western Pacific–Indonesian subduction system. The boundary between these two zones has been gradational and may have changed position during Cenozoic time, but its present position lies within Plate 1.

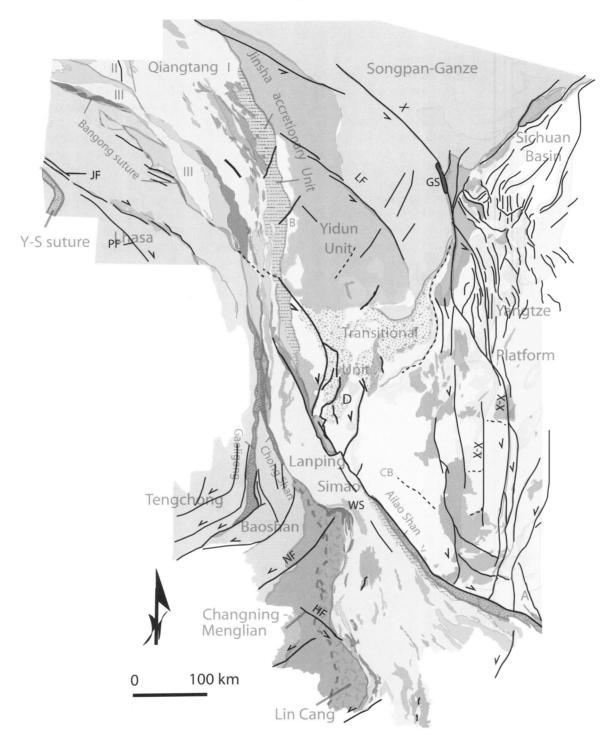

although some faults in the southwestern part of the Xianshuihe-Xiaojiang system follow old faults along the Kungdian high, some of which may be as old as Precambrian. The splaying of the faults to the southeast was accompanied by an increase in abundance of pull-apart basins that indicate an increasing component of east-west extension. Both systems contain numerous small faults with northeast strike between the major through-going faults and have right-slip oblique components that are probably related to the counterclockwise rotation between the major faults (England and Molnar, 1990). In some places the small fault segments have thrust components and have produced elevated areas as constricting bends.

Figure 16-10. Distribution of structures of late Cenozoic age. The Xianshuihe (X) and Xiaojiang (X-X) faults mark the eastern boundary of crustal fragments that rotate clockwise around the eastern Himalayan syntaxis (at Y-S suture). The similar but smaller Dali fault system (D) is also shown. Both of these major left-lateral fault systems end at the northwest-striking, right-lateral Red River fault and its continuation to the northwest. The late Cenozoic folds and faults in the Longmen Shan and Sichuan basin trend to the south and become more numerous where they cross the northeast-trending Early Cretaceous structures in the South China fold belts to the east (cf. Fig. 16-5). The structures from the Sichuan basin continue southward, where they are parallel and merge into north-south–trending folds and faults of Paleogene age in the southwestern part of the Yangtze Unit (cf. Fig. 16-7), making it difficult to determine the age of any given structure. The Gonga Shan pluton (GS) is interpreted to have intruded into a constricting bend in the Xianshuihe fault at ca. 12 Ma. Compare these structures to the global positioning system (GPS) velocities in Figure 16-11. The right-lateral Jiali (JF) and Pugu (PF) faults and the left-lateral Nantinghe (NF) and Heihe (HF) faults are also shown. LF—Litang fault zone; GS—Gonga Shan pluton; WS—Wuliang Shan; Y-S—Yarlung-Zangbo suture; CB—Chuxiong basin; A—Day Nui Con Voi.

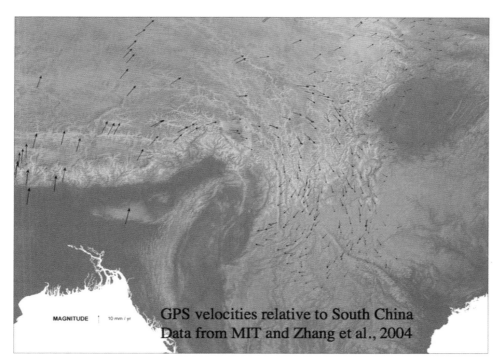

Figure 16-11. GPS velocity vectors from Zhang et al. (2004) relative to South China clearly show the present-day clockwise rotation around the eastern Himalayan syntaxis and the deviation of motion to the northeast, west of the Sichuan basin. The eastern velocity gradient in the clockwise motion is marked by the late Cenozoic Xianshuihe and Xiaojiang fault zones (Fig. 2-10). The clockwise motion sheared material across the Red River fault, and all tectonic units to its west as crustal fragments are moved into the subduction system in the Andaman Ranges of western Burma (Myanmar) and eastern Bangladesh. There is only a small component of convergence in the Longmen Shan west of the Sichuan basin, and this small part of the Tibetan Plateau moves slowly with the rest of South China. South of the Sichuan basin the late Cenozoic deformation spreads farther east, and the Tibetan Plateau rises just south of the basin and east of the major strike-slip faults. MIT—Massachusetts Institute of Technology.

Unusual is the termination of the north-south–striking left-slip faults at the Red River, which is marked by one of the rare late Cenozoic right-slip faults in the region. Wang et al. (1998) showed that the Red River fault was bent by ~60 km of counterclockwise rotation on the Xianshuihe-Xiaojiang fault system, interpreted as a result of the left-lateral shear passing through the Red River fault and the Ailao Shan Unit without displacing them, an interpretation supported by the GPS data (Fig. 16-11). Active left-slip faults are not present as continuous faults in the Lanping-Simao Unit, but within or at the western margin of the Lanping-Simao Unit, left-slip faults reappear and continue to the southwest across all the tectonic units to the west (Figs. 2-10 and 16-10; see Buschfiel, 2004). Total displacement on these western faults is ~70 km, and coupled with the GPS data, this suggests that the clockwise rotational shear around the eastern Himalayan syntaxis continues into the area southwest of Plate 1 and is absorbed in the rollback of the east-dipping Andaman subduction zone (Molnar and Tapponnier, 1975; Tapponnier and

Molnar, 1976), which has a major component of right-lateral shear as India continues to move northward relative to Indochina.

Why the fault systems terminate at the Red River fault, and why it is an active right-slip fault, are unclear. The Red River fault, a fault zone of parallel brittle faults, is along one of the major structural discontinuities in the region, the southern boundary of the Yangtze and Transitional Units and the Qiangtang I unit, indicating that this northwest-striking zone that served as a major left-slip boundary during Paleogene extrusion, was reactivated in late Cenozoic time as a right-slip zone during the change in tectonic regimes from early to late Cenozoic time (Tapponnier et al., 1982). The fault has a slow slip-rate of only a few millimeters per year and a maximum total displacement of ~50 km (Wang et al., 1998; Schoenbohm et al., 2006b), and it has not had an earthquake during recorded history. This must require further examination in light of the 2008 Wenchuan catastrophic earthquake that had no previous history of major earthquakes prior to 2008 (Densmore et al., 2007). The origin of the Red River fault remains somewhat of a problem, for if the left-lateral shear on the Xianshuihe-Xiaojiang fault system passes through it why do the Red River fault and three shorter parallel faults to its north have right-slip? Many Chinese geologists make the Red River fault the southern boundary of a wedge-shaped crustal fragment, the Chuan Dian fragment (Huchon et al., 1994), which is extruding to the southeast and is bounded on the north by the left-slip Xianshuihe-Xiaojiang fault. Another interpretation is that the Red River fault slips right-laterally along a preexisting extrusional fault zone reactivated by the more recent clockwise rotation of the whole region around the eastern Himalayan syntaxis. Toward the southeast the Red River fault appears to become less active. It must be pointed out that this interpretation is not shared by all the contributors to this work.

When the clockwise rotation of crustal fragments and their related faults began is not well constrained. The last movement on the early Cenozoic extrusion-related faults was ca. 17 Ma; thus the late Cenozoic fault zones are younger and may have begun by then or at least by ca. 12 Ma during the tectonic regime that caused uplift of the eastern part of the plateau (Clark et al., 2005, 2006; Kirby et al., 2002). The left-slip faults within the Tengchong Unit are transtensional faults that bound associated pull-apart basins containing late Cenozoic and Pliocene to Recent sediments and which are also related to the transtensional region with numerous active volcanoes. The initiation of this volcanism was ca. 17 Ma, which may support the older age for the beginning of this late Cenozoic rotation and associated system of faults.

The active left-slip faults south of the Lanping-Simao Unit were initiated during the Paleogene extrusion as right-lateral faults (Lacassin et al., 1998) related to clockwise rotation of smaller scale extruded fragments (Wang and Burchfiel, 1997). They also caused deformation of the boundary between the Lanping-Simao and Lin Cang Units by folding around a vertical axis at the northeastern end of the Nantinghe fault (Figs. 16-7 and 16-10). This deformation we interpret to have been responsible for the uplift, the folding of the Wuliang Shan, and the superposed deformation within the Lanping-Simao Unit east of the Wuliang Shan. Similar smaller scale, vertical axis folding is also present at the eastern end of the Heihe fault where the fault dies out by vertical axis distortion of older folds in the Lanping-Simao Unit. The distortion of the western boundary of the Lanping-Simao Unit is in contrast to its linear eastern boundary, which is marked by the Red River fault and the Ailao Shan Unit.

To the west the active right-lateral Jiali-Parlung and Pugu faults have yielded $^{40}Ar/^{39}Ar$ ages of at least 11–17 Ma (Lee et al., 2003), which suggests that the present system of late Cenozoic faults that accommodate the rotation around the eastern Himalayan syntaxis may have been initiated just after the early Cenozoic extrusion terminated. Some workers have suggested that the Jiali fault continues eastward into the Red River fault zone (e.g., Lin et al., 2009; Searle et al., 2007, 2010), but it is clear from Plate 1 that this is not possible.

We tentatively interpret that the time of change from the early to late Cenozoic patterns of tectonism was ca. 17–15 Ma, but the change was probably not abrupt. The cause of this change may have been related to the position of the eastern Himalayan syntaxis and the rheological changes in the Tibetan lithosphere. Prior to this time the northeastern corner of the India indenter lay ~700 km south of its present position. At the time of initial collision the India-Eurasia subduction zone had a general WNW–ESE trend, and with the continued advance of the Indian continental lithosphere the corner of the indenter moved northeastward, but the subduction zone beneath the Indoburman ranges progressively became more north-south oriented, and the subduction of oceanic lithosphere became progressively more right-lateral, strike-slip, and the subduction more eastward. We suggest that at ca. 17–15 Ma, with increasingly more eastward subduction beneath the Indoburman Ranges, the rollback of the subducting lithosphere caused the lithosphere to the east to extrude westward, beginning the present pattern of rotation around the Eastern Himalayan syntaxis. The westward motion of the Indoburman Ranges continued as the northeastern corner of the India indenter moved past them so that now they lie west of the track taken by the northeastern corner of India. This would have been also the time at which the lower Tibetan crust became ductile enough to flow under the influence of gravitational potential energy, beginning with the extensional tectonism in central Tibet and augmenting the clockwise rotation around the eastern Himalayan syntaxis, and ending the early Cenozoic southeastward extrusion. Although it must be pointed out that in the western part of the late Cenozoic rotational system, crustal rocks are still being extruded to the east, and it is only east of the syntaxis that the late Cenozoic movements deviated and began to cross the early Cenozoic structures in the southeastern part of the Tibetan Plateau and its foreland. It is at this time, with the onset of significant lower crustal flow, that the change began to take place to the present pattern of deformation and motions shown by the GPS data (Fig. 16-11).

The deformation in the Longmen Shan and the southern Sichuan basin lies northeast of the Xianshuihe-Xiaojiang fault system and contrasts with the region of clockwise rotation (Figs. 16-10 and 16-12; Plate 2). The steep eastern mountain front of the Longmen Shan is a result of major but slow uplift, exposing Precambrian basement at an elevation of >4900 m from the floor of the Sichuan basin at 500–600 m. Although recent work in the Pengguan massif has suggested a possible early Cenozoic uplift (Wang et al., 2012), the present-day Longmen Shan mountain front was formed by a late Cenozoic structure that can be interpreted to be a fault-propagation fold with a moderate to steeply west-dipping lystric thrust fault on its steep eastern limb, becoming near vertical at the surface (Burchfiel et al., 1995, 2008c; Zhang et al., 2010a). To the east are a series of imbricate faults and folds in its footwall that underlie the lower eastern Longmen Shan slope that connects to a horizontal décollement thrust that underlies the Sichuan basin father east and which is responsible for shortening in the Longquan arcuate anticline along the eastern side of the basin (Burchfiel et al., 1995, 2008a; Plate 4). This active anticline trends northward obliquely into the Longmen Shan mountain front (Fig. 16-12 and Plate 2) and has ponded Pleistocene sediments to the west, forming the Chengdu plain. In the southern part of the Chengdu plain the deformation above the décollement east of the Longmen Shan becomes broader and more complex. Southward, a few active north- to northeast-plunging folds appear within the southern part of the Chengdu plain west of the Longquan anticline and which involve only sedimentary Mesozoic strata, but when traced southward,

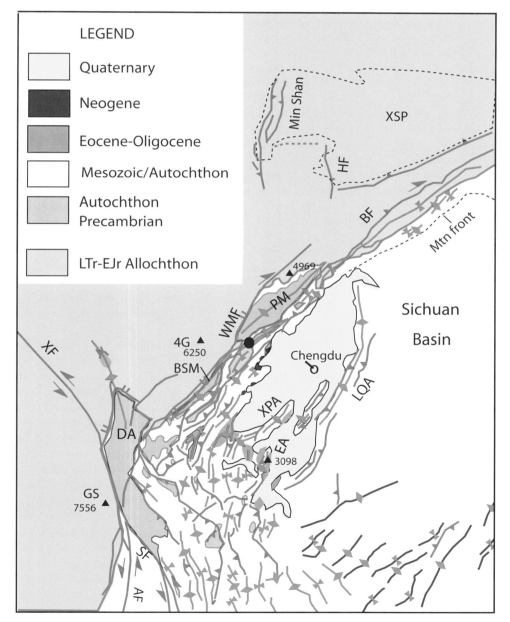

Figure 16-12. Generalized map of the Longmen Shan and Southern Sichuan basin area, showing structures of late Cenozoic age. Faults in red are decorated with barbs for thrust faults, arrows for strike-slip faults, and double ticks for normal faults. In blue are the folds of late Cenozoic age, except in the southeast, where they are of middle Cretaceous age (dark blue). These structures overprint the early Mesozoic Longmen Shan thrust faults that moved the Songpan Ganze Unit (pale red) eastward above the Yangtze Unit (white). The area of the southwestern Sichuan basin, shown in yellow, has a thin cover of Pleistocene sediments derived from the Longmen Shan and ponded behind the active Longquan anticline (LQA). The large black dot at the south end of the Pengguan massif marks the epicenter of the 2008 Wenchuan earthquake. 4G—"Four Girls" peaks; AF—Anninghe fault; BF—Beichuan fault; BSM—Baoxing massif; DA—Danba antiform; EA—Emei Shan anticline; GS—Gonga Shan; HF—Huya fault; PM—Pengguan massif; SF—Shimian fault; WMF—Wenchuan-Maowen fault; XSP—Xushan platform; XPA—Xiong Po anticline; LTr–EJr (in Legend)—Late Triassic to Early Jurassic.

the folds increase in amplitude and involve basement rocks south of the basin (Fig. 16-12 and Plate 2). These north-northeast– to north-south–trending folds and associated thrust faults continue southward, where the elevation rises abruptly where they cross the older northeast-trending Cretaceous folds and faults of the western South China fold belt to form dome and basin interference patterns (Plate 2). Similar folds with basement involvement continue south into the northern part of the area of Plate 1; however, in most of this area are fewer interference patterns because the older northeast-trending folds trend more north-south and become more parallel to the active folds from the southern Sichuan basin. The late Cenozoic folds interact with northwest-striking, south-dipping thrust faults that cause the topography to elevate to plateau heights; this area is part of the plateau, with its associated low relief at a high elevation, and it lies northeast of the Xianshuihe fault. From here southward the north-south–trending folds and faults continue into similar trending folds and faults, which in the area within and north of the Chuxiong basin are considered to be of Paleogene age (cf. Figs. 16-7 and 16-10). Trying to assign an age to a given structure in this area thus is difficult and results in apparent ambiguous age relations. The structures in the southern Sichuan basin are late Cenozoic, perhaps ranging from Miocene to Recent, and the folds and faults within and around the Chuxiong basin are, at least in part, Paleogene. Furthermore, the folds and faults from both areas are cut by the Xianshuihe-Xiaojiang fault system, which may have begun at ca. 12–17 Ma. It may be that whereas there is presently little evidence for Paleogene deformation within the Longmen Shan, the structures in the area south of the Sichuan basin may be of both early and late Cenozoic ages.

Regional structure in the southern Sichuan basin area may explain why the 2008 Wenchuan earthquake had its epicenter at the south end of the Pengguan Precambrian cored antiform along the eastern margin of the Longmen Shan (Fig. 16-12) and propagated to the northeast, but not to the south or southwest. North of the latitude of the earthquake's epicenter, GPS data show that the crustal motions within the eastern Tibetan Plateau relative to South China trend more to the northeast rather than normal to the Longmen Shan mountain front in the epicentral area (Fig. 16-11; Burchfiel et al., 1995, 2008c). But south of the epicenter an increasing number of structures absorb the east-west convergence and distribute the strain more broadly east of the Longmen Shan mountain front. This is supported by the distribution of earthquakes in this region that show a few epicenters within the eastern part of the plateau north of the 2008 epicenter, but with a broader distribution of epicenters in the area of the structures in the southern part of the Sichuan basin.

Along the Longmen Shan to the south of the epicenter the mountains increase in elevation, so that at Siguniang Shan ("four girls") the peaks reach a height of >6250 m. The mountain front east of the Siguniang Shan is very steep, suggesting again rapid uplift similar to that of the Pengguan massif; however, although it appears similar to the Longmen Shan to the north the stresses appear to be more widely distributed to the east in this region. A second steep mountain front is also present east of Emei Shan (EA in Fig. 16-12), and where an active fault is present at its base. In this region south of the Sichuan basin and east of the Xianshuihe fault the mountains reach a height of almost 4000 m, preserving a gentle topographic relief as a part of the Tibetan Plateau.

The development of the plateau in this region east of the Xianshuihe fault is interpreted to be related to the geometry of the Xianshuihe fault. Around the area of Gonga Shan the fault curves more sharply than either to the NW or the SE and forms a restraining bend. The compression across this part of the fault is where the large Cenozoic pluton of Gonga Shan is at the surface within the restraining bend (Plate 2; Fig. 16-10). The high elevation of the Gonga Shan pluton (7556 m) is interpreted to be similar to the upward extrusion of deeper crustal rocks that we propose for the Ailao Shan Unit and the Day Nui Con Voi in southeastern China (see above). East of the restraining bend the Tibetan Plateau was elevated, and the folds that continue southward from the southern part of the Sichuan basin involve Precambrian basement rocks. There are cross structures here: WNW–trending thrust faults involve Precambrian rocks that cut across the N-S–trending folds from the southern Sichuan basin that also involve Precambrian rocks. The décollement here is in the basement beneath the folds and is probably beneath the huge, NW-trending Danba antiform, cored by Precambrian rocks east of Gonga Shan. This raises the questions of how deep the décollement is, and how it interacts with the Xianshuihe fault, major questions to be resolved.

In the western part of the area of Plate 1 are fewer faults of known late Cenozoic age, with the exception of the Parlung-Jiali and Pugu faults. However, this may be, at least partly, the result of less study and fewer earthquakes in this area. A few west-northwest–trending faults cut Neogene strata (Fig. 16-10), but they are poorly studied (Wang et al., 1998). Within the Lhasa Unit similar faults cut Neogene strata, but little is known of their kinematics, although their trends might suggest they involve a component of strike-slip.

In the western part of the area of Plate 1 are fewer of the large through-going faults (such as the Xianshuihe and Ganze faults), and west of the area of Plate 1, and in the central part of the Tibetan Plateau, active faults are characterized by conjugate strike-slip faults and north-south–trending grabens (Zhang et al., 2004; Taylor and Yin, 2009). This pattern of faults shortens the plateau north-south and lengthens it east-west. Wang et al. (2008a, 2008b) showed that the Ganze fault, the western offset continuation of the Xianshuihe fault, gradually loses displacement westward and ends in the area where the conjugate fault systems begin. It appears that the east-west extension within the central plateau is transferred from the conjugate pattern of faults to eastward movement of plateau rocks on large strike-slip faults as the crust in the eastern part of the plateau moves toward and around the eastern syntaxis. The Parlung-Jiali and Pugu faults are right-lateral and have been interpreted to be the southern boundary for the extrusion of the central part of the plateau, and the Ganze fault may be the northern boundary, although the beginning of extrusion is more diffuse than it becomes farther east. If so, it suggests that the Xianshuihe fault system begins where the Ganze fault first appears in the west. The termination of the Kun Lun fault, studied by Harkins et al. (2010), could be interpreted to be the reverse in that it shows how one of the large strike-slip faults within the plateau can begin and increase in displacement. Thus the major fault systems that show so well the rotation of crust in the eastern part of the Tibetan Plateau may have begun in the west by transfer of the crustal motion from the conjugate strike-slip and extensional fault systems of the central plateau.

## MAJOR UNRESOLVED PROBLEMS

### Geological Reconstruction Prior to Early Cenozoic Extrusion

The paleogeography and structural configuration of all the tectonic units at the time of the India-Eurasia collision, a time that still remains poorly documented in the eastern part of the region, and prior to extrusion, remain unresolved problems. The geology indicates that in pre–Late Triassic time the collage elements south of the Yangtze Unit and west of the Yidun Unit were separated from it and from each other along the mélange belts in the eastern Lanping-Simao, the Changning-Menglian, Chong Shan, Bangong-Nujiang, and Jinsha sutures. Also, the Ailao Shan and possibly the Lancang were likewise shear zones across which tectonic units were displaced and were not in their present relative positions. Where these units were following their accretion together remains unknown and is dependent upon the whether these tectonic elements were further displaced relative to one another by early Cenozoic extrusion. Thus it must be determined if early Cenozoic extrusion has occurred, and if so, what is its magnitude, and what was the paleogeography at that time? Late Cenozoic overprints on early Cenozoic tectonic elements can be approximately removed but not quantitatively. The displacement on the major late Cenozoic faults that are the main features of late Cenozoic tectonism can be qualitatively reconstructed, and because their displacements are measured in tens of kilometers, they did not greatly disrupt early Cenozoic tectonic elements whose displacements appear to be much larger. However, some models of Cenozoic deformation within Tibet require almost no extrusion at all (England and Houseman, 1988; Houseman and England, 1993), and the question of whether large-scale, early Cenozoic extrusion has taken place needs to be addressed.

The key evidence is the paleomagnetic studies of J.W. Geissman and B.C. Burchfiel (unpublished report) and Tanaka et al. (2008) that show that a large part of the western Lanping-Simao Unit was translated southward between 500 and 1000 km relative to South China during early Cenozoic time. Because the transport for extrusion is to the southeast the magnitude of displacement must be larger, perhaps 750–1500 km. These data strongly support the southeastward transport of at least the Lanping-Simao Unit and probably also the units to its southwest that are still the focus of continuing paleomagnetic investigation. Geological relations indicate that the extrusion of crustal-lithospheric units has taken place along major faults and shear zones of early Cenozoic age; this was first suggested by Tapponnier et al. (1982) but has been greatly modified in our extrusion hypothesis.

Major early Cenozoic faults bound at least several continental fragments across which rock units do not match and which truncate major geological elements. Whereas these relations alone do not imply major strike-slip displacement and extrusion, they do point out where major disruption of the crust has occurred. Such disruption occurs across the Gaoligong, Chong Shan, Lancang, and both sides of the Ailao Shan shear zones in the southern part of the area of Plate 1. Their continuations to the north are less well known. These shear zones are generally steeply dipping, and the well-documented ones in the south exhibit horizontal shear senses, ages consistent with early Cenozoic strike-slip displacements, and a component of thrust displacement where higher grade rocks, commonly mylonitized, are present on one side.

An excellent review of existing data on the evolution of the Ailao Shan shear zone was presented by Leloup et al. (1995, 2001), who concluded that the shear zone was perhaps active prior to 36 Ma (late Eocene) and that it ceased left-lateral motion at 16 Ma. But more importantly they tried to determine its displacement during early Cenozoic extrusion. In their analysis they pointed out correctly that piercing points across shear zones of large displacement are coarse and difficult to recognize, so the uncertainties are large. As we have shown here, the geology indicates truncations along the southern margin of the Yangtze and Transitional Units with no obvious recognition of the displaced features. The exception is the offset of the thick Emei Shan basalt and a plutonic complex of uncertain age in the southern Ailao Shan Unit, with the rocks near Erhai Lake yielding an offset of ~300–350 km, an offset that Leloup et al. (1995) also identified. These authors suggested a range of displacements of several other less constrained features across the shear zone from South China into Vietnam that yielded magnitudes up to 1000 km, showing that there likely was a large early Cenozoic displacement across the Ailao Shan shear zone, a conclusion with which we agree. However, our interpretation of the extrusion differs significantly from theirs. The differences include the following: (1) The Ailao Shan shear zone is only one of several known and potential shear zones that were active during the extrusion and may not have been the shear zone with the largest displacement, and (2) in relating the motion on the Ailao Shan shear zone to the opening of the South China Sea, where we interpret the extrusion to have been related to the interaction of two different dynamic systems, the India-Eurasia intercontinental convergence and the subduction systems of the Western Pacific and Indonesia (see below).

On the first point (above), Leloup et al. (1995) show that the active left-lateral Dien Bien Phu fault, along the Uttaradit suture zone that closed in late Triassic time, to be correlative with the poorly defined Jinsha suture along the west side of the Yidun and Transitional Units. The Dien Bien Phu fault, a north-striking fault just east of the area of Plate 1, appears to be a major boundary (see below), but it does not offset the plutonic complex at the southern end of the Ailao Shan within China, as these plutons, such as the Phang Si Phen complex (Chung et al., 2011), continue into northern Vietnam. The fault ends to the north, where it meets the mélange zone along the west side of the Ailao Shan. There are important differences across the fault, such as a thick sequence of Upper Triassic to Cretaceous rocks to the west, which contrasts with Cretaceous strata resting unconformably on Upper Triassic rocks, commonly highly deformed, metamorphosed, and intruded by plutons to the east. The Uttaradit suture, where it is best defined in Thailand, separates Cathaysian and Gondwanan upper Paleozoic rocks to the east and west, respectively, whereas the Dien Bien Phu fault has Cathaysian biota on both sides. The more probable northward continuation of the Uttaradit suture is the Changning-Menglian suture, which separates the two different realms (Plate 1) (Wu et al., 1995; Metcalfe, 2006). This does not negate the difference in rock units across the Dien Bien Phu fault but suggests that the displacement on this feature is not at the Ailao Shan shear zone but along the fault zone that we propose that lies along the mélange zone southwest of the Ailao Shan shear zone. In our interpretation a major displacement along the mélange zone of early Cenozoic age is expressed by the oblique termination of the folds in the Lanping-Simao Unit, and this displacement continues southward along the Dien Bien Phu fault, but the extrusion fault or fault zone has been modified by younger fault activity. Thus we envision two major parallel zones of displacement of early Cenozoic age on both sides of the Ailao Shan metamorphic rocks. The fault zone, recognized from paleomagnetic data by J.W. Geissman and B.C. Burchfiel (unpublished report) is yet another zone of possible major displacement that would continue south into Thailand, but no work has been done on elucidating this fault zone.

Trying to match rock units within Vietnam with those in China has not been very successful. In northern Vietnam, north of the Day Nui Con Voi, the rocks appear to incorporate rocks that are the continuation of the southern part of the South China deformed belt of both Paleozoic and Mesozoic ages. South of the Day Nui Con Voi, most of Vietnam was strongly affected by Triassic deformation (Carter et al., 2001; Lepvrier et al., 2004), the type area of the Indosinian orogeny (Deprat, 1914; Fromaget, 1927), and these mainly highly deformed rocks are overlain in many places unconformably by Upper Triassic strata. Rocks of this character are not generally present below the Upper Triassic strata of the Lanping-Simao Unit to the west. The description given by Lepvrier et al. (2004) for rocks in northern Vietnam is of numerous fragments of continental rocks separated by dated Triassic right-lateral shear zones that contain mafic and ultramafic rocks. Carter et al. (2001) and Lepvrier et al. (2004) consider all these rocks to be part of a collage of fragments assembled during their accretion to South China along the Song Ma–Song Da suture zone in Triassic time. This is a similar scenario that we have outlined for the rocks below the Lanping-Simao and Qiangtang Units (Plate 1), except that the rocks in Vietnam are much more highly deformed, metamorphosed, and intruded, and they can be shown to be separated by one or more major faults (see below).

We agree with previous models calling for early Cenozoic extrusion, but we caution that many reconstructions calling for large amounts of extrusion are based on poorly constrained geological arguments. Even displacements on the Ailao Shan shear zone, currently the best studied shear zone, but still with many unresolved problems, have major ambiguities. For example, large magnitude extrusion of ~700 km would align a belt of Cretaceous intrusions along the southeastern coast of China with those in Vietnam (Leloup et al., 1995), but it would place the deformed and metamorphosed rocks of northern Vietnam adjacent to the Yangtze Unit, which has no rocks that would be their correlatives either in rock type or in time of deformation. It cannot go unnoticed that the edge of the continental shelf of Vietnam juts out to the east south of the probable extrusion faults, but we interpret that as a result of eastward displacement of the Sunda Shelf toward the Western Pacific subduction system relative to the South China coast. Regions both to the north and south of the extrusion boundary were and still are subjected to extension that began in Late Cretaceous time (north) or early Cenozoic time (south), but are related to two different geometries of rollback within the subduction system: To the north, subducted slabs only descend to the 660 km discontinuity, whereas they reach depths of >1200 km with a crudely defined boundary along the northern margin of the Sunda Shelf (Fig. 16-13).

Accepting the large but poorly defined magnitude of early Cenozoic extrusion, the problem within the area of Plate 1 is, first, to locate from where the extruded fragments came, and second, to determine whether a reconstruction can be supported or tested by the geology. In our interpretation, as presented above, we tentatively suggest that the fragments at the latitude of the northern Lanping-Siamo Unit were positioned north and west of the eastern Himalayan syntaxis along a broad shear zone that lies between the Lhasa Unit and the northern Qiangtang I unit, but not necessarily along the Bangong suture. The shear zone is marked by residual fragments left behind as the main bodies of the Baoshan to Lanping-Simao Units passed eastward. This is different from the reconstructions presented by Leloup et al. (1995) and Replumaz and Tapponnier (2003), who placed the fragments north of the Yarlung-Zangbo suture and involved the Lhasa Unit in the extrusion. As we have shown, there are possible effects of the extrusion west of the Gaoligong shear zone, but at present they are poorly understood. Reconstructing the fragments cannot be done quantitatively, because the amount of their internal deformation and the displacements on the many faults active during extrusion remain unknown. Such field studies as comparing the Mesozoic red-bed basins of the Lanping-Siamo Unit to those of the Qiangtang I unit, determining if the lens-bounding faults have strike-slip components, and matching the geology in the lenses within the hypothesized broad shear zone with their possible protolith units are necessary to test the hypothesis.

In addition, large-scale southeastward extrusion of the tectonic units south of the Yangtze, Transitional, and Yidun Units,

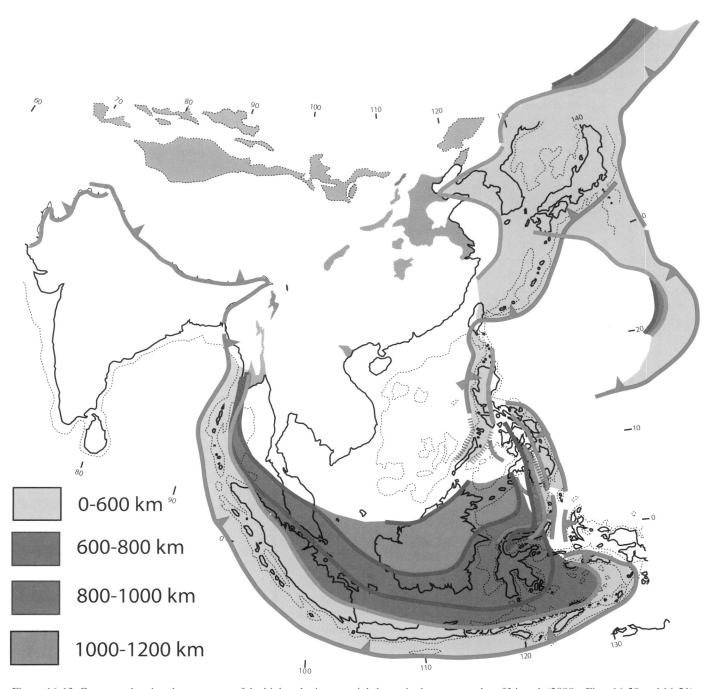

Figure 16-13. Contours showing the geometry of the high-velocity material shown in the tomography of Li et al. (2008a; Figs. 16-20 and 16-21). The high-velocity material extends to only ~660 km in eastern Asia, but south of the Philippines in the Indonesian subduction system it extends to >1200 km. It is in this southern area where we hypothesize that the trench rollback was greatest in early Cenozoic time and that it is related to the southeastward extrusion of crust-lithosphere from the area of India-Eurasia intracontinental convergence.

as shown within Plate 1, would indicate that many of the displaced parts of the tectonic units do not lie within the area of Plate 1 but should be sought farther southeast. Such an investigation is under way, but the data are not currently available to resolve this problem.

## Late Cenozoic Tectonism

The elevation of the eastern part of the plateau began in middle Miocene time (Clark et al., 2005; Ouimet et al., 2010). Uplift of the plateau resulted in a plateau topography with low relief, except in local areas, that is uniform across all the tectonic units and their structures, even though these tectonic units probably had very different crustal-lithospheric structures prior to plateau formation (Fig. 16-14). The area of low relief ends abruptly at the east-facing Longmen Shan topographic front, one of the steepest topographic slopes anywhere at the margins of the plateau. The topographic front ends to the south, where it merges into the southeastern part of the plateau that has a very gentle slope to the southeast, extending into Indochina (Fig. 16-14). These relations were explained by Royden et al. (1997) as the result of eastward flow of ductile lower crust driven by gravitational potential energy from an elevated central Tibet, where the crust was thickened by earlier Cenozoic shortening (England and Molnar, 1977). The eastward flow from central Tibet thickened the crust to the east by injection of the ductile crust from below, resulting in the elevation of the upper crust and the topographic surface without significant shortening. The flow was impeded by a colder and stronger lithosphere below the Sichuan basin, causing a sharp boundary to the elevation of the crust at the Longmen Shan. Tomographic data from the eastern part of the plateau shows a distinct contrast in the crust and lithosphere across the Longmen Shan, with a cold (high-velocity) lithosphere that extends to a depth >200 km below the Sichuan basin, and a thicker and warmer lithosphere to the west beneath the eastern part of the plateau (Fig. 16-15). Support for this interpretation is provided in a number of recent geophysical studies; for example, see the Pn velocity study by Xu and Song (2010). Eastward flow of lower crust to the south was not impeded, and it spread broadly eastward, forming a long, gentle slope. Flow in the lower crust evened out the differences in the upper crustal–lithospheric structure such that these differences were not expressed in the topography, with local exceptions. Support for this hypothesis comes from the geophysical studies of Yao et al. (2010), which show that the lower crust in the region is characterized by a layer of low velocity, interpreted to be the lower ductile flowing crust. The presence of a lower crustal ductile layer would decouple deformation within the upper crust from that in the mantle.

An objection this hypothesis is that lower crustal flow cannot spread very far eastward, because the crust is not weak enough to permit the flow to penetrate it. However, prior to the beginning of eastward flow, a broad belt of early Cenozoic alkalic magmatism extended from the central plateau into Indochina in the region of a gentle eastward slope (Fig. 16-16). Geochemical studies by Holbig and Grove (2008) on such alkalic rocks indicates that they were the product of melting at high temperatures (1400 °C) at relatively shallow depths in the mantle (~80–100 km). The intrusion of these magmatic rocks into the crust may have prefigured the rheology of the crust-lithosphere to permit the eastward flow of lower crust into the southeastern plateau area. The belt of alkalic rocks also lies in the area where extrusion took place in early Cenozoic time, and the magmatism and extrusion may be related. Extrusion may have begun in Paleogene time, and during rapid southeastward movement the lithosphere was stretched in the direction of motion, thinning it and elevating the temperatures in the upper mantle at shallow depth along the direction of movement so that magmatism followed. Both the extrusion and magmatism ended ca. 20 Ma, although it is poorly dated, particularly in southeastern China. Following these events the late Cenozoic pattern of tectonism was superposed across the early Cenozoic features.

## Crustal and Lithospheric Evolution: Broader Implications of the Early Cenozoic Extrusion Hypothesis

Extensive new geophysical studies in the area of eastern Tibet and adjacent areas have provided new information on the nature of its crust and lithosphere. These studies yielded mainly information on the late Cenozoic tectonics; however, it remains to be determined how much information they contain relating to early Cenozoic tectonism. Most studies of the tectonics of the southeastern part of the Tibetan Plateau and adjacent areas to the southeast focus on the India-Eurasia zone of convergence as the driving force for the tectonism. However, in our interpretation this is only half of the story; the subduction systems of the Western Pacific and Indonesia play an equal if not greater role. These two dynamic systems together are the main drivers for the tectonics (Fig. 16-9).

The large magnitude of early Cenozoic extrusion of continental fragments as shown within Plate 1 requires that large areas to the southeast must also be involved. Our working hypothesis thus extends into Indochina, and in our interpretation to the subduction systems of the Western Pacific and Indonesia (Fig. 16-9). This broad region during Paleogene time was dominated by extensional basins that follow the coastal areas of China in the north, but they extended farther east in the area of the Sunda Shelf, a region underlain by continental crust (Fig. 16-17), south of a boundary that trends south-southeast along the eastern edge of the Sunda Shelf. In the northern area, subsidence began in Late Cretaceous time within northeast-trending basins within and along the margin of southeastern China, and continues to the present. In the southern area, basin subsidence began in Eocene time, approximately the time when extrusion began. Basins in both areas began with mainly continental sedimentation, but sedimentation became more dominantly marine in the northern area in early Cenozoic time, but it became progressively more marine in Miocene time in the southern area (Figs. 16-18 and 16-19), the result of thermal subsidence. The basins in both regions are dominated by extensional tectonism, a condition that does not support the earlier hypothesis of Tapponnier et al. (1982) where extrusion in the southern area would be by forcefully extruding crust from SE Tibet into a region undergoing extension. Rather, we support the hypothesis that basin development in both areas was related to a pull from the subduction systems of the Western Pacific and Indonesia by trench rollback. In such a tectonic setting, crustal fragments were pushed from a region of compression north of the India indenter, but were pulled within the southern area by trench rollback, forming a source-sink relationship. Depending upon the nature of the coupling between the lithosphere and deeper

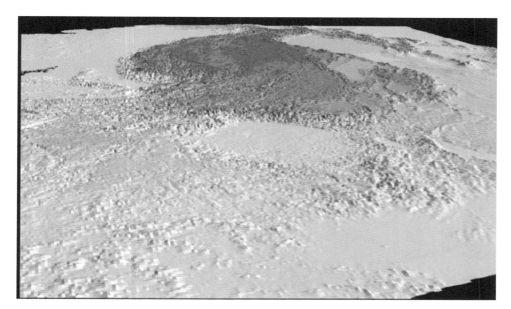

Figure 16-14. View westward across the Tibetan Plateau and Sichuan basin, showing the steep topographic front of the Longmen Shan west of the Sichuan basin, and the gentle topographic slope of the plateau to the southeast into Indochina. The gentle relief in the eastern plateau and the gentle slope into Indochina are not marked by evidence for crustal shortening during their elevation. From Burchfiel et al. (2008).

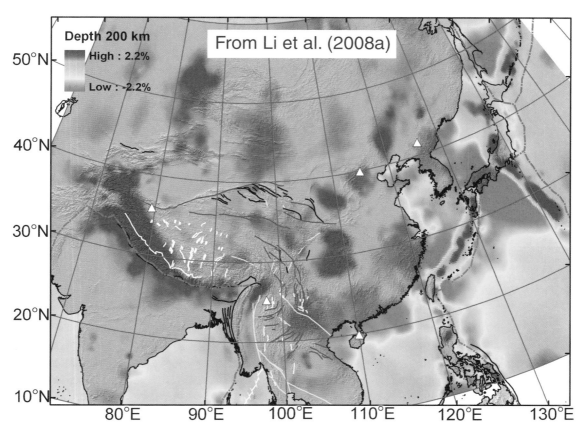

Figure 16-15. Relative P-wave velocities in eastern Asia, from Li et al. (2008a). Relatively high-velocity material is shown in blue, and low-velocity material in red, at 200 km depth. The basis for the two dynamic systems that controlled Cenozoic tectonism is expressed by low-velocity mantle (hot) beneath Tibet, related to the India-Eurasia convergent zone, and all along the eastern part of Asia, related to the subduction systems in the Western Pacific and Indonesia. Note the apparent separation of the two regions northeast of the eastern Himalayan syntaxis by a belt of somewhat higher velocity that lies within the area of Plate 1. Also shown is the higher velocity mantle below the Sichuan basin around which the lower crustal flow from Tibet has taken place.

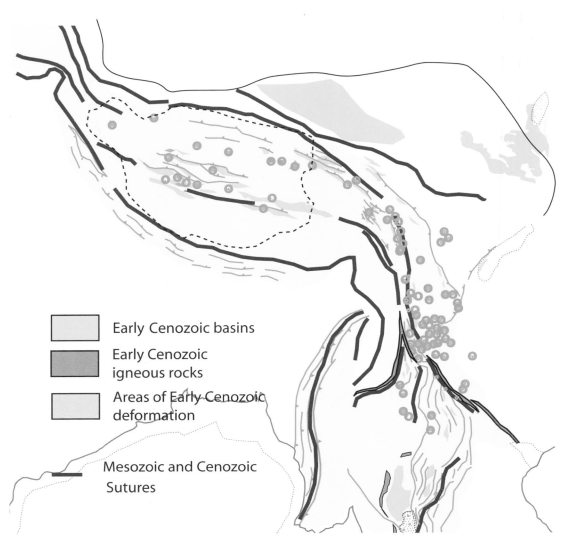

Figure 16-16. Early to early late Cenozoic alkalic volcanic rocks form a broad belt from central Tibet southeastward into Indochina (orange dots show outcrops of these rocks). This magmatism indicates melting conditions that are hot and shallow, implying a thinning of the lithosphere below this area. This figure also encompasses the area where extrusion occurred in early Cenozoic time, which may have also prefigured a ductile lower crust so that lower crustal flow could have penetrated the area southeast of the thick Tibetan crust. The dashed black line shows the area of internal drainage within the plateau. The broad area of early Cenozoic deformation and the areas of sedimentary basins are shaded. Some of the shear zones active during early Cenozoic extrusion are shown where they follow major sutures.

mantle, there is no particular relation between how far southeast the Sunda Shelf has moved nor the magnitude of basin extension. In the northern area there is no major eastward extrusion of crustal material from China, and the extension is due to trench rollback alone.

Recent tomographic studies by Li et al. (2006 and 2008; Figs. 16-20 and 16-21) show high mantle velocity zones (both Vs and Vp) in the northern area that can be followed from present subduction zones westward down to about the 660-km discontinuity, where they are overlain by low mantle velocities well beneath eastern China (Fig. 16-20). In contrast, in the southern area the high velocity zones extend to depths of >1200 km (Figs. 16-13 and 16-21), with low velocity regions above. The high velocity zones are interpreted by Li et al. (2006 and 2008) to be the subducted mantle and that the geometry of the subducted slabs supports the interpretation of rollback of the subducted slabs. The geometry of the slabs at depth is generalized from the work of Li et al. (2008) in Figure 16-13, which shows the contrast between the subducted lithosphere in the northern and southern areas; the dip is west down to 660 km, and then subhorizontal in the north, but dipping beneath the Sunda Shelf, forming a dish-shaped geometry to the west beneath the Sunda Shelf down to

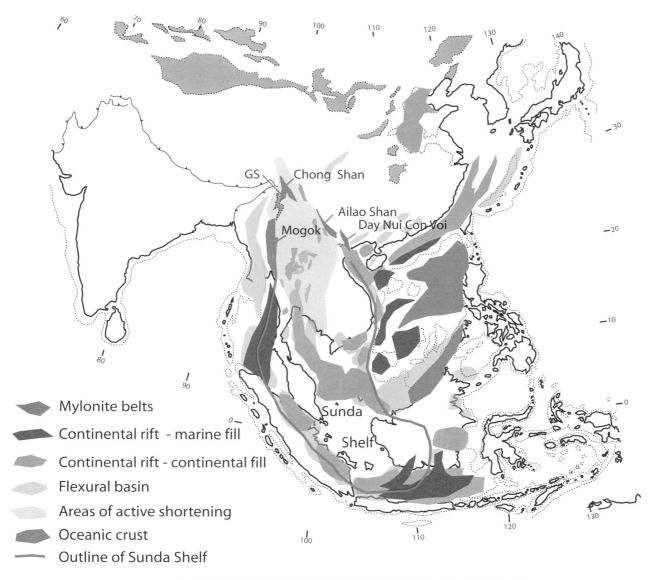

Figure 16-17. Paleogene basins of eastern China, Indochina, and the Sunda Shelf region of SE Asia. Basin development has been mainly extensional and began in Southeastern China in late Cretaceous time, continuing to Recent. Within the Sunda Shelf and the Indochina basin, development began in Eocene time, with most of the basins having been filled with continental strata. The gray area is where crust-lithosphere has been extruded to the southeast and was the site of, first, shortening, followed by extensional tectonism, as rocks moved from the India-Eurasia convergent area into the region dominated by trench rollback within the Indonesian subduction system. Major mylonitic shear zones are shown in red. GS—Gaoligong shear zone. After Hall and Morley (2004) and Sibuet et al. (2004).

1200 km in the south. The extensional basins in both areas lie above the subducted slabs. The boundary between the two different regions of subducted lithosphere trends NW-SE along the projected trend of the extruded crustal fragments in SE China, and also along the northern boundary of the Sunda Shelf; this may partly explain why the Sunda Shelf lies so far to the SE. Although the extensional basins in the northern area show a consistent trend approximately parallel to the China coast, the basins in the southern area have more variable trends, indicating that slab rollback and extension within the crust were more complex.

The regions of low velocity in the mantle above the subducted slabs extend beneath eastern China (Fig. 16-20) and indicate that the influence of the subduction systems of the Western Pacific and Indonesia extend well into eastern China, defining the general, almost certainly broad and shifting boundary between the India-Eurasia collision system and the Western

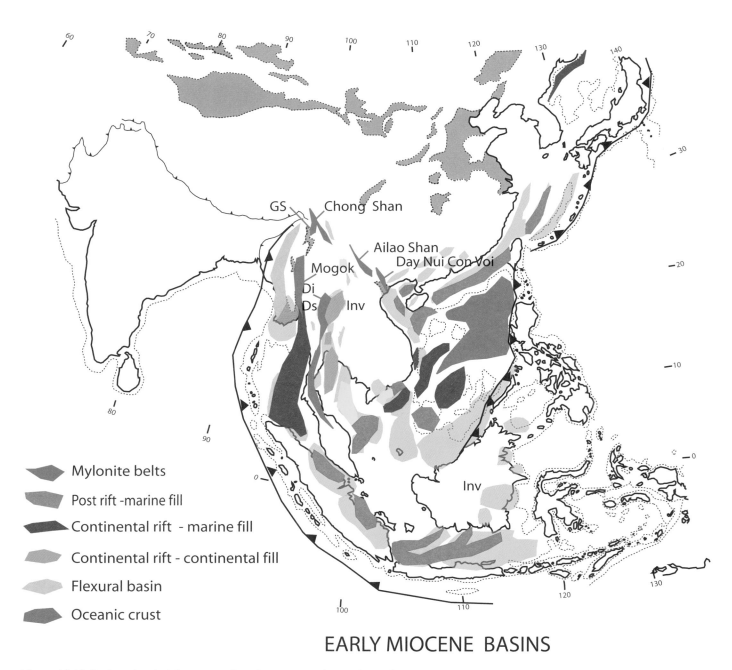

Figure 16-18. Basins of early Miocene age have begun to receive marine sediments, whereas others continue to receive continental sediments. This is interpreted to indicate thermal subsidence of thinned lithosphere. GS—Gaoligong shear zone; Di—Doi Inthanon complex; Ds—Uttaradit suture zone; Inv—inversion. After Hall and Morley (2004) and Sibuet et al. (2004).

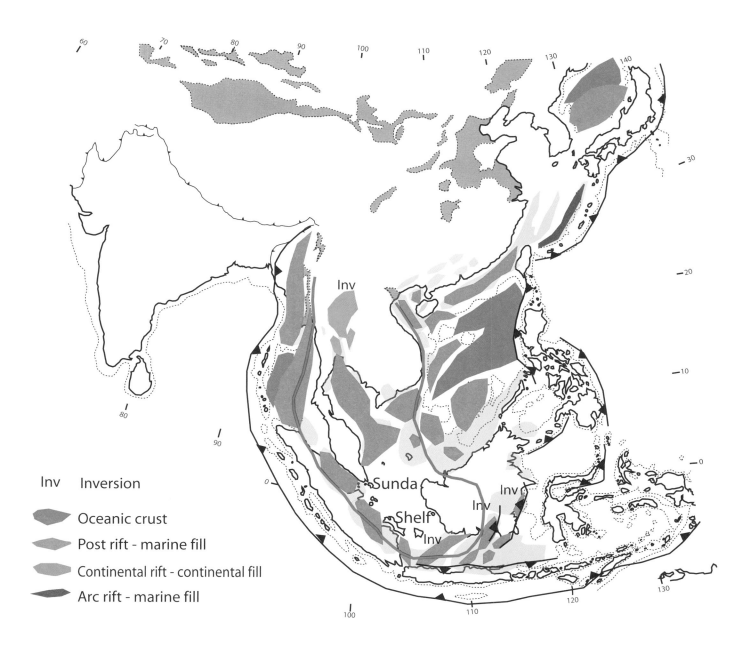

**LATE MIOCENE BASINS**

Figure 16-19. By Late Miocene time, most of the basins on the Sunda Shelf were receiving marine sediments interpreted to indicate continued thermal subsidence of thinned lithosphere. This time postdated the main extrusion of fragments from east of southeastern Tibet. After Hall and Morley (2004) and Sibuet et al. (2004).

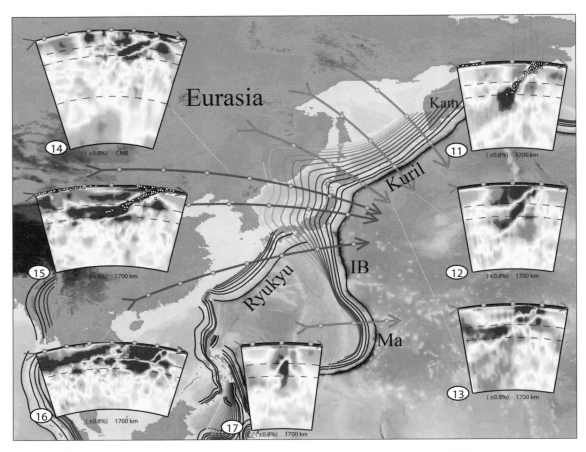

Figure 16-20. P-wave velocities in vertical section across southeastern Asia, from Li et al. (2006). High-velocity material is shown in blue and is the basis for contours in three dimensions in Figure 16-13. IB—Izu-Bonhin subduction zone; CMB—core-mantle boundary.

Pacific–Indonesian systems (Figs. 16-9, 16-15, and 16-21). The boundary between these two systems passes through the area of Plate 1 (Fig. 16-9). It is important to note that this is the same place where Argand (1922) drew a boundary between two regions of SE Asia that underwent two different patterns of deformation in responding to two different areas of tectonism.

**Mass Balance of Intracontinental Deformation**

Within the India-Eurasian collisional system, seafloor spreading data can track the path of India with respect to Eurasia, showing that, accepting the time of collision at ca. 50 Ma at the eastern Himalayan syntaxis, India has moved northward ~3600 km relative to Eurasia (Molnar and Stock, 2009). Assuming significant shortening within northern India, now in the Himalaya, of ~1000 km, the remaining 2600 km must have been absorbed within Eurasia. In taking a doubling of the Tibetan crust related to thickening, this still leaves >1000 km of convergence unaccounted for. However at present, doubling of the Tibetan crust by shortening cannot be supported, as most studies of the magnitude of shortening, albeit in local areas, is determined to be <100% shortening, so the amount of mass unaccounted for is probably >1000 km of north-south motion. The extrusion of crustal-lithospheric fragments from southeastern Tibet can account for most of the mass, as first proposed by Tapponnier et al. (1982), although our hypothesis, as outlined above, differs significantly from their original hypothesis by the addition of major shortening and rotation of the extruded fragments.

The dynamics of extrusion require looking also at the effects of the Western Pacific–Indonesian subduction systems. Tapponnier et al. (1982, and later papers) interpreted the India-Eurasian convergent system as the driving force for all the Cenozoic tectonics of Southeastern Asia. This includes the events such as the opening of the South China Sea and the southeastward movement of Sundaland. In our interpretation the Cenozoic tectonic events partly within and east of the southeastern part of the Tibetan Plateau are the product of retreat within the Western Pacific–Indonesian subduction system (Royden et al., 2008). The entire region of coastal and offshore Southeastern Asia and Indonesia is a region of Cenozoic extension as the result of trench rollback in the subduction zones to the east and south. This area is one of extensive development of extensional

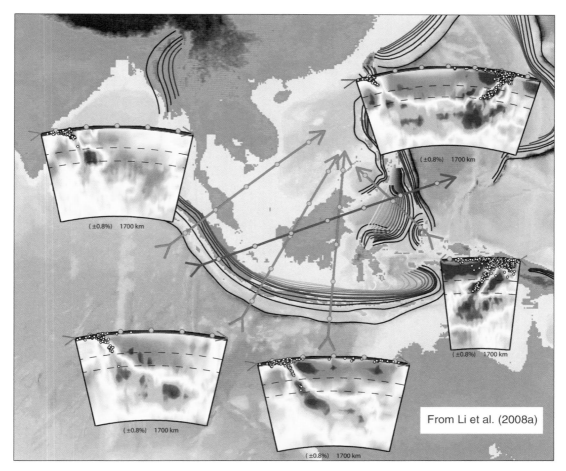

Figure 16-21. P-wave velocities in vertical section across southeastern Asia, from Li et al. (2006). High-velocity material is shown in blue and is the basis for contours in three dimensions in Figure 16-13.

sedimentary basins, beginning in Late Cretaceous time in southeastern China and by Eocene time in Indochina and on the Sunda Shelf.

Thus the Cenozoic extrusion of crustal fragments from southeastern Tibet is the result of the convergence in the Indian-Eurasia system, but east of the syntaxis the southeastward motion is increasingly controlled by extension within the rollback at the subduction zones to the east and south. The boundary between these two systems is gradational and probably shifts temporally, but it must lie within the area of Plate 1. In our hypothesis it lies along a broad zone from the Longmen Shan southward into the Tengchong Unit. As pointed out above, what is remarkable is that this is the same boundary that was identified by Emil Argand in his extraordinary synthesis *Tectonique de l'Asie,* published in 1922, which he regarded as a major tectonic boundary between the same two systems. In the north, GPS studies show that the eastern part of the Tibetan Plateau is moving eastward slowly, with only a small magnitude of convergence within the Longmen Shan (but enough over a long period of time to have caused the disastrous 2008 earthquake in eastern China). Tomographic data show a slight increase in seismic velocity in the upper mantle, which connects the high velocity regions in the Sichuan basin to high velocities at the syntaxis (Li et al., 2008a). This boundary also separates two areas of lower upper mantle velocities, one beneath the Tibetan Plateau, and the other that extends along the entire eastern margin of southeastern Asia (Fig. 16-9), the expression of the two dynamic systems that controlled the Cenozoic tectonics in SE Asia.

If the source-sink relations hypothesized above are correct, the following remain to be determined: (1) whether extrusion solves the mass balance problem north of the India indenter; (2) whether extrusion from southeastern Tibet can be confirmed by paleogeographic reconstructions; (3) the dynamics between India-Eurasia, the Western Pacific, and Indonesian subduction systems; (4) whether the magnitude of SE extrusion from Tibet is balanced by the southeastward motion of the Sunda Shelf, and the magnitude of extension in the crustal material above the Indonesian subducted lithosphere; and (5) the nature, depth, and volume of southeastward flow of mantle from Tibet. These are questions currently being investigated.

# *Epilogue*

Our study has attempted to cover the evolution of a highly complex region in southeastern China from the late Paleozoic and Mesozoic accretion, followed by the late Mesozoic to Cenozoic intracontinental deformation related to the India-Eurasia convergence and related tectonism to the east, associated with the Western Pacific and Indonesian subduction systems. This can be considered only a work in progress. It is an exciting and wonderful area in which to do geology. The discussions and interpretations presented above not only try to describe the geology as best we understand it, but also to point out numerous important unknown relations, differences in interpretations, and ambiguities. The maps and cross sections that accompany this Memoir also present a new type of format for regional geology that we hope will be useful for workers, not only to understand the geological discussion but also for use in future updates and compilations. We hope that this report will stimulate directions for future research.

# APPENDIX 1

## Notes on map file formats

The PDF files (Plate 1 and Plate 2) can be opened with any PDF viewer (for example, Adobe Acrobat, Adobe Reader, Preview) and also in Adobe Illustrator, the software used to compile the digital maps. The files were saved from Adobe Illustrator CS5 as PDF compatible with Acrobat 8 (PDF v. 1.7), with Illustrator files embedded, and they can thus be opened, viewed, printed, and modified by any software capable of opening files in either format. Additionally, Plates 1 and 2 have been combined and exported to ESRI Shapefile format for use with GIS software.

## MAP SPECIFICATIONS

The maps are Lambert Conformal Conic projections with standard parallels at 25°N and 29°N (WGS84 datum). Plate 1 has a central meridian of 99.5°E, and Plate 2 is centered at 104.5°E. Graticule ticks are arranged along the border of the maps at 1° intervals. Both plates share the same scale and projection, but Plate 2 has been rotated to fit the page layout. The plate can still be positioned into the NE corner of Plate 1, although the labels on the map will appear at an angle. The two map plates have scale bars but no numeric scale, as scale will vary depending on how the map is displayed. If printed at the original resolution of the Adobe Illustrator file, the map has a scale of 1:453,250.

Since both Plates 1 and 2 were compiled from manuscript maps, there is some positional error. Registering known points to topographic and hydrologic features, and to lines of latitude and longitude, provided some measure of ground-truth, but the overall goal of the project was the compilation of the map data rather than attaining strict positional accuracy standards. Users of the GIS data will undoubtedly discover areas where substantial positional errors remain. No attempt to further georectify the GIS data was made, such that the GIS files are essentially identical to the original Adobe Illustrator maps in terms of the linework. Because the GIS file format is limited to straight line segments, the number of points on each line was increased prior to export in order to reproduce the smooth Bézier curves of the linework in the Adobe Illustrator maps.

The maps were designed for printing, and the colors in the PDF files are CMYK, with primarily black linework and labels set to overprint.

Both Plate 1 and Plate 2 are presented on pages that are 82 in. wide to allow for printing at 100% in two adjoining N-S strips on a 42 in. plotter. Plate 1 is 82 in. wide × 122 in. tall. Plate 2 is 82 in. wide × 70 in. tall.

## VIEWING AND PRINTING THE MAPS IN ADOBE ILLUSTRATOR CS5

Open the PDF file for Plate 1 or Plate 2 from the file menu in Illustrator, or drop the file on the application icon. In Illustrator, you have full access to the vector linework and to all of the sublayers, whereas Adobe Reader and other PDF viewers only allow you to turn top-level layers on and off. For example, the Folds layer contains sublayers (synclines, overturned synclines, anticlines, etc.) each of which is organized by the type of artwork (lines, approximate lines, hidden lines, and symbols). While, in many cases it is possible to search for these features by their line styles, the nested layers provide an easy, intuitive way to customize the map view, and hence to focus on particular aspects of the geology. The full listing of layers for Plate 1 (the file for Plate 2 follows the same organizational structure) is provided in Appendix 2: List of layers on Plates 1 and 2.

Colors for the geologic units are unique, and each of the unit colors appears in the Swatches palette, as do the lithologic patterns that overlay some regions of the map.

## VIEWING AND PRINTING THE MAPS WITH PDF SOFTWARE

The PDF files can be viewed in a variety of applications that can open, display and print PDF files, including Adobe Acrobat, Adobe Reader, and Preview. The files were saved with layers, so that the top-most layers can be turned on and off to customize the map display. In Adobe Acrobat (v. 9), clicking on the Layers icon reveals a list of the layers that can be turned on and off. For example, the color layers that fill the geologic units (lowermost layers in the stacking order) can be turned off to create a black-and-white linework version of the map.

## WORKING WITH THE MAPS IN GIS SOFTWARE

The shapefiles provided along with the PDF files combine Plates 1 and 2 in a seamless dataset. The projection is the same Lambert Conformal Conic as the Plate 1 PDF file, with standard parallels at 25°N and 29°N, centered at 0°N 99.5°E (WGS84 datum). ESRI .prj files accompany each of the shapefiles. The GIS data is provided for users performing additional analysis of the data, and does not include some artwork that appears on the plates, such as the inset relief map on Plate 1, and the map titles. The database files for each shapefile contain styling information (fill color, line style, as well as the layer name of the corresponding feature in the Adobe Illustrator file). In this way the map can be styled to match the original published map, but can also be modified so it is useful for other analyses and further study.

**PLATE 3. UNCONFORMITY MAP**

Plate 3 is a PDF file that can be opened in Illustrator. The layers show the different unconformities and are highlighted by a subdued map of Plate 1 that shows only the linework. Layers are color-differentiated unconformities that can be turned on or off individually.

**PLATE 4. GEOLOGICAL CROSS SECTIONS**

Plate 4 is a PDF file that can be opened in Illustrator. It has only two layers. One layer contains colored cross sections reproduced from the Chinese 1:200,000-scale map sheets. The second layer shows the interpretation at depth. Unfortunately there is no seismic-reflection data to support the interpretations. The two layers can be viewed separately by turning a layer on or off individually.

## APPENDIX 2

## *List of layers on Plates 1 and 2*

Top-level layers, in bold, can be manipulated when viewing the PDF files in Adobe Acrobat and other software that is layer-aware. The full hierarchy of layers can be manipulated in Adobe Illustrator. Plate 2 contains only a subset of the unit labels and colors of Plate 1; layers that are specific to Plate 2 are indicated in this listing.

**Map title**

**Map layout**
 Map neatline
 Country boundaries
 Graticule
 Scale bar

**Inset map**

**Legend**
 Unit legends
 Stratigraphy legend
 Master legend

**Cross-sections**

**Feature labels and symbols**
 Basin
 Faults
 Mountain
 Physical

Continuation of **Feature labels and symbols**
 Pluton
 Populated place
 Units

**Map labels**
 Unit labels
  E
  J
  J2
  J3
  Jm
  K1
  Mw (Plate 2)
  Mx (Plate 2)
  Mz
  Mz1
  Mz1a
  Mz1b
  Mz1s
  Mz2
  Mz2a

*(Continued)*

*Continuation of* **Map labels,** Unit labels

Mz2a II

Mz2a?

Mz2b

Mz2b II

Mz2b1

Mz2b2

Mz2bx

Mz2β

Mz3

Mz3/E

Mz3?

Mz3c

Mzf

N

N1

N1-2

N2

NBd

NBp

NBz

Ni

Op (white)

Op

Pz

Pz3ma

Pz3mb

Pzl

Pzl2

Pzlm

*Continuation of* **Map labels,** Unit labels

Pzm

Pzm3

Pzu

Pzu1

Pzu2

Pzub

Pzum

Pzβ

Pβ

Q

Qal

Qb

Qgl

Ql

Qm

Qsb (Plate 2)

Qβ

Sa

T

T1

T1-2

T2

T3

T31

T32

T3b

T3m

T3sc

Ta1

*(Continued)*

*Continuation of* **Map labels,** *Unit labels*

Ta2
Tt
WNj
Wc
Wj
Wlb
Wm
Wo
Wp
Ws
Wt
Z1 (Plate 2)
Z2
Za
amp
m
mLB
mch
me
mg
ms
pt
s
scp
$v_5^1$
w
$\Sigma$
$\alpha_6$
$\beta$

*Continuation of* **Map labels,** *Unit labels*

$\beta_4$
$\beta_5$
$\beta_5^1$
$\beta_6$
$\beta\mu_5^1$
$\beta\upsilon$
$\beta\upsilon_5^1$
$\gamma$
$\gamma_2$
$\gamma_{3\text{-}4}$
$\gamma_4$
$\gamma_4^1$
$\gamma_4^2$
$\gamma_4^3$
$\gamma_5$
$\gamma_5^1$
$\gamma_5^{1\text{-}2}$ (Plate 2)
$\gamma_5^2$
$\gamma_5^{2\text{-}3}$
$\gamma_5^3$
$\gamma_6$
$\gamma_6^1$
$\gamma m$
$\gamma\beta_5^1$
$\gamma\beta_5^2$
$\gamma\beta_5^3$
$\gamma\delta_4$
$\gamma\delta_5^1$
$\gamma\delta_5^2$

*(Continued)*

*Continuation of* **Map labels,** *Unit labels*

$\gamma\delta_5^{2-3}$

$\gamma\delta_5^3$

$\gamma\delta_6$

$\gamma\delta\pi_5^1$

$\gamma\delta\pi_6$

$\gamma\epsilon$ (Plate 2)

$\gamma\epsilon_6$

$\gamma\eta$ (Plate 2)

$\gamma\eta_{51}$ (Plate 2)

$\gamma\eta_5^{1-2}$ (Plate 2)

$\gamma\eta_5^2$ (Plate 2)

$\gamma\eta 6$

$\gamma o_4^2$

$\gamma o_5^1$ (Plate 2)

$\gamma o_5^2$

$\gamma\pi_5^1$

$\gamma\pi_6$

$\gamma\pi_6^1$

$\gamma\sigma\pi_6$

$\gamma\omega_5$

$\delta$

$\delta_4$

$\delta_4^1$

$\delta_4^3$

$\delta_5$

$\delta_5^1$

$\delta_5^2$

$\delta_5^3$

$\delta_6$

*Continuation of* **Map labels,** *Unit labels*

$\delta o_5^1$

$\delta\beta_5^1$

$\delta\beta_5^2$

$\delta\beta o_5^1$

$\delta\gamma_5^1$

$\delta\epsilon o_5^3$

$\delta\epsilon\sigma_5^3$

$\delta\mu_5^1$

$\delta\mu_5^3$

$\delta\mu_6$

$\delta\eta_5^1$ (Plate 2)

$\delta\eta o_5^1$ (Plate 2)

$\delta o$ (Plate 2)

$\delta o_5^1$

$\delta o_5^2$

$\delta o_5^3$

$\delta o\mu_5^1$

$\delta o\mu_5^2$

$\delta o\mu_6$

$\delta\pi_5^1$

$\delta\pi_6$

$\delta\rho_5^1$

$\delta\rho o_5^1$

$\delta\sigma_5^1$

$\epsilon$ (Plate 2)

$\epsilon_4^3$

$\epsilon_5^1$

$\epsilon_6$

$\epsilon_6^1$

*(Continued)*

*Layers on Plates 1 and 2*

Continuation of **Map labels,** Unit labels

$\varepsilon o_{53}$
$\varepsilon \gamma_5^1$
$\varepsilon \gamma_5^3$
$\varepsilon \gamma_6$
$\varepsilon \delta \pi_5^3$
$\varepsilon \delta \pi_6$
$\varepsilon \pi_5^1$
$\varepsilon \pi_5^3$
$\varepsilon \pi_6$
$\varepsilon \sigma \pi_6$
$\eta$
$\eta_5^1$
$\eta_5^2$
$\eta_5^3$
$\eta_6$
$\eta \gamma_4$
$\eta \gamma_5^1$
$\eta \gamma_5^2$
$\eta \gamma_5^{2\text{-}3}$
$\eta \gamma_5^3$
$\eta \gamma \pi_5^1$
$\eta \gamma \pi_5^3$
$\eta \delta_4$
$\eta o_5^1$ (Plate 2)
$\eta o_6$
$\eta o \pi_5^3$
$\eta o \pi_6$
$\eta \pi_6$
$\kappa_6$

Continuation of **Map labels,** Unit labels

$\kappa \gamma_5^2$
$\kappa \tau_6$
$\lambda \gamma_6$
$\lambda \pi$
$\lambda \pi_6$
$\nu$
$\nu_4$
$\nu_4^3$
$\nu_5$
$\nu_5^1$
$\nu_6^1$
$\nu \pi_6^1$
$\xi_5^1$ (Plate 2)
$\xi o_5^1$ (Plate 2)
$\xi o_5^{1\text{-}2}$ (Plate 2)
$\pi$
$\pi_6$
$\pi \eta \gamma_5^1$
$\rho \mu_5^1$
$\tau_6$
$\chi_6$
$\chi \gamma_5^3$
$\chi \nu_6$
$\omega_4^3$
$\omega_5^1$

Isograd labels

Leader lines

*(Continued)*

**Active volcanoes**

**Blocks**

    Blocks

    Blocks, exotic

    Blocks in sediments

**Metamorphic isograds**

**Foliations**

    .2 pt

    .4 pt

**Folds**

    Synclines

    Synclines, overturned

    Anticlines

    Anticlines, overturned

    Folds, approximate

    Folds, hidden

**Faults**

    Faults

    Dip

    Thrust

    Normal

    Strike-slip

    Detachment syntaxial unit

**Unconformities**

    Symbol

    Neogene   N

    Paleogene   E

    Upper Cretaceous   K2

    Upper Jurassic   J3

    Middle Jurassic   J2

    Lower Jurassic   J1

    Upper Triassic   T3

    Middle Triassic   T2

    Lower Triassic   T1

    Paleozoic   Intra Pz

        Basal Uppermost Proterozoic   Z2, Zb

    Intra Lower Proterozoic   Za

**Contacts**

    Contacts

    Sedimentary

    Intrusive

    Igneous

    Igneous (approximate)

    Igneous basalt

    Basalt

    Serpentine

    Ultramafic

    Approximate

    Glacial

    Lake

    Non-geologic contacts

*(Continued)*

Layers on Plates 1 and 2 — 217

**Lithologic patterns**

    Stroke

    Fill

**Unit boundaries**

**Unit patterns**

    Pzl2 Baoshan

    Pzl2 Changning

    Mz Tengchong

    Mz2a? Lhasa

    Pzu Sedimentary Tengchong

    T/Tt Transitional

    J2

    Jm

    Pzu1

    Pzu2

    Lin Cang Pzm

    Wuliang Shan

    Jinsha suture

    Syntaxis core

    Syntaxis carapace

**Color**

    E

    J

    J2

    J3

    Jm

    K1

*Continuation of* **Color**

    Mw (Plate 2)

    Mx (Plate 2)

    Mz

    Mz1

    Mz1a

    Mz1b

    Mz1s

    Mz2

    Mz2a

    Mz2a - Lhasa

    Mz2a II

    Mz2a QII

    Mz2b

    Mz2b II

    Mz2b1

    Mz2b2

    Mz2β

    Mz3

    Mz3c

    Mzf

    N

    N1

    N1-2

    N2

    NBd

    NBp

    NBz

    Ni

    Op

*(Continued)*

*Continuation of* **Color**

- Pz
- Pz s
- Pz scp
- Pz3ma
- Pz3mb
- Pzl
- Pzl2
- Pzlm
- Pzm
- Pzm3
- Pzu
- Pzu1
- Pzu2
- Pzub
- Pzum
- Pzuβ
- Pzβ
- Pβ
- Q
- Qal
- Qb
- Qgl
- Ql
- Qm
- Qβ
- Sa
- Serpentine
- T
- T1

*Continuation of* **Color**

- T1 Changning Menglian
- T1-2
- T2
- T3
- T31
- T32
- T3b
- T3m
- T3sc
- Ta1
- Ta2
- Tt
- Ultramafic
- Ultramafic Ailao Shan
- WNj
- Wc
- Wj
- Wlb
- Wm
- Wo
- Wp
- Ws
- Wt
- Z2
- Za
- m
- m - Ailao Shan
- m - Lin Cang
- mLB

*(Continued)*

*Continuation of* **Color**

    me

    ms, mg

    pt

    w

**Color, igneous**

    Mafic rocks

        Σ   serpentinite

        υ   basalt

        ω51   gabbro

    β

    $β_4$

    $β_5$

    $β_5^1$

    $β_6$

    βυ

    $βυ_5^1$

    Felsic

        γ

        $γ_2$

        $γ_{3-4}$

        $γ_4$

        $γ_4^1$

        $γ_4^2$

        $γ_4^3$

        $γ_5$

        $γ_5^1$

        $γ_5^{1-2}$ (Plate 2)

        $γ_5^2$

*Continuation of* **Color, igneous**

    $γ_5^{2-3}$

    $γ_5^3$

    $γ_6$

    $γ_6^1$

    $α_6$

Metamorphosed and mylonitized igneous rocks

    γm

Intermediate

    $γβ_5 1$

    $γβ_5^2$

    $γβ_5^3$

    $γδ_4$

    $γδ_5^1$

    $γδ_5^2$

    $γδ_5^{2-3}$

    $γδ_5^3$

    $γδ_6$

    η

    $η_5$

    $η_5^1$

    $η_5^2$

    $η_5^3$

    $η_6$

    $ηγ_4$

    $ηγ_5^1$

    $ηγ_5^2$

    $ηγ_5^{2-3}$

    $ηγ_5^3$

    $ηγπ_5^1$

*(Continued)*

*Continuation of* **Color, igneous**

$\eta\gamma\pi_5^3$

$\eta\delta_4$

$\gamma\delta_5^1$ (Plate 2)

$\gamma\delta_5^2$ (Plate 2)

$\gamma\delta\pi_5^1$

$\gamma\delta\pi_6$

$\gamma\epsilon$ (Plate 2)

$\gamma\epsilon_6$

$\gamma\eta$ (Plate 2)

$\gamma\eta_5^1$ (Plate 2)

$\gamma\eta_5^{1-2}$ (Plate 2)

$\gamma\eta_5^2$ (Plate 2)

$\gamma\eta_6$

$\gamma\omega_5$

Alkalic

$\gamma o_4^2$

$\gamma o_5^1$ (Plate 2)

$\gamma o_5^2$

$\gamma\pi_5^1$

$\gamma\pi_6$

$\gamma\pi_6^1$

$\gamma\sigma\pi_6$

$\epsilon$ (Plate 2)

$\epsilon_4^3$

$\epsilon_5^1$

$\epsilon_6$

$\epsilon_6^1$

$\epsilon o_5^3$

$\epsilon\gamma_5^1$

*Continuation of* **Color, igneous**

$\epsilon\gamma_5^3$

$\epsilon\gamma_6$

$\epsilon\delta\pi_5^3$

$\epsilon\delta\pi_6$

$\epsilon\pi_5^1$

$\epsilon\pi_5^3$

$\epsilon\pi_6$

$\epsilon\sigma\pi_6$

$\eta o_5^1$ (Plate 2)

$\eta o_6$

$\eta o\pi_5^3$

$\eta o\pi_6$

$\eta\pi_6$

$\kappa_6$

$\kappa\gamma_5^2$

$\kappa\tau_6$

$\lambda\gamma_6$

$\lambda\pi$

$\lambda\pi_6$

$\xi_5^1$ (Plate 2)

$\xi o_5^1$ (Plate 2)

$\xi o_5^{1-2}$ (Plate 2)

$\pi$

$\pi_6$

$\pi\eta\gamma_5^1$

$\rho\mu_5^1$

$\tau_6$

$\chi_6$

$\chi\nu_6$

*(Continued)*

*Continuation of* **Color, igneous**
    Diabase

        $\delta$

        $\delta_4$

        $\delta_4^1$

        $\delta_4^3$

        $\delta_5$

        $\delta_5^1$

        $\delta_5^2$

        $\delta_5^3$

        $\delta_6$

## Color, volcanic

    Basalt

    $\nu_4$

    $\nu_4^3$

    $\nu_5^1$

    $\nu_6^1$

    $\delta\beta_5^1$

    $\delta\beta_5^2$

    $\delta\beta o_5^1$

*Continuation of* **Color, volcanic**

    $\delta\gamma_5^1$

    $\delta\epsilon o_5^3$

    $\delta\epsilon\sigma_5^3$

    $\delta\eta_5^1$ (Plate 2)

    $\delta\eta o_5^1$ (Plate 2)

    $\delta\mu_5^1$

    $\delta\mu_5^3$

    $\delta\mu_6$

    $\delta o$ (Plate 2)

    $\delta o_5^1$

    $\delta o_5^2$

    $\delta o_5^3$

    $\delta o\mu_5^1$

    $\delta o\mu_5^2$

    $\delta o\mu_6$

    $\delta\pi_5^1$

    $\delta\pi_6$

    $\delta\rho_5^1$

    $\delta\rho o_5^1$

    $\delta\sigma_5^1$

## GLOSSARY OF IGNEOUS ROCK SYMBOLS AND NAMES

| Symbol | English | Chinese |
|---|---|---|
| $\alpha$ | Andesite | 安山岩 |
| $\kappa$ | Alkaline rock | 碱性岩 |
| $\beta$ | Basalt | 玄武岩 |
| $\beta\upsilon$ | Diabase | 玄武质玻璃 |
| $\gamma$ | Granite | 花岗岩 |
| $\gamma m$ | Muscovite granite | 白云母花岗岩 |
| $\gamma\beta$ | Biotite granite | 黑云母花岗岩 |
| $\gamma\delta$ | Granodiorite | 花岗闪长岩 |
| $\gamma\delta\pi$ | Granodiorite-porphyry | 花岗闪长斑岩 |
| $\gamma\varepsilon$ | Granosyenite | 花岗正长岩 |
| $\gamma\eta$ | Granite-monzonite | 花岗二长岩 |
| $\gamma o$ | Plagioclase granite | 斜长花岗岩 |
| $\gamma\pi$ | Granite-porphyry | 花岗斑岩 |
| $\gamma\omega$ | Granite-picrite | 花岗橄榄岩 |
| $\delta$ | Diorite | 闪长岩 |
| $\delta\beta$ | Biotite diorite | 黑云母闪长岩 |
| $\delta\beta o$ | Biotite quartz-diorite | 黑云母石英闪长岩 |
| $\delta\gamma$ | Diorite-granite | 闪长花岗岩 |
| $\delta\varepsilon o$ | Quartz syenite diorite | 石英正长闪长岩 |
| $\delta\varepsilon\sigma$ | Syenite diorite peridotite | 闪长正长橄榄岩 |
| $\delta\mu$ | Diorite-porphyrite | 闪长玢岩 |
| $\delta o$ | Quartz-diorite | 石英闪长岩 |
| $\delta o\mu$ | Quartz diorite porphyrite | 石英闪长玢岩 |
| $\delta\pi$ | Diorite-porphyry | 闪长斑岩 |
| $\delta\rho$ | Diorite-pegmatite | 伟晶质闪长岩 |
| $\delta\rho o$ | Quartz diorite-pegmatite | 伟晶质石英闪长岩 |
| $\delta\sigma$ | Plagioclase peridotite | 斜长橄榄岩 |
| $\varepsilon$ | Nepheline syenite | 霞石正长岩 |
| $\varepsilon o$ | Alkalic intrusive | 石英正长岩 |
| $\varepsilon\gamma$ | Potash feldspar-granite | 钾长花岗岩 |
| $\varepsilon\delta\pi$ | Syenite diorite porphyry | 正长闪长斑岩 |
| $\varepsilon\pi$ | Syenite-porphyry | 正长斑岩 |
| $\varepsilon\sigma\pi$ | Alkalic porphyry | 正长橄榄斑岩 |
| $\eta$ | Monzonite | 二长岩 |
| $\eta\gamma$ | Monzonitic granite | 二长花岗岩 |
| $\eta\gamma\pi$ | Monzonitic granite-porphyry | 二长花岗斑岩 |
| $\eta\delta$ | Monzodiorite | 二长闪长岩 |

*(Continued)*

## GLOSSARY OF IGNEOUS ROCK SYMBOLS AND NAMES (Continued)

| Symbol | English | Chinese |
|---|---|---|
| $\eta o$ | Quartz-monzonite | 石英二长岩 |
| $\eta o\pi$ | Quartz-monzonite-porphyry | 石英二长斑岩 |
| $\eta\pi$ | Monzonite-porphyry | 二长斑岩 |
| $\kappa\gamma$ | Alkaline granite | 碱性花岗岩 |
| $\kappa\tau$ | Alkali trachyte | 碱性粗面岩 |
| $\lambda\gamma$ | Rhyolitic granite | 流纹质花岗岩 |
| $\lambda\pi$ | Rhyolite | 流纹岩 |
| $\lambda\pi$ | Rhyolite-porphyry | 流纹斑岩 |
| $\nu$ | Gabbro | 辉长岩 |
| $\pi$ | Porphyry | 斑岩 |
| $\pi\eta\gamma$ | Porphyritic monzonitic granite | 斑状二长花岗岩 |
| $\rho\mu$ | Pegmatitic porphyrite | 伟晶玢岩 |
| $\tau$ | Trachyte | 粗面岩 |
| $\upsilon$ | Glasstone, porphyrite | 玻璃岩 隐晶岩 |
| $\chi$ | Lamprophyre | 煌斑岩 |
| $\chi\nu$ | Lamprophyric gabbro | 煌斑辉长岩 |
| $\omega$ | Picrite | 苦橄岩 |

# References cited

Aikman, A.B., Harrison, M.T., and Lin, D., 2008, Evidence for early (>44 Ma) Himalayan crustal thickening, Tethyan Himalaya, southeastern Tibet: Earth and Planetary Science Letters, v. 274, p. 14–23.

Akciz, S., 2004, Structural and geochronological constraints on the ductile deformation observed along the Gaoligong Shan and Chong Shan shear zone, Yunnan (China) [Ph.D. thesis]: Cambridge, Massachusetts Institute of Technology, 211 p.

Akciz, S., Burchfiel, B.C., Crowley, J.L., Yin, J., and Chen, L., 2008, Geometry, kinematics and regional significance of the Chong Shan Shear Zone, eastern Himalayan syntaxis, Yunnan, China: Geosphere, v. 4, p. 292–314, doi:10.1130/GES00111.1.

Ali, J.R., and Aitchison, J.C., 2004, Problem of positioning Paleogene Eurasia: A review; efforts to resolve the issue, implications for the India-Asia collision, in Clift, P.D., Hayes, D.E., Khunt, W., and Wang, P.Y., eds., Continent-Ocean Interactions within the East Asian Marginal Seas: Washington, American Geophysical Union Monograph Series, p. 23–35.

Allen, C.R., Gillespie, A.R., Han, Y., Sieh, K.E., Zhang, B., and Zhu, C., 1984, Red River and associated faults, Yunnan Province, China: Quaternary geology, slip-rates, and seismic hazard: Geological Society of America Bulletin, v. 95, p. 686–700, doi:10.1130/0016-7606(1984)95<686:RRAAFY>2.0.CO;2.

Allen, C.R., Luo, A., Qian, H., Wen, X., Zhou, H., and Huang, W., 1991, Field study of a highly active fault zone: The Xianshuihe fault of southwestern China: Geological Society of America Bulletin, v. 103, p. 1178–1199, doi:10.1130/0016-7606(1991)103<1178:FSOAHA>2.3.CO;2.

Argand, E., 1922, La tectonique de l'Asie: Brussels, International Geological Congress, 171 p.

Armijo, R., Tapponnier, P., Mercier, J.L., and Lin, H.T., 1986, Quaternary extension in southern Tibet: Field observation and tectonic implications: Journal of Geophysical Research, v. 91, p. 13,803–13,872.

Armijo, R., Tapponnier, P., and Han, T.-L., 1989, Late Cenozoic right-lateral strike-slip faulting in southern Tibet: Journal of Geophysical Research, v. 94, p. 2787–2838, doi:10.1029/JB094iB03p02787.

Besse, J., and Courtillot, V., 1991, Revised and synthetic apparent polar wander paths for the African, Eurasian, North American, and Indian plates, and true polar wander since 200 Ma: Journal of Geophysical Research, v. 96, no. B3, p. 4029–4050.

Booth, A.L., Zietler, P.K., Kidd, W.S.F., Wooden, J., Liu, Y., Idelman, B., Hren, M., and Chamberlain, C.P., 2004, U-Pb zircon constraints on the tectonic evolution of southeastern Tibe Namche Barwa area: American Journal of Science, v. 304, p. 889–929, doi:10.247/ajs.304.10.889.

Booth, A.L., Chamberlain, C.P., Kidd, W.S.F., and Zeitler, P.K., 2009, Constraints on the metamorphic evolution of the eastern Himalayan syntaxis from geochronologic and petrologic studies of Namche Barwa: Geological Society of America Bulletin, v. 121, p. 385–407, doi:10.1130/B26041.1.

Burchfiel, B.C., 2004, New technology, new challenges: GSA Today, v. 14, no. 2, p. 4–9, doi:10.1130/1052-5173(2004)014<4:PANNGC>2.0.CO;2.

Burchfiel, B.C., and Wang, E., 2003, Northwest-trending, middle Cenozoic, left-lateral faults in southern Yunnan, China, and their tectonic significance: Journal of Structural Geology, v. 25, p. 781–792, doi:10.1016/S0191-8141(02)00065-2.

Burchfiel, B.C., Chen, Z., Hodges, K.V., Liu, Y., Royden, L.H., Deng, C., and Xu, J., 1992, Extension Contemporaneous with and Parallel to Shortening in the Himalaya: Geological Society of America Special Paper 269, 41 p.

Burchfiel, B.C., Chen, Z., Liu, Y., and Royden, L.H., 1995, Tectonics of the Longmen Shan and adjacent regions: International Geology Review, v. 37, p. 661–735, doi:10.1080/00206819509465424.

Burchfiel, B.C., Chen, L., Wang, E., and Swanson, E., 2008a, Preliminary investigation into the complexities of the Ailao Shan and Day Nui Con Voi shear zones in SE Yunnan and Vietnam, in Burchfiel, B.C., and Wang, E., eds., Investigations into the Tectonics of the Tibetan Plateau: Geological Society of America Special Paper 444, p. 45–58.

Burchfiel, B.C., Royden, L.H., Van der Hilst, R.D., Hager, B., Chen, Z., King, R.W., Li, C., Lu, J., Yao, H., and Kirby, E., 2008b, A geological and geophysical context for the Wenchuan earthquake of 12 May 2008, Sichuan, People's Republic of China: GSA Today, v. 18, no. 7, p. 4–11, doi:10.1130GSATG18A1.

Burchfiel, B.C., Studnicki-Gizbert, C., Geissman, J.W., King, R.W., Chen, Z., Chen, L., and Wang, E., 2008c, How much strain can continental crust accommodate without developing through-going faults?, in Sears, J.W., Harms, T.A., and Evenchick, C.A., eds., Whence the Mountains? Inquiries into the Evolution of Orogenic Systems: A Volume in Honor of Raymond A. Price: Geological Society of America Special Paper 433, p. 51–61, doi:10.1130/2007.2433(03).

Bureau of Geology and Mineral Resources of Sichuan Province, 1991, Regional Geology of Sichuan Province: Beijing, Geological Publishing House, 610 p. (in Chinese with English abstract).

Bureau of Geology and Mineral Resources of Xizang Autonomous Region, 1993, Regional Geology of Xizang (Tibet) Autonomous Region: Beijing, Geological Publishing House, 146 p.

Bureau of Geology and Mineral Resources of Yunnan Province, 1990, Regional Geology of Yunnan Province: Beijing, Geological Publishing House, 178 p. (in Chinese with English abstract).

Burg, J.-P., and Podladchikov, Y., 2000, From buckling to asymmetric folding of the continental lithosphere: Numerical modeling and application to the Himalayan syntaxes, in Kahn, M.A., Treloar, P.J., Searle, M.P., and Jan, M.Q., eds., Tectonics of the Nanga Parbat Syntaxis and the Western Himalaya: Geological Society [London] Special Publication 170, p. 219–236.

Burg, J.-P., Davy, P., Nieverget, P., Oberli, F., Seward, D., Dial, Z., and Meier, M., 1997, Exhumation during crustal folding in the Namche Barwa syntaxis: Terra Nova, v. 9, p. 53–56, doi:10.1111/j.1365-3121.1997.tb00001.x.

Burg, J.-P., Nieverget, P., Oberli, F., Seward, D., Davy, P., Maurin, H.C., Diao, Z., and Meier, M., 1998, The Namche Barwa syntaxis: Evidence for exhumation related to compressional crustal folding: Journal of Asian Earth Sciences, v. 16, p. 239–252, doi:10.1016/S0743-9547(98)00002-6.

Calassou, S., 1994, Étude tectonique d'une chaîne de décollement-tectonique Triasique et Tertiaire de la chaine de Songpan-Ganze (Est Tibet) [Ph.D. thesis]: Université de Montpellier II.

Cao, R.G., 1986a, Palaeogeography and tectonic evolution of West Yunnan: Bulletin of the Chinese Academy of Sciences, v. 13, p. 37–49.

Cao, R.G., 1986b, The discovery of Late Carboniferous glacial-marine strata in West Yunnan: Geological Review, v. 32, p. 236–242.

Cao, S.-H., Liao, L.-G., and Deng, S.-Q., 2005, Sequences, geochemistry and genesis of the Bangong Lake ophiolites in Xizang: Sedimentary Geology and Tethyan Geology, v. 25, p. 101–110.

Carter, A., Roques, D., Bristow, C., and Kinny, P., 2001, Understanding Mesozoic accretion in Southeast Asia: Significance of Triassic thermotectonism (Indosinian orogeny) in Vietnam: Geology, v. 29, p. 211–214, doi:10.1130/0091-7613(2001)029<0211:UMAISA>2.0.CO;2.

Chen, B., Wang, K., Liu, W., Cai, Z., Zhang, Q., Peng, X., Qiu, Y., and Zheng, Y., 1987, Geotectonics of the Nujiang-Lancang-Jiang-Jinshajiang Region: Beijing, Geological Publishing House (in Chinese).

Chen, H., Dobson, J., Heller, F., and Hao, J., 1995, Palaeomagnetic evidence for clockwise rotation of the Simao region since the Cretaceous: A consequence of India-Asia collision: Earth and Planetary Science Letters, v. 134, p. 203–217, doi:10.1016/0012-821X(95)00118-V.

Chen, J., 1987, Ages of granitoids in western Yunnan and discussion on their isotopic datings: Yunnan Geology, v. 6, p. 101–113.

Chen, L.-C., Wang, H., Ran, Y.-K., Sun, X.-Z., Su, G.-W., Wang, J., Tan, X.-B., Li, Z.-M., and Zhang, X.-Q., 2010, The Ms 7.1 Yushu earthquake surface rupture and large historical earthquakes on the Garze-Yushu Fault: Chinese Science Bulletin, v. 55, p. 3504–3509, doi:10.1007/s11434-010-4079-2.

Chen, Z., and Chen, X., 1987, On the Tectonic Evolution of the West Margin of the Yangzi Block: Chengdu, China, Chengdu Institute of Geology and Mineral Resources, 171 p.

Chen, Z., Burchfiel, B.C., Liu, Y., King, R.W., Royden, L.H., Tang, W., Wang, E., Zhao, J., and Zhang, X., 2000, Global Positioning System measurements from eastern Tibet and their implications for India/Eurasia intercontinental deformation: Journal of Geophysical Research, v. 105, p. 16,215–16,227, doi:10.1029/2000JB900092.

Cheng, Y., and Mao, J., 2012, Geochronology of the gabbro-mafic microgranular enclaves-granite associates in the Gejiu District, Yunnan Province and their geodynamic significance: Acta Geologica Sinica, v. 86, no. 3, p. 748–761.

Chengdu Institute of Geology and Mineral Resources, 2002, Map sheet F47C001002–F47C001003: scale 1:250,000 (unpublished).

Chiu, H.-Y., Chung, S.-L., Wu, F.-Y., Liu, D., Liang, Y.-H., Lin, I.-J., Iizika, Y., Xie, L.-W., Wang, Y., and Chu, M.-F., 2009, Zircon U-Pb and Hf isotopic constraints from eastern Transhimalayan batholiths on the precollisional magmatic and tectonic evolution in southern Tibet: Tectonophysics, v. 477, p. 3–19, doi:10.1016/j.tecto.2009.02.034.

Chung, S.-L., Lee, T.-Y., Lo, C.-H., Wang, P.-L., Chen, C.-Y., Yem, N.T., Hoa, T.T., and Wu, G., 1997, Intraplate extension prior to continental extrusion along the Ailao Shan–Red River shear zone: Geology, v. 25, p. 311–314, doi:10.1130/0091-7613(1997)025<0311:IEPTCE>2.3.CO;2.

Chung, S.-L., Chu, M.-F., Zhang, Y., Xie, X., Lo, C.-H., Lee, T.-Y., Lan, C.-Y., Li, X., Zhang, Q., and Wang, Y., 2005, Tibetan tectonic evolution inferred from spatial and temporal variations in post-collisional magmatism: Earth-Science Reviews, v. 68, p. 173–196, doi:10.1016/j.earscirev.2004.05.001.

Clark, M., Schoenbohm, L.M., Royden, L.H., Whipple, K.X., Burchfiel, B.C., Zhang, X., Tang, W., Wang, E., and Chen, L., 2004, Surface uplift, tectonics, and erosion of eastern Tibet from large-scale drainage patterns: Tectonics, v. 23, TC1006, doi:10.1029/2002TC001402.

Clark, M.K., House, M.A., Royden, L.H., Whipple, K.X., Burchfiel, B.C., Zhang, X., and Tang, W., 2005, Late Cenozoic uplift of eastern Tibet: Geology, v. 33, p. 525–528, doi:10.1130/G21265.1.

Clark, M.K., Royden, L.H., Whipple, K.X., Burchfiel, B.C., Zhang, X., and Tang, W., 2006, Use of a regional, relict landscape to measure vertical deformation of the eastern Tibetan Plateau: Journal of Geophysical Research, v. 111, F03002, doi:10.1029/2005JF000294.

Cong, B., and Zhai, M., 2000, Metamorphic geology and metamorphism in western Yunnan, in Zhong, D., ed., Paleotethysides in West Yunnan, China: Beijing, VSP Science Press, p. 26–38.

Densmore, A.L., Ellis, M.A., Li, Y., Zhou, R., Hancock, G.S., and Richardson, N., 2007, Active tectonics of the Beichuan and Pengguan faults at the eastern margin of the Tibetan Plateau: Tectonics, v. 26, TC4005, doi:10.1029/2006TC001987.

Deprat, J., 1914, Étude des plissements et des zones d'écrasement de la moyenne et de la basse Rivière Noire: Mémoires du Service géologique de l'Indochine, v. 3: Imprinte d'Extrême-Orient, 59 p.

Ding, L., and Zhong, D., 1999, Metamorphic characteristics and geotectonic implications of the high-pressure granulites from Namjagbarwa, eastern Tibet: Science in China, v. 42, p. 491–505.

Ding, L., Zhong, D., Yin, A., Kapp, P., and Harrison, T.M., 2001, Cenozoic structural and metamorphic evolution of the eastern Himalayan syntaxis (Namche Barwa): Earth and Planetary Science Letters, v. 192, p. 423–438, doi:10.1016/S0012-821X(01)00463-0.

Dirks, P.H.G.M., Wilson, C.J.L., Chen, S., Luo, Z.L., and Liu, S., 1994, Tectonic evolution of the NE margin of the Tibetan Plateau; evidence from the central Longmen Mountains, Sichuan Province, China: Journal of Southeast Asian Earth Sciences, v. 9, p. 181–192, doi:10.1016/0743-9547(94)90074-4.

England, P., and Houseman, G.A., 1988, The mechanics of the Tibetan Plateau, in Shackleton, R.M., Dewey, J.F., and Windley, B.F., eds., Tectonic Evolution of the Himalayas and Tibet: Royal Society of London, v. 326, p. 301–319.

England, P., and Molnar, P., 1990, Right-lateral shear and rotation as the explanation for strike-slip faulting in east Tibet: Nature, v. 344, p. 140–142, doi:10.1038/344140a0.

England, P., and Molnar, P., 1997, Active deformation of Asia: From kinematics to dynamics: Science, v. 278, p. 647–650, doi:10.1126/science.278.5338.647.

Enkelmann, E., Weislogel, A., Ratschbacher, L., Eide, E., Renno, A., and Wooden, J., 2007, How was the Triassic Songpan-Ganze basin filled? A provenance study: Tectonics, v. 26, TC4007, doi:10.1029/2006TC002078.

Enkin, R., Courtillot, V., Xing, L., Zhang, Z., Zhuang, Z., and Zhang, J., 1991, The stationary Cretaceous paleomagnetic pole of Sichuan (South China Block): Tectonics, v. 10, p. 547–559.

Enkin, R., Yang, Z., Chen, Y., and Courtillot, V., 1992, Paleomagnetic constraints on the geodynamic history of the major blocks of China from the Permian to the present: Journal of Geophysical Research, v. 97, p. 13,953–13,989.

Fan, C., and Zhang, Y., 1994, The structure and tectonics on western Yunnan: Journal of Southeast Asian Earth Sciences, v. 9, p. 355–361, doi:10.1016/0743-9547(94)90047-7.

Fromaget, J., 1927, Études geologiques sur le Nord de l'Indochine centrale: Bulletin du Service Géologique de l'Indochine, v. 16, 368 p.

Gan, W., Zhang, P., Shen, Z.-K., Niu, Z., Wang, M., Wan, Y., Zhou, D., and Cheng, J., 2007, Present-day crustal motion within the Tibetan Plateau inferred from GPS measurements: Journal of Geophysical Research, v. 112, B08416, doi:10.1029/2005JB004120.

Garzanti, E., and Van Haver, T., 1988, The Indus clastics: Forearc basin sedimentation in the Ladakh Himalaya (India): Sedimentary Geology, v. 59, p. 237–249, doi:10.1016/0037-0738(88)90078-4.

Geng, Q., Pan, G., Zheng, L., Chen, Z., Fisher, R.D., Sun, Z., Ou, C., Dong, H., Wang, X., Li, S., Lou, X., and Fu, H., 2005, The Eastern Himalayan syntaxis: Major tectonic domains, ophiolitic mélanges and geologic evolution: Journal of Asian Earth Sciences, v. 25, p. 1–21, doi:10.1016/j.jseaes.2005.03.009.

Gilley, L.D., Harrison, T.M., Leloup, P.H., Ryerson, F.J., Lovera, O., and Wang, J.-H., 2003, Direct dating of left-lateral deformation along the Red River shear zone: China and Vietnam: Journal of Geophysical Research, v. 108, p. 14–21, doi:10.1029/2001JB001726.

Green, O.R., Searle, M.P., Corfield, R.I., and Corfield, R.M., 2008, Cretaceous–Tertiary carbonate platform evolution and the age of the India-Asia collision along the Ladakh Himalaya (northwest India): Journal of Geology, v. 116, p. 331–353, doi:10.1086/588831.

Harkins, N., Kirby, E., Shi, X., Wang, E., Burbank, D., and Chun, F., 2010, Millennial slip rates along the eastern Kunlun fault. Implications for the dynamics of intracontinental deformation in Asia: Lithosphere, v. 2, p. 247–266, doi:10.1130/L85.1.

Harrowfield, M.J., and Wilson, C.H.L., 2005, Indosinian deformation of the Songpan Garze fold belt, northeast Tibet Plateau: Journal of Structural Geology, v. 27, p. 101–117, doi:10.1016/j.jsg.2004.06.010.

He, B., Xu, Y.G., Chung, S.-L., Xiao, O., and Wang, Y., 2003, Sedimentary evidence for a rapid, kilometer-scale crustal doming prior to the eruption of the Emeishan flood basalts: Earth and Planetary Science Letters, v. 213, p. 391–405, doi:10.1016/S0012-821X(03)00323-6.

Hennig, D., Lehmann, B., Frei, D., Belyatsky, B., Zhao, X.F., Cabral, A.R., Zeng, P.S., Zhou, M.F., and Schmidt, J., 2009, Early Permian seafloor to continental arc magmatism in the eastern Paleo-Tethys: U-Pb age and Nd-Sr isotope data from the southern Lancangjiang zone, Yunnan, China: Lithos, v. 113, p. 408–422, doi:10.1016/j.lithos.2009.04.031.

Heppe, K., 2004, Plate tectonic evolution and mineral resource potential of the Lancang River zone, Southwestern Yunnan, People's Republic of China [Ph.D. thesis]: Georg-August-Universität zu Göttingen, 115 p.

Heppe, K., Helmcke, D., and Wemmer, K., 2007, The Lancang River Zone of southwestern Yunnan, China: A questionable location for the active continental margin of Paleotethys: Journal of Asian Earth Sciences, v. 30, p. 706–720, doi:10.1016/j.jseaes.2007.04.002.

Holbig, E.S., and Grove, T.L., 2008, Mantle melting beneath the Tibetan Plateau: Experimental constraints on the generation of ultra-potassic lavas from Qiangtang, Tibet: Journal of Geophysical Research, v. 113, B04210, doi:10.1029/2007JB005149.

Horton, B.K., Yin, A., Spurlin, M., Zhou, J., and Wang, J., 2002, Paleocene–Eocene syncontractional sedimentation in narrow, lacustrine-dominated basins of east-central Tibet: Geological Society of America Bulletin, v. 114, p. 771–786, doi:10.1130/0016-7606(2002)114<0771:PESSIN>2.0.CO;2.

Houseman, G., and England, P., 1993, Crustal thickening versus lateral expansion in the India-Asia continental collisions: Journal of Geophysical Research, v. 98, p. 12,233–12,249, doi:10.1029/93JB00443.

Hsü, K.J., 1980, Thin-skinned plate-tectonic model for collision-type orogeneses: Scientia Sinica, v. 24, p. 100–110.

Hsü, K.J., Sun, S., and Li, J., 1987, Mesozoic suture in the Huanan Alps and the tectonic assembly of South China: Scientia Sinica (Ser. B), p. 1107–1115.

Hsü, K.J., Shu, S., Li, J., Chen, H., Haihong, C., Haipo, P., and Şengör, A.M.C., 1988, Mesozoic overthrust tectonics in south China: Geology,

v. 16, p. 418–421, doi:10.1130/0091-7613(1988)016<0418:MOTISC>2.3.CO;2.

Hu, D., Wu, Z., Wan, J., et al., 2005, Shrimp zircon U-Pb ages and Nd isotopic study on the Nyainqentanglha Group in Tibet: Science in China, Ser. Earth Science, v. 48, p. 1377–1386.

Hubbard, J., and Shaw, J.H., 2009, Uplift of the Longmen Shan and Tibetan plateau, and the 2008 Wenchuan (M = 7.9) earthquake: Nature, v. 458, p. 194–197, doi:10.1038/nature07837.

Huchon, P., LePichon, X., and Rangin, C., 1994, Indochina peninsula and the collision of India and Eurasia: Geology, v. 22, p. 27–30, doi:10.1130/0091-7613(1994)022<0027:IPATCO>2.3.CO;2.

Jagoutz, O., Muntener, O., Manatschal, G., Rubattos, D., Peron-Pinvidic, G., Turrin, B.D., and Villa, I.M., 2007, The rift to drift transition in the North Atlantic: A shuttering start to the MORB machine: Geology, v. 35, p. 1087–1090, doi:10.1130/G23613A.1.

Ji, J.Q., 1998, Petrology and Cenozoic lithospheric tectonic evolution of Tengchong-Yingjinag-Nagang area, west Yunnan, southwest China [Ph.D. thesis]: Institute of Geology, Chinese Academy of Sciences, p. 33–57.

Jian, P., Wang, X., He, L., and Wang, C., 1998, U–Pb zircon dating of the Shuanggou ophiolite from Xinping County, Yunnan Province: Acta Petrologica Sinica, v. 58, p. 1–17.

Jian, P., Liu, D., Kroner, A., Zhang, Q., Wang, Y., Sun, X., and Zhang, W., 2009a, Devonian to Permian plate tectonic cycle of the Paleo-Tethys Orogen in southwest China (I): Geochemistry of ophiolites, arc/back-arc assemblages and within-plate igneous rocks: Lithos, v. 113, p. 748–766, doi:10.1016/j.lithos.2009.04.004.

Jian, P., Liu, D., Kroner, A., Zhang, Q., Wang, Y., Sun, X., and Zhang, W., 2009b, Devonian to Permian plate tectonic cycle of the Paleo-Tethys Orogen in southwest China (II): Insights from zircon ages of ophiolites, arc/back-arc assemblages and within-plate igneous rocks and generation of Emeishan CFB province: Lithos, v. 113, p. 767–784, doi:10.1016/j.lithos.2009.04.006.

Jin, X., Hao, H., Shi, Y., and Zhang, L., 2011, Lithologic boundaries in Permian post-glacial sediments of the Gondwana-affinity regions of China: Typical sections, age range and correlation: Acta Geologica Sinica, v. 85, p. 373–386, doi:10.1111/j.1755-6724.2011.00406.x.

Jolivet, L., Beyssac, O., Goffe, G., Avigad, D., Lepvrier, C., Maluske, H., and Thang, T.T., 2001, Oligo-Miocene midcrustal subhorizontal shear zone in Indochina: Tectonics, v. 20, p. 46–57, doi:10.1029/2000TC900021.

Kan, R., Zhao, J., and Kan, D., 1996, The tectonic evolution and volcanic eruption in Tengchong volcanic and geothermal region (in Chinese): Seismology: Geomagnetism Observations Research, v. 17, p. 28–33.

Kapp, P., Yin, A., Manning, C.E., Harrison, T.M., Taylor, M.H., and Lin, D., 2003, Tectonic evolution of the early Mesozoic blueschist-bearing Qiangtang metamorphic belt, central Tibet: Tectonics, v. 22, p. 1043–1059, doi:10.1029/2002TC001383.

Kapp, P., Yin, A., Harrison, M.T., and Ding, L., 2005a, Cretaceous–Tertiary shortening, basin development, and volcanism in central Tibet: Geological Society of America Bulletin, v. 117, p. 865–878, doi:10.1130/B25595.1.

Kapp, J.I.D., Harrison, T.M., Kapp, P., Grove, M., Lovera, O.M., and Ding, L., 2005b, Nyainqentanglha Shan: A window into the tectonic, thermal, and geochemical evolution of the Lhasa Block, southern Tibet: Journal of Geophysical Research, v. 110, B08413, doi:10.1029/2004JB003330.

Kapp, P., DeCelles, P.G., Gehrels, G.E., Heizler, M., and Lin, D., 2007, Geological records of the Lhasa-Qiangtang and Indo-Asian collisions on the Nima area of central Tibet: Geological Society of America Bulletin, v. 119, p. 917–932, doi:10.1130/B26033.1.

King, R.W., Shen, F., Burchfiel, B.C., Chen, Z., Li, Y., Liu, Y., Royden, L.H., Wang, E., Zhang, X., and Zhao, J., 1997, Geodetic measurement of crustal motion in southwest China: Geology, v. 25, p. 179–182, doi:10.1130/0091-7613(1997)025<0179:GMOCMI>2.3.CO;2.

Kirby, E., Whipple, K.X., Burchfiel, B.C., Tang, W., Berger, G., Sun, Z., and Chen, Z., 2000, Neotectonics of the Min Shan, China: Implications for mechanisms driving Quaternary deformation along the eastern margin of the Tibetan Plateau: Geological Society of America Bulletin, v. 112, p. 375–393, doi:10.1130/0016-7606(2000)112<375:NOTMSC>2.0.CO;2.

Kirby, E., Reiners, P., Krol, M., Hodges, K., Whipple, K., Farley, K., Tang, W., and Chen, Z., 2002, Late Cenozoic evolution of the eastern margin of the Tibetan Plateau: Inferences from $^{40}Ar/^{39}Ar$ and U-Th/He thermochronology: Tectonics, v. 21, p. 1001, doi:10.1029/2000TC001246.

Lacassin, R.F., Scharer, U., Leloup, P.H., Arnaud, N., Tapponnier, P., Liu, X., and Zhang, L., 1996, Tertiary deformation and metamorphism SE of Tibet; the folded Tiger-leap décollement of NW Yunnan, China: Tectonics, v. 15, p. 605–622, doi:10.1029/95TC03749.

Lacassin, R., Replumaz, A., and Leloup, H.P., 1998, Hairpin river loops and slip-sense inversion on southeast Asian strike-slip faults: Geology, v. 26, p. 703–706, doi:10.1130/0091-7613(1998)026<0703:HRLASS>2.3.CO;2.

Lee, H.-Y., Chung, S.-L., Wang, J.-R., Wen, D.-J., Lo, C.-H., Yang, T.F., Zhang, Y., Xie, Y., Lee, T.-Y., Wu, G., and Ji, J., 2003, Miocene Jiali faulting and its implications for Tibetan tectonic evolution: Earth and Planetary Science Letters, v. 205, p. 185–194, doi:10.1016/S0012-821X(02)01040-3.

Leloup, P.H., Harrison, T.M., Ryerson, F.J., Chen, W., Li, Q., Tapponnier, P., and Lacassin, R., 1993, Structural, petrological and thermal evolution of a Tertiary ductile strike-slip shear zone, Diancang Shan, Yunnan: Journal of Geophysical Research, v. 98, p. 6715–6743, doi:10.1029/92JB02791.

Leloup, P.H., Lacassin, R., Tapponnier, P., Scharer, U., Zhong, D., Liu, X., Zhang, L., Ji, S., and Trinh, P.T., 1995, The Ailao Shan–Red River shear zone (Yunnan, China), Tertiary transform boundary of Indochina: Tectonophysics, v. 251, p. 3–84, doi:10.1016/0040-1951(95)00070-4.

Leloup, P., Arnaud, N., Lacassin, R., Kienast, J., Harrison, T., Trong, T., Replumaz, A., and Tapponnier, P., 2001, New constraints on the structure, thermochronology, and timing of the Ailao Shan–Red River shear zone, SE Asia: Journal of Geophysical Research, v. 106, p. 6683–6732, doi:10.1029/2000JB900322.

Lepvrier, C., Maluski, H., Tich, V.V., Leyreloup, A., Thi, P.T., and Vuong, N.V., 2004, The Early Triassic Indosinian orogeny in Vietnam (Tuong Son Belt and Kontum Massiv): Implications for the geodynamic evolution of Indochina: Tectonophysics, v. 393, p. 87–118, doi:10.1016/j.tecto.2004.07.030.

Li, C., Van der Hilst, R.D., and Toksöz, M.N., 2006, Constraining spatial variations in P-wave velocity in the upper mantle beneath SE Asia: Physics of the Earth and Planetary Interiors, v. 154, p. 180–195, doi:10.1016/j.pepi.2005.09.008.

Li, C., Van der Hilst, R.D., Engdahl, E.R., and Burdick, S., 2008a, A new global model for P wave speed variations in Earth's mantle: Geochemistry Geophysics Geosystems, v. 9, Q05018, doi:10.1029/2007GC001806.

Li, C., Van der Hilst, R.D., Meltzer, A.S., and Engdahl, E.R., 2008b, Subduction of the Indian lithosphere beneath the Tibetan Plateau and Burma: Earth and Planetary Science Letters, v. 274, p. 157–168, doi:10.1016/j.epsl.2008.07.016.

Li, X.-X., Wu, Y.-M., and Fu, Z.-B., 1985, Preliminary study on a mixed Permian flora from Xiagangjiang of Gerze district, Xizang, and its palaeobiogeographic significance: Acta Palaeontologica Sinica, v. 24, p. 150–170.

Li, X., Liu, C., and Pan, G., 1996, Geology and Tectonics of the Hengduan Mountains: International Geological Congress, 30th, Field Trip Guide T116: Beijing, Geological Publishing House, 60 p.

Li, Z.X., Li, X.H., Zhou, H.W., and Kinny, P.D., 2002, Grenvillian continental collision in South China: New SHRIMP U-Pb zircon results and implications for the configuration of Rodinia: Geology, v. 30, p. 163–166, doi:10.1130/0091-7613(2002)030<0163:GCCISC>2.0.CO;2.

Li, Z., Li, X.H., Kinny, P.D., Wang, J., Zhang, X., and Zhou, H., 2003, Geochronology of Neoproterozoic syn-rift magmatism in the Yangtze Craton, South China, and correlations with other continents: Evidence for a mantle superplume that broke up Rodinia: Precambrian Research, v. 122, p. 85–109, doi:10.1016/S0301-9268(02)00208-5.

Liang, X., Li, X., and Qiu, Y., 2004, Intracontinental collisional orogeny during Late Permian–Middle Triassic in South China: Sedimentary records of the Shiwandashan basin: Acta Geologica Sinica, v. 78, p. 756–762.

Lin, A., Jia, D., Rao, G., Yan, B., Wu, X., and Ren, Z., 2011, Recurrent morphogenic earthquakes in the past millennium along the strike-slip Yushu Fault, Central Tibetan Plateau: Bulletin of the Seismological Society of America, v. 101, p. 2755–2764, doi:10.1785/0120100274.

Lin, T.-H., Lo, C.-H., Chung, S.-L., Hsu, F.-J., Yeh, M.-W., Lee, T.-Y., Ji, J.-Q., Wang, Y.-Z., and Liu, D., 2009, $^{40}Ar/^{39}Ar$ dating of the Jaili and Gaoligong shear zones: Implications for crustal deformation around the eastern Himalayan Syntaxis: Journal of Asian Earth Sciences, v. 34, p. 674–685, doi:10.1016/j.jseaes.2008.10.009.

Lingzhi, G., Yangshen, S., Huafu, L., Honggren, D., and Shufeng, Y., 1989, The pre-Devonian tectonic pattern of South China: Journal of Southeast Asian Earth Sciences, v. 3, p. 87–93, doi:10.1016/0743-9547(89)90012-3.

Lister, G.S., and Davis, G.A., 1989, The origin of metamorphic core complexes and detachment faults formed during Tertiary continental extension in the northern Colorado River region, U.S.A.: Journal of Structural Geology, v. 11, p. 65–94, doi:10.1016/0191-8141(89)90036-9.

Liu, B., Feng, Q., Fang, N., et al., 1993, Tectonic evolution of Palaeo-Tethys poly-island-ocean in Changning-Menglian and Lancangjiang belts, southwestern Yunnan, China: Earth Science Journal of the University of Geosciences, v. 18, p. 529–538.

Liu, C., Zhu, J., and Xu, X., 1989, The Hercynian collision type granites of West Yunnan and their tectonic significance: Journal of Southeast Asian Earth Sciences, v. 3, p. 263–270, doi:10.1016/0743-9547(89)90031-7.

Liu, J., Liu, F., He, J., Chen, H., and You, Q., 2001, Study of seismic tomography in Panxi paleorift area of southwestern China: Science in China, Ser. D, v. 44, p. 277–288, doi:10.1007/BF02882262.

Liu, Y., and Zhong, D., 1997, Petrology of high-pressure granulite from the Eastern Himalaya: Journal of Metamorphic Geology, v. 15, p. 451–466, doi:10.1111/j.1525-1314.1997.00033.x.

Liu, Y., and Zhong, D., 1998, Petrology of high-pressure granulite from Eastern Himalaya: Implications to tectonic significance: Scientia Geologica Sinica, v. 33, p. 267–281.

Luo, H., and Jiang, Z., 1996, The Sinian-Cambrian boundary section and the Meishucun and Chenjiang fauna in Yunnan: International Geological Congress, 30th, Field Trip Guide T118/T381: Beijing, Geological Publishing House.

MacDonald, A.S., Barr, S.M., Dunning, G.R., and Yaowanoiyothin, W., 1993, The Doi Inthanon metamorphic core complex: Age and significance: Journal of Southeast Asian Earth Sciences, v. 8, p. 117–125, doi:10.1016/0743-9547(93)90013-F.

Maluski, H., Lepvrien, C., Jolivet, L., Carter, Z., Rogues, D., Beyssac, O., Trong Tang, T., Duc Thang, N., and Avigad, D., 2001, Ar/Ar fission-track ages in the Song Chay Massif: Early Triassic and Cenozoic tectonics in northern Vietnam: Journal of Asian Earth Sciences, v. 19, p. 233–248, doi:10.1016/S1367-9120(00)00038-9.

Mattauer, M., Matte, P., Malavieille, J., Tapponnier, P., Maliski, H., Qin, Z.Z., Lun, L.Y., and Qin, T.Y., 1985, Tectonics of the Qinling Belt: Build-up and evolution of eastern Asia: Nature, v. 317, p. 496–500, doi:10.1038/317496a0.

Mattauer, M., Malavieille, J., Calassou, S., Lancelot, J., Roger, R., Hao, Z., Xu, Z., and Hou, L., 1992, Le chaine Triassique de Songpan-Garze (ouest Sichuan et est Tibet): Une chaine de pissement-décollement sur marge passive: Paris, Comptes Rendus de l'Académie des sciences, v. 314, Ser. II, p. 619–626.

McCaffrey, R., 1982, Lithospheric deformation within the Molucca Sea arcarc collision: Evidence from shallow and intermediate earthquake activity: Journal of Geophysical Research, v. 87, p. 3663–3678, doi:10.1029/JB087iB05p03663.

Meng, Q., and Zhang, G., 1999, Timing of collision of the North and South China blocks: Controversy and reconciliation: Geology, v. 27, p. 123–126, doi:10.1130/0091-7613(1999)027<0123:TOCOTN>2.3.CO;2.

Metcalfe, I., 1986, Late Palaeozoic palaeogeography of Southeast Asia: Some stratigraphical, palaeontological and palaeomagnetic constraints: Bulletin of the Geological Society of Malaysia, v. 19, p. 153–164.

Metcalfe, I., 1996, Pre-Cretaceous evolution of SE Asian terranes, in Hall, R., and Blundell, D.J., eds., Tectonics of SE Asia: Geological Society [London] Special Publication 106, p. 97–122.

Metcalfe, I., 1998, Palaeozoic and Mesozoic geological evolution of the SE Asian region: Multidisciplinary constraints and implications for biogeography, in Hall, R., and Holloway, J.D., eds., Biogeography and Geological Evolution of SE Asia: Leiden, Netherlands, Backhuys Publishers, p. 25–41.

Metcalfe, I., 2006, Palaeozoic and Mesozoic tectonic evolution and palaeogeography of East Asian crustal fragments: The Korean Peninsula in context: Gondwana Research, v. 9, p. 24–46, doi:10.1016/j.gr.2005.04.002.

Mo, X., Shen, S., Zhu, Q., et al., 1998, Volcanics-Ophiolite and Mineralization of Middle-Southern Part in Sanjiang Area of Southwestern China: Beijing, Geological Publishing House, 128 p. (in Chinese with English abstract).

Molnar, P., and Stock, J.M., 2009, Slowing of India's convergence with Eurasia since 20 Ma and its implications for Tibet mantle dynamics: Tectonics, v. 28, TC3001, doi:10.1029/2008TC002271.

Molnar, P., and Tapponnier, P., 1975, Cenozoic tectonics of Asia: Effects of a continental collision: Science, v. 189, p. 419–426, doi:10.1126/science.189.4201.419.

Mu, Z.G., Tong, W., and Garniss, H.C., 1987, Times of volcanic activity and origin of magma in Tengchong Geothermal Area, West Yunnan Province: Acta Geophysica Sinica, v. 30, p. 261–270.

Otofuji, Y.-I., My, C.L., Tanaka, K., Miura, D., Inokuchi, H., Kamei, R., Tamai, M., Takemoto, K., Zaman, H., and Yokoyama, M., 2007, Spatial gap between Lhasa and Qiangtang blocks inferred from Middle Jurassic to Cretaceous paleomagnetic data: Earth and Planetary Science Letters, v. 262, p. 581–593, doi:10.1016/j.epsl.2007.08.013.

Ouimet, W., Whipple, K., Royden, L.H., Reiners, P., Hodges, K., and Pringle, M., 2010, Regional incision of the eastern margin of the Tibetan Plateau: Lithosphere, v. 2, p. 50–63, doi:101130/L57.1.

Pan, G., Liu, Y., Zhong, I., Geng, Q., Wag, L., Li, G., Yin, F., Liao, Z., and Zhu, D., 2005, Qinghai-Tibet plateau: Collisional tectonics and their effects: Guangdong Science and Technology, p. 12–153 (in Chinese with English abstract).

Peng, X., and Luo, W., 1982, Discovery and tectonic significance of glaucophane schist from the southern section of the Lancang River in western Yunnan: Regional Geology of China, v. 2, p. 69–75.

Péron-Pinvidic, G., and Manatschal, G., 2009, The final rifting evolution at deep magma-poor passive margins from Iberia-Newfoundland: A new point of view: International Journal of Earth Sciences, v. 98, p. 1581–1597, doi:10.1007/s00531-008-0337-9.

Pullen, A., Kapp, P., Gehrels, G.E., Vervoort, J.D., and Ding, L., 2008, Triassic continental subduction in central Tibet and Mediterranean-style closure of the Paleo-Tethys Ocean: Geology, v. 36, p. 351–354, doi:10.1130/G24435A.1.

Qin, J., Qian, X., and Huangpu, G., 1996, The seismicity feature of the volcanic area in Tengchong: Seismology: Geomagnetism Observations Research, v. 17, p. 19–27 (in Chinese).

Ratschbacher, L., Frisch, W., Liu, G., and Chen, C., 1994, Distributed deformation in southern and western Tibet during and after the India–Asia collision: Journal of Geophysical Research, v. 99, p. 19,917–19,945, doi:10.1029/94JB00932.

Ratschbacher, L., Frisch, W., Chen, C., and Pan, G., 1996, Cenozoic deformation, rotation, and stress patterns in eastern Tibet and western Sichuan, China, in Yin, A., and Harrison, T.M., eds., The Tectonic Evolution of Asia: Cambridge, UK, Cambridge University Press, p. 227–249.

Reid, A., Wilson, C.J.L., Phillips, D., and Liu, S., 2005a, Mesozoic cooling across the Yidun Arc, central-eastern Tibetan Plateau: A reconnaissance $^{40}$Ar/$^{39}$Ar study: Tectonophysics, v. 398, p. 45–66, doi:10.1016/j.tecto.2005.01.002.

Reid, A., Wilson, C.J.L., and Liu, S., 2005b, Structural evidence for the Permo-Triassic tectonic evolution of the Yidun Arc, eastern Tibetan Plateau: Journal of Structural Geology, v. 27, p. 119–137, doi:10.1016/j.jsg.2004.06.011.

Reid, A., Wilson, C.J.L., Liu, S., Pearson, N., and Belousova, E., 2007, Mesozoic plutons of the Yidun Arc, SW China: U/Pb geochronology and Hf isotopic signature: Ore Geology Reviews, v. 31, p. 88–106, doi:10.1016/j.oregeorev.2004.11.003.

Replumaz, A., and Tapponnier, P., 2003, Reconstruction of the deformed collision zone between India and Asia by backward motion of lithospheric blocks: Journal of Geophysical Research, v. 108, 2285, doi:10.1029/2001JB000661.

Richardson, N.J., 2008, Extraordinary denudation in the Sichuan Basin: Insights from low-temperature thermochronology adjacent to the eastern margin of the Tibetan Plateau: Journal of Geophysical Research, v. 113, BO4409, doi:10.1029/2006JB004739.

Roger, F., Calassou, S., Lancelot, J., Malavieille, J., Mattauer, M., Xu, Z., Hao, Z., and Hou, L., 1995, Miocene emplacement and deformation of the Konga Shan granite (Xianshi He fault zone, west Sichuan, China): Geodynamic implications: Earth and Planetary Science Letters, v. 130, p. 201–216, doi:10.1016/0012-821X(94)00252-T.

Roger, F., Malavieille, M., Leloup, P.H., Calassou, S., and Xu, Z., 2004, Timing of granite emplacement and cooling in the Songpan Garze Fold Belt (eastern Tibetan Plateau with tectonic implications): Journal of Asian Earth Sciences, v. 22, p. 465–481, doi:10.1016/S1367-9120(03)00089-0.

Roger, F., Jolivet, M., and Malavieille, J., 2008, Tectonic evolution of the Triassic fold belts of Tibet: Comptes Rendus Geoscience, v. 340, p. 180–189, doi:10.1016/j.crte.2007.10.014.

Roger, F., Jolivet, M., and Malavieille, J., 2010, The tectonic evolution of the Songpan-Garze (North Tibet) and adjacent areas from Proterozoic to

Present: A synthesis: Journal of Asian Earth Sciences, v. 39, p. 254–269, doi:10.1016/j.jseaes.2010.03.008.

Rowley, D.B., 1996, Age of initiation of collision between India and Asia: A review of stratigraphic data: Earth and Planetary Science Letters, v. 145, p. 1–13, doi:10.1016/S0012-821X(96)00201-4.

Rowley, D.B., 1998, Minimum age of initiation of collision between India and Asia north of Everest based on the subsidence history of the Zhepure Mountain section: Journal of Geology, v. 106, p. 220–235, doi:10.1086/516018.

Royden, L.H., Burchfiel, B.C., King, R.W., Wang, E., Chen, Z., Shen, F., and Liu, Y., 1997, Surface deformation and lower crustal flow in eastern Tibet: Science, v. 276, p. 788–790, doi:10.1126/science.276.5313.788.

Royden, L.H., Burchfiel, B.C., and van der Hilst, R.D., 2008, The geological evolution of the Tibetan Plateau: Science, v. 321, p. 1054–1058, doi:10.1126/science.1155371.

Schoenbohm, L.M., Burchfiel, B.C., Chen, L., and Yin, J., 2005, Exhumation of the Ailao Shan shear zone recorded by Cenozoic sedimentary rocks, Yunnan Province, China: Tectonics, v. 24, TC6015, doi:10.1029/2005TC001803.

Schoenbohm, L.M., Burchfiel, B.C., and Chen, L., 2006a, Propagation of surface uplift, lower crustal flow, and Cenozoic tectonics of the southeastern margin of the Tibetan Plateau: Geology, v. 34, p. 813–816, doi:10.1130/G22679.1.

Schoenbohm, L.M., Burchfiel, B.C., Chen, L., and Yin, J., 2006b, Miocene to present activity along the Red River fault, China, in the context of continental extrusion, upper crustal rotation, and lower crustal flow: Geological Society of America Bulletin, v. 118, p. 672–688, doi:10.1130/B25816.1.

Scrotese, C.R., and Golonka, J., 1992, Paleogeographic Atlas: Paleomap Program Report 20: Department of Geology, University of Texas at Arlington.

Searle, M.P., Noble, S.R., Cottle, J.M., Waters, D.J., Mitchell, A.H.G., Hlaing, T., and Horstworrd, M.S.A., 2007, Tectonic evolution of the Mogok metamorphic belt, Burma (Myanmar) constrained by U-Th-Pb dating of metamorphic and magmatic rocks: Tectonics, v. 26, TC3014, doi:10.1029/2006TC002083.

Searle, M.P., Yeh, M.W., Lin, T.H., and Chung, S.L., 2010, Structural constraints on the timing of left-lateral shear along the Red River shear zone in the Ailao Shan and Diancang Shan Ranges, Yunnan, SW China: Geosphere, v. 6, p. 316–338, doi:10.1130/GES00580.1.

Şengör, A.M.C., 1984, The Cimmeride Orogenic System and the Tectonics of Eurasia: Geological Society of America Special Paper 195, 82 p.

Şengör, A.M.C., 1990, Plate tectonics and orogenic research after 25 years: A Tethyan perspective: Earth-Science Reviews, v. 195, p. 1–82.

Şengör, A.M.C., and Hsü, K.J., 1984, The Cimmerides of eastern Asia: History of the eastern end of the Tethys: Mémoires de la Société Géologique de France, Nouvelle Serie, v. 147, p. 139–167.

Şengör, A.M.C., Cin, A., Rowley, D.B., and Nie, S.-Y., 1993, Space-time patterns of magmatism along the Tethysides: Journal of Geology, v. 101, p. 51–84, doi:10.1086/648196.

Sha, S.-H., Xie, Y.-W., Chen, Y.-M., Xilou, L.-J., Liu, X.-L., Zhang, N., and Jiang, C.-X., 2009, Tongka garnet pyroxenite in eastern Tibet and Songduo blue schist in Lhasa block: Geology in China, v. 36, p. 1302–1311.

She, Z., Ma, C., Mason, R., Li, J., Wang, G., and Lei, Y., 2006, Provenance of the Triassic Songpan-Ganzi flysch, west China: Chemical Geology, v. 231, p. 159–175, doi:10.1016/j.chemgeo.2006.01.001.

Shellnutt, J.G., and Jahn, B.-M., 2010, Formation of the Late Permian Panzhihua plutonic-hypabyssal-volcanic igneous complex: Implications for the genesis of Fe-Ti oxide deposits and A-type granites of SW China: Earth and Planetary Science Letters, v. 289, p. 509–519, doi:10.1016/j.epsl.2009.11.044.

Shellnutt, J.G., Zhou, M.-F., Yan, D.-P., and Wang, Y., 2008, Longevity of the Permian Emeishan mangle plume (SW China): 1 Ma, 8 Ma, or 18 Ma?: Geological Magazine, v. 145, p. 373–388, doi:10.1017/S0016756808004524.

Shen, F., Royden, L.H., and Burchfiel, B.C., 2001, Large scale crustal deformation of the Tibetan plateau: Journal of Geophysical Research, v. 106, p. 6793–6816, doi:10.1029/2000JB900389.

Shi, R., Yang, J., Xu, Z., and Qi, X., 2008, The Bangong Lake ophiolite (NNW Tibet) and its bearing on the tectonic evolution of the Bangong-Nujiang suture zone: Journal of Asian Earth Sciences, v. 32, p. 438–457, doi:10.1016/j.jseaes.2007.11.011.

Silver, E.A., and Moore, J.C., 1978, The Molucca Seas Collision Zone, Indonesia: Journal of Geophysical Research, v. 83, p. 1681–1691, doi:10.1029/JB083iB04p01681.

Socquet, A., and Pubellier, M., 2005, Cenozoic deformation in western Yunnan (China-Myanmar border): Journal of Asian Earth Sciences, v. 24, p. 495–515, doi:10.1016/j.jseaes.2004.03.006.

Song, X.-Y., Zhou, M.-F., Cao, Z.-M., and Robinson, P.T., 2004, Late Permian rifting of the South China Craton caused by the Emeishan mantle plume: Journal of the Geological Society [London], v. 161, p. 773–781, doi:10.1144/0016-764903-135.

Stewart, R.J., Hallet, B., Zeitler, P.K., Malloy, M.A., Allen, C.M., and Trippett, D., 2008, Brahmaputra sediment flux dominated by highly localized rapid erosion from the easternmost Himalaya: Geology, v. 36, p. 711–714, doi:10.1130/G24890A.1.

Studnicki-Gizbert, C.T., 2006, Deformation, erosion and sedimentation in collisional orogens: Case studies from eastern Tibet and southwestern China [Ph.D. thesis]: Cambridge, Massachusetts Institute of Technology, 296 p.

Studnicki-Gizbert, C., Burchfiel, B.C., Li, Z., and Chen, Z., 2008, Early Tertiary Gonjo basin, eastern Tibet: Sedimentary and structural record of the early history of India-Asia collision: Geosphere, v. 4, p. 713–735, doi:10.1130/GES00136.1.

Sun, W.-H., Zhou, M.-F., Gao, J.-F., and Yang, Y.-H., 2009, Detrital zircon U-Pb geochronological and Lu-Hf isotopic constraints on the Precambrian magmatic and crustal evolution of the western Yangtze Block, SW China: Precambrian Research, v. 172, p. 99–126.

Tanaka, K., Mu, C., Sato, K., Takemoto, K., Miura, D., Liu, Y., Zaman, H., Yang, Z., Yokoyama, M., Iwamoto, H., Uno, K., and Otofuji, Y.-I., 2008, Tectonic deformation around the eastern Himalayan syntaxis: Constraints from the Cretaceous paleomagnetic data of the Shan-Thai Block: Geophysical Journal International, v. 175, p. 713–728, doi:10.1111/j.1365-246X.2008.03885.x.

Tapponnier, P., Peltzer, G., Le Dain, A.Y., Armijo, R., and Cobbold, P., 1982, Propagating extrusion tectonics in Asia: New insights from simple experiments with plasticine: Geology, v. 10, p. 611–616, doi:10.1130/0091-7613(1982)10<611:PETIAN>2.0.CO;2.

Tapponnier, P., Peltzer, G., and Armijo, R., 1986, On the mechanics of the collision between India and Asia, in Coward, M.P., and Ries, A.C., eds., Collision Tectonics: Geological Society [London] Special Publication 19, p. 115–157.

Tapponnier, P., Lacassin, R., Leloup, P.H., Scharer, U., Zhong, D., Liiu, X., Ji, S., Zhang, L., and Zhong, J., 1990, The Ailao Shan/Red River metamorphic belt: Tertiary left-lateral shear between Indochina and South China: Nature, v. 343, p. 431–437, doi:10.1038/343431a0.

Tapponnier, P., Xu, Z., Roger, F., Meyer, B., Arnaul, N., Wittlinger, G., and Yang, J., 2001, Oblique stepwise rise and growth of the Tibet plateau: Science, v. 294, p. 1671–1677, doi:10.1126/science.105978.

Taylor, M., and Yin, A., 2009, Active structures of the Himalayan-Tibetan orogen and their relationships to earthquake distribution, contemporary strain field, and Cenozoic volcanism: Geosphere, v. 5, p. 199–214, doi:10.1130/GES00217.1.

Ueno, K., 2003, The Permian fusulinoidean faunas of the Sibumasu and Baoshan blocks, their implications for the paleogeographic and paleoclimatologic reconstruction of the Cimmerian continent: Palaeogeography, Palaeoclimatology, Palaeoecology, v. 193, p. 1–24.

Wang, E., and Burchfiel, B.C., 1997, Interpretation of Cenozoic tectonics in the right-lateral accommodation zone between the Ailao Shan shear zone and the eastern Himalayan syntaxis: International Geology Review, v. 39, p. 191–219, doi:10.1080/00206819709465267.

Wang, E., and Burchfiel, B.C., 2000, Late Cenozoic to Recent deformation in SW Sichuan and adjacent Yunnan, China, and its role in formation of the SE part of the Tibetan plateau: Geological Society of America Bulletin, v. 112, p. 413–423, doi:10.1130/0016-7606(2000)112<413:LCTHDI>2.0.CO;2.

Wang, E., Burchfiel, B.C., Royden, L.H., Chen, L., Chen, J., and Li, W., 1998, Late Cenozoic Xianshuihe-Xiaojiang and Red River Fault Systems of Southwestern Sichuan and Central Yunnan, China: Geological Society of America Special Paper 327, 108 p.

Wang, E., Kirby, E., Furlong, K.P., van Soest, M., Xu, G., Shi, X., Kamp, P.J.H., and Hodges, K.V., 2012, Two-phase growth of high topography in eastern Tibet during the Cenozoic: Nature Geoscience Letters, v. 5, p. 640–645.

Wang, G., Wan, J., Wang, E., Zheng, D., and Li, F., 2008, Late Cenozoic to recent transtensional deformation across the southern part of the Gaoligong shear zone between the Indian plate and SE margin of the

Tibetan plateau and its tectonic origin: Tectonophysics, v. 460, p. 1–20, doi:10.1016/j.tecto.2008.04.007.

Wang, J., and Li, Z.-X., 2003a, History of Neoproterozoic rift basins in South China: Implications for Rodinia break-up: Precambrian Research, v. 122, p. 141–158, doi:10.1016/S0301-9268(02)00209-7.

Wang, J., and Li, Z.-X., 2003b, History of Neoproterozoic rift basins, in Wang, X., Metcalfe, I., Jian, P., He, L., and Wang, C., eds., 2000, The Jinshajiang Suture Zone: Tectono-Stratigraphic Subdivision and Revision of Age: Science in China (Ser. D), v. 43, p. 10–22.

Wang, K., Long, B.L., Yanlong, Z.K., Yang, Z., and Wen, D., 2002, Discovery of marine Jurassic strata at Muli, Sichuan, and its geological implications: Geological Bulletin of China, v. 21, p. 421–427.

Wang, S., Fan, C., Wang, G., and Wang, E., 2008a, Late Cenozoic deformation along the northwestern continuation of the Xianshuihe fault system, Eastern Tibetan Plateau: Geological Society of America Bulletin, v. 120, p. 312–327, doi:10.1130/B25833.1.

Wang, S., Wang, E., Fang, X., and Fu, B., 2008b, Late Cenozoic systematic left-lateral stream deflections along the Ganze-Yushu Fault, Xianshuihe Fault System, eastern Tibet: International Geology Review, v. 50, p. 624–635, doi:10.2747/0020-6814.50.7.624.

Wang, X.-L., Zhou, J.-C., and Griffin, W.L., 2007, Detrital zircon geochronology of Precambrian basement sequences in the Jiangnan orogen: Dating the assembly of the Yangtze and Cathaysia Blocks: Precambrian Research, v. 159, p. 117–131, doi:10.1016/j.precamres.2007.06.005.

Wang, Y.Z., 1983, Characteristics and significance of the pebbly Carboniferous strata in the Tengchong and Baoshan areas: Contributions to the Geology of Qinghai-Tibet Plateau, v. 11, p. 71–77.

Wang, Y., Fan, W., Zjang, Y., Peng, T., Chen, X., and Xu, Y., 2006, Kinematics and $^{40}$Ar/$^{39}$Ar geochronology of the Gaoligong and Chongshan shear systems, western Yunnan, China: Implications for early Oligocene tectonic extrusion of SE Asia: Tectonophysics, v. 418, p. 235–254, doi:10.1016/j.tecto.2006.02.005.

Wang, Y., Fan, W., Zhao, G., Ji, S., and Peng, T., 2007, Zircon U-Pb geochronology of gneissic rocks in the Yunkai massif and its implications on the Caledonian event in the south China block: Gondwana Research, v. 12, p. 404–416, doi:10.1016/j.gr.2006.10.003.

Wang, Z., Zhang, J., Li, T., Xie, G., and Ma, Z., 2010, Structural analysis of the multi-layer detachment folding in Eastern Sichuan Province: Acta Geologica Sinica, v. 84, p. 497–514.

Weislogel, A.L., Graham, S.A., Chang, E.Z., Wooden, J.L., Gehrels, G.E., and Yang, H., 2006, Detrital zircon provenance of the Late Triassic Songpan-Ganzi complex: Sedimentary record of collision of the North and South China blocks: Geology, v. 34, p. 97–100, doi:10.1130/G21929.1.

Weislogel, A.L., 2008, Tectonostratigraphic and geochronologic constraints on evolution of the northeast Paleotethys from the Songpan-Ganzi complex, central China: Tectonophysics, v. 451, p. 331–345, doi:10.1016/j.tecto.2007.11.053.

Weislogel, A.L., Graham, S.A., Chang, E.Z., Wooden, J.L., and Gehrels, G.E., 2010, Detrital zircon provenance from three turbidite depocenters of the Middle–Upper Triassic Songpan-Ganzi complex, central China: Record of collisional tectonics erosional exhumation and sediment production: Geological Society of America Bulletin, v. 122, p. 2041–2062, doi:10.1130/B26606.1.

Wernicke, B., 1981, Low-angle normal faults in the Basin and Range Province: Nappe tectonics in an extending orogen: Nature, v. 291, p. 645–648, doi:10.1038/291645a0.

Wu, G.Y., 1993, Permian basalts in Lijiang and Jinping, western Yunnan: A comparative study and its geological significance: Acta Petrologica Sinica, v. 9, Supplement, p. 63–69.

Wu, H., Boulter, C.A., Ke, B., Stow, D.A.V., and Wang, Z., 1995, The Changning-Menglian suture zone: A segment of the major Cathaysian-Gondwana divide in Southeast Asia: Tectonophysics, v. 242, p. 267–280, doi:10.1016/0040-1951(94)00210-Z.

Xiao, S., Zhang, Y., and Knoll, A.H., 1998, Three dimensional preservation of algae and animal embryos in a Neoproterozoic phosphorite: Nature, v. 391, p. 553–558, doi:10.1038/35318.

Xiao, W., and He, H., 2005, Early Mesozoic thrust tectonics of the northwest Zhejiang region (Southeast China): Geological Society of America Bulletin, v. 117, p. 945–961, doi:10.1130/B25417.1.

Xing, Y., and Luo, H., 1984, Precambrian-Cambrian boundary candidate, Meishucun, Jinning, Yunnan, China: Geological Magazine, v. 121, p. 133.

Xu, G., and Kamp, P.J.J., 2000, Tectonics and denudation adjacent to the Xianshuihe Fault, eastern Tibetan Plateau: Constraints from fission track thermochronology: Journal of Geophysical Research, v. 105, p. 19,231–19,251, doi:10.1029/2000JB900159.

Xu, G.X., He, B., Chung, S.-L., Menzies, M.A., and Frey, F.A., 2004, Geologic, geochemical and geophysical consequences of plume involvement in the Emeishan flood-basalt province: Geology, v. 32, p. 917–920, doi:10.1130/G20602.1.

Xu, X., Zhang, Y., Jia, D., and Shu, L., 2011, U-Pb dating of volcanic rocks and granites along the Wuyishan Belt: Constraints on timing of Late Mesozoic tectonic events in Southeast China: Acta Geologica Sinica, v. 85, p. 130–144, doi:10.1111/j.1755-6724.2011.00385.x.

Xu, Y.-G., and Chung, S.-L., 2001, The Emeishan large igneous province: Evidence for mantle plume activity and melting conditions: Geochimica, v. 30, p. 1–9 (in Chinese with English abstract).

Xu, Y.-G., and He, B., 2007, Thick, high-velocity crust in the Emeishan large igneous province, southwestern China: Evidence for crustal growth by magmatic underplating or intraplating, in Foulger, G.R., and Jurdy, D.M., eds., Plates, Plumes, and Planetary Processes: Geological Society of America Special Paper 430, p. 841–858, doi:10.1130/2007.2430(39).

Xu, Y.-G., Mei, H.J., Xu, J.F., Huang, X.L., Wang, Y.J., and Chung, S.L., 2003, Origins of two differentiation trends in the Emeishan flood basalts: Chinese Science Bulletin, v. 48, p. 390–394, doi:10.1360/03tb9082.

Xu, Z.J., and Song, X., 2010, Joint inversion for crustal and Pn velocities and Moho depth in Eastern Margin of the Tibet Plateau: Tectonophysics, v. 491, p. 185–193, doi:10.1016/j.tecto.2009.11.022.

Xua, J.-F., and Castillob, P.R., 1994, Geochemical and Nd-Pb Isotopic characteristics of the Tethyan asthenosphere: Implications for the origin of the Indian Ocean mantle domain: Tectonophysics, v. 393, 1–4, p. 9–27.

Yan, D., and Wang, X., 2000, Analysis of fold style and its formation mechanism in the area of boundary among Sichuan, Hubei and Hunan: Geoscience (Journal of Graduate School: China University of Geoscience), v. 14, p. 37–43.

Yan, D.-P., Zhou, M.-F., Song, H.-L., Wang, X.-W., and Malpas, J., 2003a, Origin and tectonic significance of a Mesozoic multi-layer overthrust system within the Yangtze Block (South China): Tectonophysics, v. 361, p. 239–254, doi:10.1016/S0040-1951(02)00646-7.

Yan, D.-P., Zhou, M.-F., Song, H., and Fu, Z., 2003b, Structural style and tectonic significance of the Jianglang dome in the eastern margin of the Tibetan Plateau, China: Journal of Structural Geology, v. 25, p. 765–779, doi:10.1016/S0191-8141(02)00059-7.

Yan, D.P., Zhou, M.F., Yan, C.Y., and Xia, B., 2006, Structural and geochronological constraints on the tectonic evolution of the Dulong-Song Chay tectonic dome in Yunnan province, SW China: Journal of Asian Earth Sciences, v. 28, p. 332–353, doi:10.1016/j.jseaes.2005.10.011.

Yang, J., and Yao, X., 1998, Giant nappe structure in the Garze-Litang Fault Zone: Acta Geologica Sichuan, v. 18, p. 81–87.

Yang, Z., and Besse, J., 2001, New Mesozoic apparent polar wander path for south China: Tectonic consequences: Journal of Geophysical Research, v. 106, p. 8493–8520, doi:10.1029/2000JB900338.

Yano, T., Wu, G., Tang, M., and Sha, S., 1994, Tectono-sedimentary development of backarc continental basin in Yunnan, southern China: Journal of Southeast Asian Earth Sciences, v. 9, p. 153–166, doi:10.1016/0743-9547(94)90072-8.

Yao, H., Van der Hilst, R.D., and Montagner, J.-P., 2010, Structure and deformation of crust and upper mantle beneath SE Tibet from surface-wave array tomography: Journal of Geophysical Research, v. 115, B12307, doi:10.1029/2009JB007142.

Yao, J., Shu, L., Santosh, M., and Li, J., 2012, Precambrian crustal evolution of the South China Block and its relation to supercontinent history: Constraints from U-Pb ages, Lu-Hf isotopes and REE geochemistry of zircons from sandstones and granodiorite: Precambrian Research, v. 208–211, p. 19–48.

Yin, A., and Nie, S., 1993, An indentation model for the North and South China collision and the development of the Tanlu and Honam fault systems, eastern Asia: Tectonics, v. 12, p. 801–813, doi:10.1029/93TC00313.

Yumul, G.P., Jr., Zhou, M.-F., Wang, C., Zhao, T.-P., and Dimalanta, C.B., 2008, Geology and geochemistry of the Shuanggou Ophiolite (Ailao Shan Ophiolitic Belt), Yunnan Province, SW China: Evidence for a slow-spreading oceanic basin origin: Journal of Asian Earth Sciences, v. 32, p. 385–395, doi:10.1016/j.jseaes.2007.11.007.